AF660809

Advanced Robust and Adaptive Control

Theory and Applications

Tsinghua University Press

NEW AGE

Advanced Robust and Adaptive Control
Theory and Applications

Edited by
Daizhan Cheng • Yuanzhang Sun
Tielong Shen • Hiromitsu Ohmori

PUBLISHING FOR ONE WORLD

NEW AGE INTERNATIONAL (P) LIMITED, PUBLISHERS
New Delhi • Bangalore • Chennai • Cochin • Guwahati • Hyderabad
Jalandhar • Kolkata • Lucknow • Mumbai • Ranchi
Visit us at www.newagepublishers.com

Copyright © 2010 by Tsinghua University Press Limited
First Indian Sub-Continent Edition : 2010
by New Age International (P) Ltd., Publishers

This special low-priced edition is for sale in **Middle East, India, Bangladesh, Bhutan, Myanmar (Burma), Maldives, Nepal, Pakistan and Sri Lanka.**

All rights reserved. No part of this book may be reproduced or transmitted in any form or by any means, electronic or mechanical, including photocopying, recording or by any information storage and retrieval system, without permission in writing from the Publisher.

Branches:

- 37/10, 8th Cross (Near Hanuman Temple), Azad Nagar, Chamrajpet, **Bangalore**-560 018. Tel.: (080) 2675 6823, Telefax: 2675 6820, E-mail: bangalore@newagepublishers.com
- 26, Damodaran Street, T. Nagar, **Chennai**-600 017. Tel.: (044) 24353401, Telefax: 24351463, E-mail: chennai@newagepublishers.com
- CC-39/1016, Carrier Station Road, Ernakulam South, **Cochin**-682 016. Tel.: (0484) 2377004, Telefax: 4051303. E-mail: cochin@newagepublishers.com
- Hemsen Complex, Mohd. Shah Road, Paltan Bazar, Near Starline Hotel, **Guwahati**-781 008. Tel.: (0361) 2513881. Telefax: 2543669, E-mail: guwahati@newagepublishers.com
- 105, 1st Floor, Madhiray Kaveri Tower, 3-2-19, Azam Jahi Road, Nimboliadda, **Hyderabad**-500 027. Tel.: (040) 24652456, Telefax: 24652457, E-mail:hyderabad@newagepublishers.com
- RDB Chambers (Formerly Lotus Cinema)106A, 1st Floor, S.N. Banerjee Road, **Kolkata**-700 014. Tel.: (033) 22273773, Telefax: 22275247, E-mail:kolkata@newagepublishers.com
- 16-A, Jopling Road, **Lucknow**-226 001. Tel.: (0522) 2209578, 4045297, Telefax: 2204098, E-mail: lucknow@newagepublishers.com
- 142C, Victor House, Ground Floor, N.M. Joshi Marg, Lower Parel, **Mumbai**-400 013. Tel.: (022) 24927869. Telefax: 24915415, E-mail: mumbai@newagepublishers.com
- 22, Golden House, Daryaganj, **New Delhi**-110 002. Tel.: (011) 23262370, 23262368, Telefax: 43551305. E-mail: sales@newagepublishers.com

Printed in India at Glorious Printers, Delhi.
C-10-05-4581

ISBN: 978-81-224-3058-5
Indian Edition Price: Rs. 495.00

PUBLISHING FOR ONE WORLD
NEW AGE INTERNATIONAL (P) LIMITED, PUBLISHERS
4835/24, Ansari Road, Daryaganj, New Delhi-110002
Visit us at **www.newagepublishers.com**

Preface

The control groups in Sophia University, Tsinghua University and Institute of Systems Science, Chinese Academy of Sciences have a long efficient collaboration in research of control theory and its applications to power systems and some other engineering problems.

The idea of organizing a joint workshop on robust and adaptive control appeared from a salon discussion on some cross problems met in different researches. Then through some personal communications it was responded warmly from the system and control community of both Chinese and Japanese sides. Finally the *China-Japan Joint Workshop on Advanced Robust Control and Adaptive Control Theory and Applications* had been held in Fragrant Hill Hotel Beijing, China. The workshop was sponsored technically by Technical Committee on Control Theory (TCCT), Chinese Association of Automation (CAA) (China), Committee of Electric Mathematics (CEM), Chinese Association of Electrical Engineering (CAEE) (China), and Committee of Adaptive Leaning Control (CALC), The Society of Instrument and Control Engineers(SICE) (Japan). Financially, the workshop was supported by Natural Science Foundation of China(NSFC) and Japan Science Promotion Society(JSPS).

The workshop provided a forum for control scientists and engineers from China and Japan to present their new research results, control techniques etc. and exchange their creative ideas. After two days discussion most of the participants found that the topics and the results presented on workshop are interesting, and a book for selected works could be beneficial to the world control community. That is the motivation for publishing this book.

The book contains four parts of different topics.

Part one is about the nonlinear systems and robust controls, which contains Hamiltonian realization, recursive design of controller, nonlinear internal model, stabilization of stochastic nonlinear systems, and control of grazing of multi-agent systems.

Part two considers the non-smooth systems and non-smooth control. Particular interest has been focused on switched systems. Non-smooth control design, non-smooth finite-time stabilization, $\mathcal{L}_2$ robust control of switched systems and robust control of discrete-time switched systems are investigated.

Part three concerns adaptive control and modelling. It discusses robust adaptive control, adaptive control of non-holonomic systems, date-based PID controller, nonparametric estimation, decoupling control using neural networks, and nonparametric estimation.

Part four consists of various applications of robust and/or adaptive controls. The applications include transfer conductance on power systems, water control under uncertainty, neural network technique, design effective FACTS controller, coordinated design for multi-machine systems, optimal cost minimization control, sound measure via wavelet analysis.

We are facing an information rich world. High technologies and complexity have put a heavy impact on control community. It is not only a challenge but also an opportunity as well. To meet such challenge we have to develop new control technologies, particularly, advanced robust and adaptive controls based on information and theory of complexity have to be further investigated and improved. New and advanced robust and adaptive control techniques have to be developed. We are expecting that this book is a meaningful contribution to these fields.

Finally, we would like to express our sincere thanks to NSFC and JSPS for their support, and for Professor Xiaohong Jiao and Ms. Shuangling Ding for helping in organizing papers.

Daizhan Cheng, Yuanzhang Sun
Tielong Shen, Hiromitsu Ohmori
Beijing, January 31, 2005

Contents

Part 1. Nonlinear Systems and Robust Control

Part 2. Nonsmooth and Switched System Control

Part 3. Adaptive Control and Estimation

Part 4. Applications of Nonlinear and Adaptive Control

Part 1

Nonlinear Systems and Robust Control

On Port Controlled Hamiltonian Systems

Daizhan Cheng[1], Romeo Ortega[2] and Elena Panteley[2]

[1] Institute of Systems Science, Chinese Academy of Sciences, Beijing 100080, P.R.China dcheng@iss03.iss.ac.cn
[2] Lab. des Signaux et Systémes, Supelec, Plateau du Moulon, 91192 Gif-sur-Yvette, France ortega@lss.supelec.fr

1 Introduction

The importance of the notion of passivity for analysis and control design can, nowadays, hardly be overestimated, see e.g. [13, 12, 8]. The central question of transforming a non–passive system into a passive system via state–feedback was elegantly settled in [2] where succinct, necessary and sufficient, geometric conditions are given. In spite of the unquestionable beauty of this result feedback equivalence to a (general) passive system has been used more as a conceptual framework to understand stabilization mechanisms than as an actual controller design procedure.

On the other hand, feedback equivalence to port controlled Hamiltonian (PCH) models, which are a class of passive systems, has attracted the attention of many researchers lately, in particular for stabilization objectives. A PCH system (with dissipation) is defined as [13]:

$$\begin{aligned} \dot{x} &= [J(x) - R(x)]\,\nabla H(x) + g(x)u, \quad x \in \mathbb{R}^n,\ u \in \mathbb{R}^m, \\ y &= g^\top(x)\nabla H(x), \quad y \in \mathbb{R}^m, \end{aligned} \tag{1}$$

where $H : \mathbb{R}^n \to \mathbb{R}$ is the total stored energy, $J(x) = -J^\top(x)$ is known as the interconnection matrix, $R(x) = R^\top(x) \geq 0$ represents the dissipation and $\nabla = \frac{\partial}{\partial x}$. The vector signals u and y are the conjugated port variables and their product $u^\top y$ has units of power. It is easy to see that, if the total energy function is non-negative, then PCH systems are passive.

As explained in [13] PCH systems constitute an extension of classical Hamiltonian and Euler–Lagrange models that naturally incorporate interaction with the environment (through power port variables) and capture the essential physical property of power conservation. Given this nice features of PCH models it is then natural to ask when is a general nonlinear system transformable into a PCH system? The investigation of this question is the topic of interest of the present work.

The rest of the paper is organized as follows: Section 2 formulates the problem. Two equivalent conditions are proved in Section 3. Section 4 considers the munimum number of linear PDEs for the solvability of PCH equivalence. Global equivalence is studied in Section 5. An illustrative example is given in Section 6. Section 7 gives some concluding remarks.

2 Problem Formulation

Given an affine system

$$\Sigma_{f,G}: \quad \dot{x} = f(x) + G(x)u, \quad x \in \mathbb{R}^n,\ u \in \mathbb{R}^m,\ m < n \tag{2}$$

and the matrix $J(x) - R(x)$, when does there exists a state feedback $\beta : \mathbb{R}^n \to \mathbb{R}^m$ and an energy function $H : \mathbb{R}^n \to \mathbb{R}$ such that

$$f(x) + G(x)\beta(x) = [J(x) - R(x)]\, \nabla H(x). \tag{3}$$

In this case, we say that the system $\Sigma_{f,G}$ is feedback equivalent to a PCH system (with given interconnection and damping matrices.)

Remark 2.1 Transforming a system to be controlled into a PCH system is the central idea of the Interconnection and Damping Assignment Passivity-based Control method firstly introduced in [9], where (3) is viewed as a set of partial differential equations (PDE's), parameterized in $J(x)$ and $R(x)$, to be solved—see Remark 3.3. A summary of some recent developments may be found in [9], see also [3, 14] for its applications to power systems. In [5], the problem is addressed from an alternative perspective. That is, in equation (3) we fix $H(x)$ and solve the resulting algebraic equations for $J(x)$, $R(x)$ and $\beta(x)$. See also [7, 1] for the case of feedback equivalence to a Lagrangian system.

Remark 2.2 As will become clear in the sequel the particular structure of the matrix $J(x) - R(x)$ does not play any role in the characterization of the class of feedback equivalent systems, therefore we will address the slightly more general problem of feedback equivalence to a pseudo–gradient system. That is, instead of the PCH model (1) we will consider pseudo–gradient systems

$$\Sigma_F \ : \ \dot{x} = F(x)\nabla H(x)$$

where $F(x)$ is a fixed, but otherwise arbitrary, $n \times n$ matrix. This yields, instead of (3), the matching equation

$$f(x) + G(x)\beta(x) = F(x)\nabla H(x), \tag{4}$$

If (4) holds, we will say that the system $\Sigma_{f,G}$ is feedback equivalent to Σ_F.

Before giving our solution to this problem a word on notation is in order.

- All vectors, including the gradient, are row vectors.
- For all vectors and matrices which are functions of x we will write explicitly this dependence only the first time they are defined.
- Throughout the paper we will assume that all functions are sufficiently smooth.
- Finally, no particular attention is given to the characterization of the domain of validity of our statements, to which the local qualifier should be attached. The global equivalence is discussed in Section 5. Where you can see that global rank condition is not enough to assure a global equivalence.

3 Two Equivalent Conditions for Feedback Equivalence

If the matrix F is full rank, Poincare's Lemma give us directly a necessary and sufficient condition for feedback equivalence. Indeed, the vector field $F^{-1}(f+G\beta)$ is a gradient vector field, that is, (4) is satisfied for some scalar function H, if and only if

$$\nabla[F^{-1}(f+G\beta)] = \left(\nabla[F^{-1}(f+G\beta)]\right)^{\top}. \tag{5}$$

The latter condition—for fixed F, f and G—translates into $\frac{n}{2}(n-1)$ PDE's in terms of β. This was the method proposed in [9]. One of the objectives of this note is to show that we can significantly reduce the number of PDE's to be solved, therefore simplifying the associated computational problem. Actually, we will identify the minimal number of PDE's that needs to be solved to achieve the feedback equivalence.

Before presenting our characterization we recall a basic linear algebra lemma.

Lemma 3.1 *Consider two linear subspaces $\mathcal{S}_1, \mathcal{S}_2 \subset \mathbb{R}^n$. If,* $\dim \mathcal{S}_1 = \dim \mathcal{S}_2$ *and $\mathcal{S}_1 \subset \mathcal{S}_2$ (or $\mathcal{S}_2 \subset \mathcal{S}_1$), then $\mathcal{S}_1 = \mathcal{S}_2$.*

We need the following standard assumption.

Assumption A.1 For all points $p \in \mathbb{R}^n$ there exists an open and simply connected neighborhood $\mathcal{N}_p$ such that for all $x \in \mathcal{N}_p$ we have $\operatorname{rank} G(x) = m$. Without loss of generality, we partition[3]

$$G(x) = \begin{bmatrix} G_1(x) \\ G_2(x) \end{bmatrix}, \quad G_2(x) \in \mathbb{R}^{m\times m}, \tag{6}$$

where $\operatorname{rank} G_2(x) = m$ for all $x \in \mathcal{N}_p$.

[3]This partition can always be locally achieved simply swapping and relabelling the state equations.

Remark 3.2 Note that under assumption A.1, the feedback control for PCH equivalence can be uniquely determined from (4) as

$$u(x) = \beta(x) = [G^\top(x)G(x)]^{-1}G^\top(x)\,[F(x)\nabla H(x) - f(x)]\,. \tag{7}$$

This control will be used through the paper.

Proposition 3.3 *Under Assumption A.1, the following statements are equivalent*

1. *$\Sigma_{f,G}$ is feedback equivalent to Σ_F.*
2. *The n PDE's*

$$[I_n - \Pi_G(x)][F(x)\nabla H(x) - f(x)] = 0 \tag{8}$$

admit a solution, where

$$\Pi_G(x) = G(x)[G^\top(x)G(x)]^{-1}G^\top(x).$$

3. *The $n-m$ PDE's*

$$[F_1(x) - G_1(x)G_2^{-1}(x)F_2(x)]\nabla H(x) = f_1(x) - G_1(x)G_2^{-1}(x)f_2(x) \tag{9}$$

admit a solution, where

$$F(x) = \begin{bmatrix} F_1(x) \\ F_2(x) \end{bmatrix}, \quad F_1(x) \in \mathbb{R}^{(n-m)\times n}, \quad f(x) = \begin{bmatrix} f_1(x) \\ f_2(x) \end{bmatrix}, \quad f_1(x) \in \mathbb{R}^{n-m}.$$

Proof. [i) $\Rightarrow$ iii)] Define the $(n-m)\times n$ matrix[4]

$$\xi(x) = \begin{bmatrix} I_{n-m} & -G_1(x)G_2^{-1}(x) \end{bmatrix}. \tag{10}$$

We have the following chain of implications

$$\begin{aligned} i) &\Leftrightarrow f + G\beta = F\nabla H \\ &\Rightarrow \xi(f + G\beta) = \xi F\nabla H \\ &\Leftrightarrow \xi(F\nabla H - f) = 0 \\ &\Leftrightarrow iii) \end{aligned} \tag{11}$$

[iii) $\Rightarrow$ i)] This will be established by contradiction. Assume iii) does not hold. From (9), or (11), we see that this is equivalent to saying that

$$F\nabla H - f \ni \text{Ker}\ \xi, \tag{12}$$

We will prove now that Ker ξ = Im G. First, note that both spaces have the same dimension, m. Consider then the chain of implications:

[4]Clearly, ξ is a left annihilator of G, that is $\xi G = 0$.

$$\begin{aligned} a \in \operatorname{Im} G \Leftrightarrow \exists b \in \mathbb{R}^n \ &: \ a = Gb \\ &\Rightarrow \xi a = \xi G b = 0 \\ &\Rightarrow a \in \operatorname{Ker} \xi \\ &\Rightarrow \operatorname{Im} G \subset \operatorname{Ker} \xi \end{aligned}$$

Finally, we can invoke Lemma 3.1 to conclude that $\operatorname{Ker} \xi = \operatorname{Im} G$. From the proof above we have that (12) is equivalent to

$$F\nabla H - f \ni \operatorname{Im} G,$$

but the latter contradicts i), that states the existence of β such that $F\nabla H - f = G\beta$.

[iii) $\Leftrightarrow$ ii)] To prove this equivalence we will establish that

$$\operatorname{Ker} \xi = \operatorname{Ker} (I_n - \Pi_G),$$

which, together with (8) and (11), shows that the set of solutions of both PDE's are the same—completing the proof. Towards this end, define two $n \times n$ matrices

$$C(x) = \left[\xi^\top(x)[\xi(x)\xi^\top(x)]^{-1} \ G(x)\right], \quad D(x) = \begin{bmatrix} \xi(x) \\ (G^\top(x)G(x))^{-1}G^\top(x) \end{bmatrix}.$$

Now, using $\xi G = 0$ we have that $DC = I_n$, and consequently $D = C^{-1}$. This also implies that $CD = I_n$, which doing the computations is equivalent to

$$\Pi_{\xi^\top}(x) = I_n - \Pi_G(x). \tag{13}$$

where we have defined the projector matrix

$$\Pi_{\xi^\top}(x) = \xi^\top(x)[\xi(x)\xi^\top(x)]^{-1}\xi(x).$$

We will prove now, by contradiction, that $\operatorname{rank} \Pi_{\xi^\top} = n - m$. Assume $\operatorname{rank} \Pi_{\xi^\top} < n - m$. We have the following set of equations, that lead to a contradiction,

$$\begin{aligned} n - m &= \operatorname{rank} \xi^\top \\ &= \operatorname{rank} \Pi_{\xi^\top}\xi^\top \\ &\leq \min\{\operatorname{rank} \Pi_{\xi^\top}, \operatorname{rank} \xi^\top\} \\ &< n - m, \end{aligned}$$

where we used $\Pi_{\xi^\top}\xi^\top = \xi^\top$ for the second identity, and to obtain the third line we invoked the fact that, for any pair of conformal matrices A, B,

$$\operatorname{rank} AB \leq \min\{\operatorname{rank} A, \operatorname{rank} B\}.$$

Therefore, $\operatorname{rank} \Pi_{\xi^\top} = n - m$ and consequently $\dim \operatorname{Ker} \Pi_{\xi^\top} = m$.

To conclude the proof we recall that $\dim \operatorname{Ker}\, \xi = m$, and given that $\operatorname{Ker}\, \xi \subset \operatorname{Ker}\, \Pi_{\xi^\top}$, we have that

$$\begin{aligned} \operatorname{Ker}\, \xi &= \operatorname{Ker}\, \Pi_{\xi^\top} \\ &= \operatorname{Ker}\, [I_n - \Pi_G] \end{aligned}$$

where we have invoked Lemma 3.1 for the first identity and (13) for the second.
□

Remark 3.4 The proposition establishes the interesting fact that the set of solutions of the n PDE's (8) exactly coincides with the set of solutions of the $n - m$ PDE's (9)—equivalently, (11). (The lack of such a formal statement was a source of some confusion on the literature, see e.g., [9].) The proposition also gives alternative parameterizations of the matching equation (4) that complements the original proposal of [9] to solve the $\frac{n}{2}(n-1)$ PDE's (5).

Remark 3.5 Although the left annihilator matrix of G is not uniquely defined the proposition remains unaffected if we choose a matrix different from (10).[5] Indeed, let us take any left annihilator, say $\bar{\xi}(x)$, of G, and partition it as

$$\bar{\xi}(x) = \left[\bar{\xi}_1(x)\; \bar{\xi}_2(x)\right], \quad \bar{\xi}_1(x) \in \mathbb{R}^{(n-m)\times(n-m)}.$$

The left annihilator condition $\bar{\xi}G = 0$ imposes the relationship $\bar{\xi}_2 = -\bar{\xi}_1 G_1 G_2^{-1}$ that leads to the factorization

$$\bar{\xi} = \bar{\xi}_1 \left[I_{n-m}\; -G_1 G_2^{-1}\right].$$

For all full–rank matrices $\bar{\xi}_1$, $\bar{\xi}$ has the same kernel as ξ, whence the proposition remains unaltered by the use of this "new" left annihilator.

Remark 3.6 The necessity of (8) for feedback equivalence can be easily established as follows. (4) implies that

$$G^\top G\beta = G^\top(F\nabla H - f),$$

which together with invertibility of[6] $G^\top G$ defines, uniquely, the control β. Equation (8) is then obtained plugging the expression of β in (4).

Remark 3.7 It is easy to prove that $\operatorname{Im}\, \xi^\top$ is the orthogonal complement of $\operatorname{Im} G$, that is

$$\operatorname{Im} G \oplus \operatorname{Im}\, \xi^\top = \mathbb{R}^n,$$

with $\oplus$ denoting direct sum. This stems from the fact that $\Pi_{\xi^\top}$ is an orthogonal projector (over the rows of ξ), whence

$$\operatorname{Im}\, \Pi_{\xi^\top} \oplus \operatorname{Ker}\, \Pi_{\xi^\top} = \mathbb{R}^n, \quad \operatorname{Im}\, \Pi_{\xi^\top} = \operatorname{Im}\, \xi^\top, \quad \operatorname{Ker}\, \Pi_{\xi^\top} = \operatorname{Ker}\, \xi$$

and that, as shown in the proof above, $\operatorname{Ker}\, \xi = \operatorname{Im} G$. See Section 5 of this paper or Section 5.8 of [6] for further details on projectors (also known as idempotent matrices.)

[5] Of course, as a space, the orthogonal complement of Im G is uniquely defined. See Remark 11.

[6] This follows from the fact that $G^\top G$ is the Gram matrix of a set of linearly independent vectors.

4 Reducing the Number of PDE's

In this section we prove that it is possible to further reduce the number of PDE's to be solved for feedback equivalence and we identify the minimal number required. Towards this end, and to simplify notation, we rewrite (9) in the form

$$W(x)\nabla H(x) = s(x) \tag{14}$$

where we have defined

$$W(x) = F_1(x) - G_1(x)G_2^{-1}(x)F_2(x), \quad s(x) = f_1(x) - G_1(x)G_2^{-1}(x)f_2(x).$$

Assumption A.2 For all points $p \in \mathbb{R}^n$ there exists an open and simply connected neighborhood $\mathcal{N}_p$ such that, for all $x \in \mathcal{N}_p$, $\operatorname{rank} W(x) = \ell$ and assume

A.2.1 $\ell < n - m$, **A.2.2** $s \in \operatorname{Im} W$

Clearly, if $\ell = n - m$ we are in the situation of iii) of Proposition 1. Hence, to reduce the number of PDE's we need Assumption A.2.1. Furthermore, if the latter holds, Assumption A.2.2 is necessary for solvability of (14).

It is well known [6] that there exists an $n \times n$ nonsingular matrix $Q(x)$ that extracts the full–rank part of W. That is, such that

$$W(x)Q(x) = \left[W_0(x) \; 0_{(n-m)\times(n-\ell)}\right], \tag{15}$$

where $W_0 \in \mathbb{R}^{(n-m)\times\ell}$ verifies $\operatorname{rank} W_0 = \ell$. Let us partition

$$Q^{-1}(x) = \begin{bmatrix} M(x) \\ N(x) \end{bmatrix}, \quad M(x) \in \mathbb{R}^{\ell \times n}. \tag{16}$$

Proposion 4.1 *Under Assumptions A.1 and A.2, the following statements are equivalent*

1. *$\Sigma_{f,G}$ is feedback equivalent to Σ_F.*
2. *The ℓ PDE's*

$$M(x)\nabla H(x) = [W_0^\top(x)W_0(x)]^{-1}W_0^\top(x)s(x) \tag{17}$$

admit a solution.

Furthermore, the equation (17) is independent of the choice of the matrix Q in the construction of (15).

Proof. First, notice that since $\operatorname{Im} W = \operatorname{Im} W_0$ we have that Assumption A.2.2 can be equivalently stated as $s \in \operatorname{Im} W_0$. Now, this condition implies that

$$\Pi_{W_0}(x)s(x) = s(x) \tag{18}$$

where

$$\Pi_{W_0}(x) = W_0(x)[W_0^\top(x)W_0(x)]^{-1}W_0^\top(x)$$

is an orthogonal projector (over the columns of W_0.)

We have established in Proposition 1 that $\Sigma_{f,G}$ is feedback equivalent to Σ_F if and only if (9), or equivalently (14), hold. We then have the following set of equivalences

$$\begin{aligned} W\nabla H = s &\Leftrightarrow WQQ^{-1}\nabla H = s \\ &\Leftrightarrow \left[W_0\ 0_{(n-m)\times(n-\ell)}\right] \begin{bmatrix} M \\ N \end{bmatrix} \nabla H = s \\ &\Leftrightarrow W_0 M \nabla H = \Pi_{W_0} s \\ &\Leftrightarrow (17), \end{aligned}$$

where we have used (15) and (16) to get the second equivalence, (18) in the third, and the full rank condition of W_0 for the latter. This establishes the equivalence between i) and ii).

Assume now that the extraction of the full–rank part of W is done with another matrix $\bar{Q}(x)$, that is, instead of (15) we have

$$W(x)\bar{Q}(x) = \left[\bar{W}_0(x)\ 0_{(n-m)\times(n-\ell)}\right], \tag{19}$$

where $\bar{W}_0 \in \mathbb{R}^{(n-m)\times\ell}$ verifies rank $\bar{W}_0 = \ell$. We note that Im W_0 = Im $\bar{W}_0$, therefore we can also construct a projector $\Pi_{\bar{W}_0}$ that leaves s invariant as in (18). It can actually be shown that the two projectors are the same, that is $\Pi_{W_0} = \Pi_{\bar{W}_0}$, see pp. 197 of [6]. Mimicking the steps of the proof with the new matrix $\bar{Q}$ we obtain in the one before the last step

$$\bar{W}_0 \bar{M} \nabla H = \Pi_{W_0} s,$$

where we have partitioned

$$\bar{Q}^{-1} = \begin{bmatrix} \bar{M} \\ \bar{N} \end{bmatrix}, \quad \bar{M} \in \mathbb{R}^{\ell\times n}.$$

Now, from (19) and (15) one obtains

$$\left[\bar{W}_0\ 0_{(n-m)\times(n-\ell)}\right]\bar{Q}^{-1} = \left[W_0\ 0_{(n-m)\times(n-\ell)}\right]Q^{-1}$$

which proves that $\bar{W}_0\bar{M} = W_0 M$, completing the proof. □

Remark 4.2 Although the construction of the matrix Q is standard, for the sake of completeness, we outline here the procedure. Referring to (16), select the $\ell \times n$ matrix M such that its ℓ rows span Im W, this can be done choosing elements proportional to the euclidian orthonormal basis of $\mathbb{R}^\ell$. Then, select and $(n-\ell)\times n$ matrix N, also from the orthonormal basis, to complete the rank. It is easy to see that this construction yields the desired decomposition (15).

5 Global Equivalence

This section considers global equivalence to PCH. We assume the following, which is a global version of A.1:

Assumption A.3 There exists an $n \times (n-m)$ matrix $\Psi(x)$ such that

$$B(x) := \left[G_(x)\ \Psi(x)\right], \quad x \in \mathbb{R}^n$$

is nonsingular.

Next, we define

$$E(x) := \left[I_n - \prod_G(x)G_(x)\right] = \left[I - G(x)(G(x)^\top G(x))^{-1}G(x)^\top\right].$$

We first claim that $E(x)$ has constant rank.

Lemma 5.1 *Assume A1 holds. Then*

$$\mathrm{rank}(E(x)) = n - m. \tag{20}$$

Proof. From A1 one sees that $g_1(x), \cdots, g_m(x)$ are linearly independent. It is easy to see that

$$E(x)\left(g_1(x)\ \cdots\ g_m(x)\right) = 0.$$

So $\mathrm{rank}(E(x)) \leq n - m$.

Denote by $\mathcal{G}(x) = \mathrm{Span}\{g_1(x), \cdots, g_m(x)\}$. Then for any vector field $Z(x) \in \mathcal{G}^\perp(x)$

$$E(x)Z(x) = Z(x), \quad \forall Z(x) \in \mathcal{G}^\perp(x).$$

That is, *Span* $\mathrm{col}\{E(x)\} \supset \mathcal{G}^\perp(x)$ Hence $\mathrm{rank}(E(x)) \geq n-m$. Therefore, (20) follows. □

Lemma 5.2 *Assume A3 holds. Then there exists an orthogonal matrix $P(x)$, $x \in \mathbb{R}^n$, such that*

$$P(x)^{-1}E(x)P(x) = \begin{bmatrix} I_{n-m} & 0 \\ 0 & 0 \end{bmatrix}, \quad x \in \mathbb{R}^n. \tag{21}$$

Proof. Using A3, we denote

$$B^{-1}(x) = \begin{bmatrix} * \\ \xi_1^T(x) \\ \vdots \\ \xi_{n-m}^T(x) \end{bmatrix}, \quad x \in \mathbb{R}^n,$$

where $*$ is the first m less important rows. Now it is clear that

$$\mathcal{G}^\perp(x) = \mathrm{Span}\{\xi_1(x), \cdots, \xi_{n-m}(x)\}, \quad x \in \mathbb{R}^n.$$

Moreover, from the proof of Lemma 5.1, we have

$$E(x)\xi_i(x) = \xi_i(x), \quad i = 1, \cdots, n-m, \quad x \in \mathbb{R}^n; \tag{22}$$

and

$$E(x)g_i(x) = 0, \quad i = 1, \cdots, m, \quad x \in \mathbb{R}^n. \tag{23}$$

Normalizing $\{\xi_1(x), \cdots, \xi_{n-m}(x)\}$ by Gram-Schmidt orthogonalization algorithm as

$$\bar{\xi}_1(x) = \xi_1(x)/\|\xi_1(x)\|;$$

$$\begin{cases} \tilde{\xi}_k(x) = \xi_k(x) - \langle \xi_k(x), \bar{\xi}_1(x) \rangle \bar{\xi}_1(x) - \cdots - \langle \xi_k(x), \bar{\xi}_{k-1}(x) \rangle \bar{\xi}_{k-1}(x), \\ \bar{\xi}_k(x) = \tilde{\xi}_k(x)/\|\tilde{\xi}_k(x)\|, \quad k = 2, \cdots, n-m. \end{cases}$$

Similarly, $g_1(x), \cdots, g_m(x)$ can be normalized as $\bar{g}_1(x), \cdots, \bar{g}_m(x)$.

Define

$$P(x) = \left(\bar{\xi}_1(x), \cdots, \bar{\xi}_{n-m}(x)\ \bar{g}_1(x) \cdots \bar{g}_m(x)\right), \quad x \in \mathbb{R}^n.$$

Then it is ready to check that $P(x)$ is an orthogonal matrix and (21) holds. □

Now left multiplying (8) by $P^{-1}(x)$ yields

$$\begin{bmatrix} I_{n-m} & 0 \\ 0 & 0 \end{bmatrix} P^{-1}(x) \left[F(x)\frac{\partial H}{\partial x} - f(x) \right] = 0. \tag{24}$$

Denote

$$P^{-1}(x)F(x) := \begin{pmatrix} W_{11}(x) & W_{12}(x) \\ W_{21}(x) & W_{22}(x) \end{pmatrix}, \tag{25}$$

where $W_{11}(x)$ is an $(n-m) \times (n-m)$ matrix, the other blocks have corresponding dimensions. Similarly, decompose

$$P^{-1}(x)f(x) := \begin{pmatrix} s(x) \\ s^c(x) \end{pmatrix}, \tag{26}$$

$s(x)$ is the first $n-m$ components.

Now we need the global version of assumption A2 as

Assumption A.4 Let

$$W := \left[W_{11}(x)\ W_{12}(x) \right].$$

Then rank $W(x) = \ell$, $x \in \mathbb{R}^n$ and assume

A.4.1 $\ell < n-m$, **A.4.2** $s(x) \in \operatorname{Im} W$;

A.4.3 there exists a nonsingular matrix $Q(x)$, such that

$$(W(x))\, Q(x) = \left(W_0(x)\ 0\right), \quad x \in \mathbb{R}^n, \tag{27}$$

where $W_0(x)$ is an $(n-m) \times \ell$ matrix of rank ℓ.

Then Proposition 4.1 becomes a global result:

Corollary 5.3 *Assume A3 and A4. Then Proposition 4.1 is globally true on* $\mathbb{R}^n$

6 An Illustrative Example

Consider global equivalence to PCH of the following system

$$\dot{x} = \begin{bmatrix} \cos x_2(\cos x_2 - \sin x_2) + 1 \\ \sin x_2(\cos x_2 - \sin x_2) \\ -\cos x_2 - \sin^2 x_2 \end{bmatrix} \phi(x) + \begin{bmatrix} \cos x_2 \\ \sin x_2 \\ \cos x_2 - \sin x_2 \end{bmatrix} u, \tag{28}$$

where $x \in \mathbb{R}^3$ and $\phi(x) = x_3^3 - x_1$. For notational compactness, we denote

$$S = \sin x_2, \quad C = \cos x_2, \quad \mu = \sqrt{2 - 2SC}.$$

Choosing

$$\Psi(x) = \begin{bmatrix} 1 & S \\ -1 & -C \\ -1 & 0 \end{bmatrix},$$

it is easy to check that $B(x) = [g(x) \quad \Psi(x)]$ is nonsingular and

$$B^{-1}(x) = \frac{1}{\mu^2} \begin{bmatrix} C & S & -S + C \\ C(C-S) & S(C-S) & -1 \\ 2S - C & S - 2C & C + S \end{bmatrix}.$$

Hence

$$\xi_1 = \frac{1}{\mu^2} \begin{bmatrix} C(C-S) \\ S(C-S) \\ -1 \end{bmatrix}, \quad \xi_2 = \frac{1}{\mu^2} \begin{bmatrix} 2S - C \\ S - 2C \\ C + S \end{bmatrix}$$

Normalizing them, we have

$$\bar{\xi}_1 = \frac{1}{\mu} \begin{bmatrix} C(C-S) \\ S(C-S) \\ -1 \end{bmatrix}, \quad \bar{\xi}_2 = \begin{bmatrix} S \\ -C \\ 0 \end{bmatrix}.$$

Then

$$P = [\bar{g} \; \bar{\xi}_1 \; \bar{\xi}_2] = \begin{bmatrix} \frac{C(C-S)}{\mu} & S & -\frac{C}{\mu} \\ \frac{S(C-S)}{\mu} & -C & \frac{S}{\mu} \\ -\frac{1}{\mu} & 0 & \frac{C-S}{\mu} \end{bmatrix},$$

and $P^{-1} = P^T$.

Now assume the structure required is

$$J = \begin{bmatrix} 0 & (-SC - S^2)/2 & S^2 - 0.5 \\ (SC + S^2)/2 & 0 & -S^2 - SC \\ 0.5 - S^2 & S^2 + SC & 0 \end{bmatrix}, \tag{29}$$

$$R = \begin{bmatrix} C^2 + S^2 - CS & S(C-S)/2 & -S^2 - 0.5 \\ S(C-S)/2 & S^2 & (3SC - S^2)/2 \\ -S^2 - 0.5 & (3SC - S^2)/2 & 2 + 2(C-S)^2 \end{bmatrix}. \tag{30}$$

Then

$$P^{-1}F(x) = P^{-1}(J - R) = \begin{bmatrix} -\mu & 0 & 2\mu \\ -S & 0 & 2S \\ 0 & -S\mu & 2(S - C)\mu \end{bmatrix}.$$

It follows that

$$W = \begin{bmatrix} -\mu & 0 & 2\mu \\ -S & 0 & 2S \end{bmatrix}.$$

$$\ell = \operatorname{rank}(W) = 1.$$

It is easy to calculate that

$$P^{-1}f(x) = \begin{bmatrix} \mu \\ S \\ 2C - 2C^2S \end{bmatrix} \phi(x)$$

So

$$s(x) = \begin{pmatrix} \mu \\ S \end{pmatrix} \phi(x) \in \operatorname{Span}\{W\}.$$

Now all the conditions of A3 and A4 are satisfied. So the only thing we have to check is the solvability of (17). Now

$$Q(x) = Q = \begin{bmatrix} 1 & 0 & 2 \\ 0 & 1 & 0 \\ 0 & 0 & 1 \end{bmatrix}, \quad W_0 = \begin{bmatrix} -\mu \\ -S \end{bmatrix}$$

$$Q^{-1} = \begin{bmatrix} 1 & 0 & -2 \\ 0 & 1 & 0 \\ 0 & 0 & 1 \end{bmatrix}, \quad M(x) = M = \begin{bmatrix} 1 & 0 & -2 \end{bmatrix}.$$

Equation (17) can be easily obtained as

$$\frac{\partial H}{\partial x_1} - 2\frac{\partial H}{\partial x_3} = x_1 - x_3^3. \tag{31}$$

An obvious solution is:

$$H(x) = \frac{1}{2}x_1^2 + x_2^2 + \frac{1}{8}x_3^4. \tag{32}$$

So system (28) has a feedback Hamiltonian equivalent form with Structure matrices J, R and Hamiltonian function H as in (29), (30) and (32) respectively.

7 Concluding Remarks

We have investigated in this paper the problem of feedback equivalence—via state–feedback control—of affine systems to PCH systems. Such a transformation lies at the core of some stabilization techniques recently reported in the literature. Necessary and sufficient conditions, expressed in terms of solvability of sets of PDE's, are given. As is well–known solving PDE's is not an easy task, therefore, particular emphasis is given to the possibility of reducing their number, which turns out to be determined by a simple rank condition.

Several variations to the feedback equivalence problem studied here are of practical interest and have been studied in the literature. For instance, in some cases it is reasonable to make the interconnection matrix J dependent on the control—e.g., in switched systems like power converters this is the natural way of modelling [4]. In this case, the matching equation (3) becomes

$$f(x) + G(x)\beta(x) = [J(x, \beta(x)) - R(x)] \nabla H(x),$$

leading to a completely different characterization of feedback equivalence. This approach has been adopted in [11] yielding some practically interesting stabilizing control laws.

One final remark concerning feedback equivalence to PCH system, which puts it in sharp contrast to feedback linearization, is the invariance vis—vis change of coordinates of the former. That is, as the PCH structure is invariant to change of coordinates [3, 5], if $\beta(x)$ is a control law that transforms the system into PCH form, then for any diffeomorphism $x = \Phi(z)$, the control $\beta(\Phi(z))$ also endows the system with a PCH structure (in the coordinates z.)

References

1. G. Blankenstein, R. Ortega and A. van der Schaft, The matching conditions of controlled lagrangians and interconnection and damping passivity–based control. *Int. J. of Control*, vol. 75, no. 9, pp. 645–665, 2002.
2. C. Byrnes, A. Isidori and J.C. Willems, Passivity, feedback equivalence, and the global stabilization of minimum phase nonlinear systems, *IEEE Trans Aut. Cont.*, Vol.36, no.11, pp. 1228–1240, 1991.
3. D. Cheng, T. Shen, T.J. Tarn, Pseudo-Hamiltonian realization and its application, *Comm. Inform. Sys.*, Vol.2, No.2, pp. 91-120, 2002.
4. G. Escobar, A. van der Schaft and R. Ortega, A Hamiltonian viewpoint in the modeling of switching power converters, *Automatica*, Vol. 1999.
5. K. Fujimoto and T. Sugie, Canonical transformations and stabilization of generalized hamiltonian systems. *Sys. Contr. Lett.*, vol. 42, no. 3, pp.217–227, 2001.
6. P. Lancaster and M. Tismenetsky, *The theory of matrices*, Academic Press, NY, 1985.
7. A. Bloch, N. Leonhard and J. Marsden, Controlled lagrangians and the stabilization of mechanical systems. *IEEE Trans. Aut. Contr.*, Vol. 45, no. 12, pp. 2253–2270, Dec. 2000.

8. R. Ortega, A. Loria, P.J. Nicklasson and H. Sira-Ramirez, *Passivity–based control of Euler–Lagrange systems*, Springer-Verlag, Berlin, Communications and Control Engineering, Sept. 1998.
9. R. Ortega, A. van der Schaft, B. Maschke and G. Escobar, Interconnection and damping assignment passivity–based control of port– controlled hamiltonian systems. *Automatica*, vol. 38, no. 4, April 2002.
10. R. Ortega, Some applications and extensions of interconnection and damping assignment passivity-based control, *Proc. IFAC Wokshop. on Lagrangian and Hamiltonian methods in nonlinear systems*, Seville, Spain, April 3-5, 2003.
11. H. Rodriguez, R. Ortega, G. Escobar and N. Barabanov, A Robustly Stable Output Feedback Saturated Controller for the Boost DC–to–DC Converter, *Sys. Contr. Lett.*, Vol. 40, No. 1, pp. 1 -8, May 2000.
12. R. Sepulchre, M. Jankovic and P.V. Kokotović, **Constructive nonlinear control**, Springer-Verlag, Berlin, 1996.
13. A. J. van der Schaft, *L_2-gain and passivity techniques in nonlinear control*, Berlin, Springer, 1999.
14. Y. Wang, D. Cheng, C. Li, and Y. Ge, Dissipative H amiltonian realization and energy-based L_2 disturbance attenuation control of multimachine power systems, *IEEE Trans. Aut. Contr.*, Vol.48, No.8, pp.1428-1433, 2003.

Recursive Design for Adaptive Stabilizing Controller of Nonlinear Time-Delay Systems

Xiaohong Jiao[1], Yuanzhang Sun[1] and Tielong Shen[2]

[1] Department of Electrical Engineering, Tsinghua University, Beijing 100084, China jiaoxh@tsinghua.edu.cn, yzsun@tsinghua.edu.cn
[2] Department of Mechanical Engineering, Sophia University,Tokyo 102-8554, Japan tetu-sin@sophia.ac.jp

1 Introduction

In this paper, we consider a class of uncertain nonlinear time-delay systems described by the form

$$\begin{cases} \dot{x}_1 = x_2 + \varphi_1(\tilde{x}_1)\theta + e_1(\tilde{x}_1)d_1(x_t) \\ \dot{x}_2 = x_3 + \varphi_2(\tilde{x}_2)\theta + e_2(\tilde{x}_2)d_2(x_t) \\ \quad \vdots \\ \dot{x}_n = u + \varphi_n(\tilde{x}_n)\theta + e_n(\tilde{x}_n)d_n(x_t) \end{cases} \tag{1}$$

where $x = [x_1 \cdots x_n]^T \in R^n$ is the state, $\tilde{x}_i = [x_1 \cdots x_i]^T$, $(i = 1, \cdots, n)$ and $u \in R$ the control input. $x_t := x(t + \tau)$, $\tau \in [-r, 0]$ describes the state-function at time t, which is the piece of trajectories between time $t - r$ and t, i.e. $x_t \in C$ for a given t, where $C = C([-r, 0], R^n)$ represents the space of continuous functions that maps the interval $[-r, 0]$ into R^n and $r > 0$ is a constant representing the largest value of delay. $x_0 =: \phi(\tau) \in C$ is the initial condition. $\theta \in R^p$ denotes unknown parameter vector. $\varphi_i(\cdot)$ and $e_i(\cdot)$ are known smooth functions with appropriate dimensions and $\varphi_i(0)=0$, $e_i(0)=0$ $(i=1, \cdots, n)$. The delay related terms $d_i(\cdot)$ are unknown functions satisfying

$$|d_i(x_t)| \leq |x_{1t}|, \qquad i = 1, \cdots, n \tag{2}$$

The main purpose of this paper is to develop a design approach to adaptive stabilizing controller for the system (1), i.e. we will provide a design method of the feedback controller in the form

$$\begin{cases} u = \alpha(x(t), \hat{\Theta}) \\ \dot{\hat{\Theta}} = \beta(x(t), \hat{\Theta}) \end{cases} \tag{3}$$

with $\alpha(0, \hat{\Theta}) = 0$, such that the closed loop system coordinated by $(x, \hat{\Theta})$ is stable and $x(t) \to 0$ as $t \to \infty$ for any given initial conditions $\phi(\cdot)$ and $\hat{\Theta}(0)$, where $\hat{\Theta}$ is parameter estimate.

As is well-known, a number of papers have investigated the adaptive control problem for nonlinear systems without time-delay (see e.g. [1, 2, 3, 4]). The most successful result for nonlinear adaptive controller design is the Lyapunov-based method in which the convergence of the system state and the boundedness of the parameter estimate are guaranteed by the Lyapunov function satisfying LaSalle-Yoshizawa theorem [1]. It has been shown that such a Lyapunov function can be constructed by recursive method if the system can be transformed to the strict parametric-feedback form under appropriate geometric condition [1, 2].

For time-delay systems, the analysis and synthesis of the stability rely on the Lyapunov functional [5, 6, 7, 8]. Simultaneously, the Lyapunov function is also used to prove the stability of time-delay systems [5, 6, 9, 10] due to the complicated construction of the functional. Therefore, the recursive method for the stabilization of the system (1) can be given based on either the Lyapunov functional or the Lyapunov function. Motivated by the Lyapunov design framework for nonlinear nondelay systems, we have proposed a Lyapunov-Razumikhin function-based design scheme for adaptive stabilization problem of nonlinear time-delay systems [11]. A main contribution of the work in [11] is the development of an essential lemma, which provides a sufficient condition for distinguishing the invariant set where the partial states converge to zero and the whole states are bounded. Using this lemma, we are able to design an adaptive stabilizing controller because the specification of the adaptive stabilization is to make the partial states, namely the system states, converge to zero and the remained states, which present the parameter estimates, be bounded. In [11], the domination design technique is employed to design the adaptive controller based on the presented condition. The key point of this design scheme is to dominate time derivative of the Lyapunov-Razumikhin function to be negative along the trajectories of the whole system whenever the Razumikhin condition holds. It has been shown in [11] that this domination design can be applied to the nonlinear time-delay systems with the cascaded structure.

Thus, a natural question is that is it possible to extend the domination design technique to the nonlinear systems with the integrator-chained structure such as the form (1). In this paper, we will first present a positive answer provided the structural condition (2) holds. The design scheme presented in this paper is really a backstepping approach, however, the significant feature of our design approach is that the subsystem forced by the virtual feedback control law at each step, which is independent of the state-delayed, is not necessary stable as one did in the usual backstepping design for stabilization. Furthermore, it should be noted that a recursive design approach has been proposed in [9] to solve the robust stabilization problem for nonlinear time-delay systems with strict-feedback form by using the Lyapunov-Krasovskii functional. However, it was later commented by [10] that the stabilizing controller proposed could not be obtained constructively. On the other hand, the

Lyapunov-Krasovskii functional is difficult in constructing than the Lyapunov function and it is not easy to be extended to the adaptive control design.

2 Technical Lemma for Adaptive Stabilization

Usually, a nonlinear time-delay system with adaptive control can be described by the form

$$\begin{cases} \dot{x} = f_c(x, x_t, \hat{\Theta}) \\ \dot{\hat{\Theta}} = \Psi(x, \hat{\Theta}) \end{cases} \tag{4}$$

with the initial conditions $\phi(\tau) = x_0(\tau),\ \tau \in [-r, 0]$ and $\hat{\Theta}(0)$, where $x \in R^n$ represents the system state and $\hat{\Theta} \in R^p$ denotes the adaptive parameter estimate.

For an adaptive control system, the specification of stabilization is the stability of the whole system in the sense of Lyapunov and the convergence of the state x, i.e. $x(t) \to 0$. The next lemma [11] provides a sufficient condition such that the system (4) satisfies this design performance.

Lemma 1 *For the system (4), if there exist continuous function $V(\tilde{x})$ with $\tilde{x} = [x^T\ \hat{\Theta}^T]^T$, $\mathcal{K}_\infty$ functions $\kappa_1(\cdot)$ and $\kappa_2(\cdot)$ and a positive definite*

$$\kappa_1(\|\tilde{x}\|) \le V(\tilde{x}) \le \kappa_2(\|\tilde{x}\|) \tag{5}$$

$$\dot{V}(\tilde{x}) \le -Q(x), \text{if} \max_{-r\le\tau\le0} V(\tilde{x}_t(\tau)) < p(V(\tilde{x}_t(0))) \tag{6}$$

then $\tilde{x}_t(\tilde{\phi})$ is bounded and $x_t(\tilde{\phi}) \to 0$ as $t \to \infty$, where $\tilde{x}_t(\tilde{\phi})(\cdot)$ denotes the state-function of the whole system with the initial condition $\tilde{\phi}$, $\tilde{\phi} = [\phi^T(\tau)\ \hat{\Theta}^T(\tau)]^T\ \tau \in [-r, 0]$. $p(\cdot)$ is a given continuous nondecreasing function and $p(s) > s$ for $s > 0$.

Remark 1. For the sake of simplicity, the function $p(\cdot)$ can be chosen as $p(s) = ps$ with the constant $p > 1$. Furthermore, if the Lyapunov function is a quadratic form, i.e. $V(\tilde{x}) = \tilde{x}^T P \tilde{x}$, P is a positive definite constant matrix, then the condition $\max_{-r\le\tau\le0} V(\tilde{x}(t+\tau)) < pV(\tilde{x}(t))$ can be replaced by the condition $\|\tilde{x}(t+\tau)\| < q\|\tilde{x}(t)\|$, $\forall \tau \in [-r, 0]$, where $q = \sqrt{p\lambda_2/\lambda_1} > 1$, λ_1 and λ_2 are the minimum and maximum eigenvalues of P, respectively. In this paper, this condition is referred to as the Razumikhin condition and $V(\tilde{x}(t))$ satisfying (5) and (6) as the Lyapunov-Razumikhin function.

3 Recursive Design Scheme

In the following, we will give a recursive design method to the adaptive stabilizing controller of the system (1).

Theorem 1 *For the system (1), an adaptive controller independent of the state-delayed, which renders the resulting closed-loop system stable in the sense* *of Lyapunov and $x(t) \to 0$ for any given initial state $\phi(\cdot)$ and estimate $\hat{\Theta}(0)$, is given by*

$$\begin{cases} u = -z_{n-1} + \sum_{j=1}^{n-1} \frac{\partial \alpha_{n-1}}{\partial x_j} x_{j+1} + \frac{\partial \alpha_{n-1}}{\partial \hat{\Theta}} \dot{\hat{\Theta}} - \frac{1}{2} q^2 z_n (e_n - \sum_{i=1}^{n-1} \frac{\partial \alpha_{n-1}}{\partial x_i} e_i)^2 \\ \quad - \frac{1}{2} q^2 z_n E_n(x, \hat{\Theta}) \|\hat{\Theta}\|^2 - \Psi_{n1} \hat{\Theta} - \frac{n+2}{2} z_n \\ \dot{\hat{\Theta}} = \beta_{n-1}(x, \hat{\Theta}) + \Gamma^{-1} \Psi_{n1}^T (x, \hat{\Theta}) z_n + \Gamma^{-1} \Psi_{n2}^T (x) \end{cases} \tag{7}$$

where $\hat{\Theta}$ is estimate of $\Theta = [\theta^T \ \rho]^T$ with $\rho := \|\theta\|$, $\Gamma \in R^{(p+1)\times(p+1)}$ is any given positive definite matrix and $q > 1$ is any given positive constant. The symbols $z_n, z_{n-1}, \alpha_{n-1}, \beta_{n-1}, \Psi_{n1}, \Psi_{n2}, E_n$ are defined in the proof.

Proof. The following design procedure is similar to the conventional back-stepping recursive procedure. It gives the explicit construction of a Lyapunov-Razumikhin function and an adaptive robust stabilizing controller in the form (7) that solves the problem.

First we define the following nonlinear coordinate transformation:

$$z_i = x_i - \alpha_{i-1}(\tilde{x}_{i-1}, \hat{\Theta}) \tag{8}$$

where $\alpha_{i-1}(\tilde{x}_{i-1}, \hat{\Theta})$ $(i = 1, 2, \cdots, n)$ are some smooth functions determined later with $\alpha_0 = 0$ and $\alpha_{i-1}(0, \hat{\Theta}) = 0$.

Note that Lyapunov-Razumikhin function of the whole system consisting of (1) with (7) will be a quadratic form on z and $\Theta - \hat{\Theta}$. Let $\tilde{z} = [z^T \ (\Theta - \hat{\Theta})^T]^T$, then $\tilde{z}_t = [z_t^T \ (\Theta - \hat{\Theta}_t)^T]^T$ and the Razumikhin condition is

$$\|\tilde{z}_t(\tau)\| < q\|\tilde{z}\|, \quad \tau \in [-r, 0] \tag{9}$$

Step 1. We first consider the x_1-subsystem of (1) with a candidate of Lyapunov-Razumikhin function as

$$V_1(z_1, \hat{\Theta}) = \frac{1}{2} z_1^2 + \frac{1}{2} (\Theta - \hat{\Theta})^T \Gamma (\Theta - \hat{\Theta}) \tag{10}$$

then, under the condition (2) the time derivative of V_1 is calculated as

$$\begin{aligned} \dot{V}_1(z_1, \hat{\Theta}) &= z_1[x_2 + \varphi_1(x_1)\theta + e_1(x_1) d_1(x_{1t})] - (\Theta - \hat{\Theta})^T \Gamma \dot{\hat{\Theta}} \\ &\le z_1[x_2 + \varphi_1(x_1)\theta] + N_1 - (\Theta - \hat{\Theta})^T \Gamma \dot{\hat{\Theta}} \end{aligned} \tag{11}$$

where $N_1 = |x_1||e_1|x_{1t}|$.

Whenever the Razumikhin condition holds, the following inequality is satisfied under the inequality $|x_{1t}| < q\|\tilde{z}\|$

$$\begin{aligned} N_1 &\le q|x_1||e_1|(\|z\| + \|\Theta\| + \|\hat{\Theta}\|) \\ &\le \frac{1}{2}q^2 x_1^2 e_1^2 + \frac{1}{2}\|z\|^2 + \frac{1}{2}q^2 x_1^2 \tilde{e}_1^2 \|\hat{\Theta}\|^2 + \frac{1}{2}x_1^2 + \frac{1}{2} q x_1^2 \tilde{e}_1^2 \|\Theta\| + \frac{1}{2} q x_1^2 \|\Theta\| \end{aligned} \tag{12}$$

Thus, it is easy to show that under the fact $\|\Theta\| \le \|\theta\| + \rho = 2\rho$ the time derivative of V_1 satisfies

$$\dot{V}_1(z_1,\hat{\Theta}) \le z_1\left[x_2+\varphi_1(x_1)\theta+\frac{1}{2}q^2x_1e_1^2+\frac{1}{2}q^2x_1\tilde{e}_1^2\|\hat{\Theta}\|^2+qx_1\tilde{e}_1^2\rho+qx_1\rho\right] +\frac{1}{2}\|z\|^2+\frac{1}{2}x_1^2-(\Theta-\hat{\Theta})^T\Gamma\dot{\hat{\Theta}} \tag{13}$$

With the choice of the virtual control law

$$\alpha_1(x_1,\hat{\Theta}) = -\frac{1}{2}q^2x_1e_1^2-\frac{1}{2}q^2x_1\tilde{e}_1^2\|\hat{\Theta}\|^2-\Psi_1(x_1)\hat{\Theta}-\frac{2n+1}{2}x_1-(n-1)qx_1\hat{\rho} \tag{14}$$

the following inequality holds

$$\dot{V}_1(z_1,\hat{\Theta}) \le x_1z_2-nx_1^2-(n-1)qx_1^2\hat{\rho}+\frac{1}{2}\|z\|^2+(\Theta-\hat{\Theta})^T\Gamma(\beta_1-\dot{\hat{\Theta}}) \tag{15}$$

where

$$\beta_1(x_1) = \Gamma^{-1}\Psi_1^T(x_1)x_1 \tag{16}$$

$$\Psi_1(x_1) = [\varphi_1(x_1) \quad qx_1(1+\tilde{e}_1^2)] \tag{17}$$

Step 2. For the (x_1, x_2)-subsystem of (1), we construct a candidate of Lyapunov-Razumikhin function as

$$V_2(\tilde{z}_2,\hat{\Theta}) = V_1(z_1,\hat{\Theta}) + \frac{1}{2}z_2^2 \tag{18}$$

Along any trajectories of the subsystem (x_1, x_2) and under the condition (2), the time derivative of V_2 is calculated as follows

$$\dot{V}_2(\tilde{z}_2,\hat{\Theta}) \le \dot{V}_1 + z_2\left[x_3+(\varphi_2-\frac{\partial\alpha_1}{\partial x_1}\varphi_1)\theta-\frac{\partial\alpha_1}{\partial x_1}x_2-\frac{\partial\alpha_1}{\partial\hat{\Theta}}\dot{\hat{\Theta}}\right]+N_2 \tag{19}$$

where $N_2 = |z_2|\left|e_2-\frac{\partial\alpha_1}{\partial x_1}e_1\right||x_{1t}|$.

Since $\alpha_1(0,\cdot) = 0$, there exists a function $\eta_{11}(\cdot)$ satisfying the decomposition

$$\alpha_1(x_1,\hat{\Theta}) = x_1\eta_{11}(x_1,\hat{\Theta}) \tag{20}$$

Also, the decompositions hold for any smooth function $e_i(\cdot)$ $(i = 1, \cdots, n)$

$$e_i(\tilde{x}_i) = \sum_{j=1}^{i} x_j\tilde{e}_{ij} \tag{21}$$

Thus, whenever the Razumikhin condition holds, we have the following inequality under (21), (8) and (20).

$$
\begin{aligned}
N_2 &\le |z_2|\left|e_2 - \frac{\partial\alpha_1}{\partial x_1}e_1\right| q(\|z\| + \|\Theta\| + \|\hat{\Theta}\|) \\
&\le \frac{1}{2}q^2z_2^2(e_2 - \frac{\partial\alpha_1}{\partial x_1}e_1)^2 + \frac{1}{2}\|z\|^2 \\
&\quad +|z_2|\left|x_1(\tilde{e}_{21} - \frac{\partial\alpha_1}{\partial x_1}\tilde{e}_{11} + \tilde{e}_{22}\eta_{11}) + z_2\tilde{e}_{22}\right| q(\|\Theta\| + \|\hat{\Theta}\|) \\
&\le \frac{1}{2}q^2z_2^2(e_2 - \frac{\partial\alpha_1}{\partial x_1}e_1)^2 + \frac{1}{2}\|z\|^2 + \frac{1}{2}q^2z_2^2E_2\|\hat{\Theta}\|^2 + \frac{1}{2}(x_1^2 + z_2^2) \\
&\quad +\frac{1}{2}qz_2^2E_2\|\Theta\| + \frac{1}{2}q(x_1^2 + z_2^2)\|\Theta\| \qquad (22)
\end{aligned}
$$

where $E_2 = \tilde{e}_{22}^2 + (\tilde{e}_{21} + \tilde{e}_{22}\eta_{11} - \frac{\partial\alpha_1}{\partial x_1}\tilde{e}_{11})^2$.
Substituting (22) and (15) into (19) and considering $\|\Theta\| \le \|\theta\| + \rho = 2\rho$, we obtain

$$
\begin{aligned}
\dot{V}_2(\tilde{z}_2, \hat{\Theta}) &\le z_2\left[x_3 + x_1 - \frac{\partial\alpha_1}{\partial x_1}x_2 - \frac{\partial\alpha_1}{\partial\hat{\Theta}}\dot{\hat{\Theta}} + \frac{1}{2}q^2z_2(e_2 - \frac{\partial\alpha_1}{\partial x_1}e_1)^2 + \frac{1}{2}q^2z_2E_2\|\hat{\Theta}\|^2\right] \\
&\quad +z_2(\varphi_2 - \frac{\partial\alpha_1}{\partial x_1}\varphi_1)\theta + qz_2^2E_2\rho + qz_2^2\rho + \frac{1}{2}z_2^2 + qx_1^2\rho + \|z\|^2 \\
&\quad -\frac{2n-1}{2}x_1^2 - (n-1)qx_1^2\hat{\rho} + (\Theta - \hat{\Theta})^T\Gamma^{-1}(\beta_1 - \dot{\hat{\Theta}}) \qquad (23)
\end{aligned}
$$

Thus, choosing the virtual control law

$$
\begin{aligned}
\alpha_2(\tilde{x}_2, \hat{\Theta}) &= -x_1 + \frac{\partial\alpha_1}{\partial x_1}x_2 + \frac{\partial\alpha_1}{\partial\hat{\Theta}}\dot{\hat{\Theta}} - \frac{1}{2}q^2z_2(e_2 - \frac{\partial\alpha_1}{\partial x_1}e_1)^2 - \frac{1}{2}q^2z_2E_2\|\hat{\Theta}\|^2 \\
&\quad -\Psi_{21}\hat{\Theta} - nz_2 - (n-2)z_2\hat{\rho} \qquad (24)
\end{aligned}
$$

renders

$$
\begin{aligned}
\dot{V}_2(\tilde{z}_2, \hat{\Theta}) &\le z_2z_3 + \|z\|^2 - \frac{2n-1}{2}(x_1^2 + z_2^2) - (n-2)q(z_1^2 + z_2^2)\hat{\rho} \\
&\quad +(\Theta - \hat{\Theta})^T\Gamma^{-1}(\beta_2(\tilde{x}_2, \hat{\Theta}) - \dot{\hat{\Theta}}) \qquad (25)
\end{aligned}
$$

where

$$
\beta_2(\tilde{x}_2, \hat{\Theta}) = \beta_1(x_1) + \Gamma^{-1}\Psi_{21}^T(\tilde{x}_2, \hat{\Theta})z_2 + \Gamma^{-1}\Psi_{22}^T(\tilde{x}_2) \qquad (26)
$$

$$
\Psi_{21}(\tilde{x}_2, \hat{\Theta}) = \left[\ \varphi_2 - \frac{\partial\alpha_1}{\partial x_1}\varphi_1 \quad qz_2(1 + E_2)\ \right] \qquad (27)
$$

$$
\Psi_{22}(\tilde{x}_2) = [\ 0 \quad qz_1^2\ ] \qquad (28)
$$

Step k. Suppose for the k–1-th subsystem of (1), there are a set of smooth virtual control laws $\alpha_i(\tilde{x}_i, \hat{\Theta})$ $(i = 1, \cdots, k-1)$ and a candidate of Lyapunov-Razumikhin function in the form with $\tilde{z}_{k-1} = [x_1 \cdots z_{k-1}]^T$

$$V_{k-1}(\tilde{z}_{k-1},\hat{\Theta}) = V_{k-2}(\tilde{z}_{k-2},\hat{\Theta}) + \frac{1}{2}z_{k-1}^2 \tag{29}$$

such that

$$\dot{V}_{k-1}(\tilde{z}_{k-1},\hat{\Theta}) \le z_{k-1}z_k+\frac{k-1}{2}\|z\|^2-\frac{2n-k+2}{2}\sum_{i=1}^{k-1}z_i^2-(n-k+1)q\sum_{i=1}^{k-1}z_i^2\hat{\rho}$$
$$+(\Theta-\hat{\Theta})^T\Gamma^{-1}(\beta_{k-1}(\tilde{x}_{k-1},\hat{\Theta})-\dot{\hat{\Theta}}) \tag{30}$$

and β_{k-1} is given in the following form

$$\beta_{k-1}(\tilde{x}_{k-1},\hat{\Theta}) = \beta_{k-2}+\Gamma^{-1}\Psi_{(k-1)1}^T(\tilde{x}_{k-1},\hat{\Theta})z_{k-1}+\Gamma^{-1}\Psi_{(k-1)2}^T(\tilde{x}_{k-1}) \tag{31}$$

$$\Psi_{(k-1)1}(\tilde{x}_{k-1},\hat{\Theta}) = \left[\varphi_{k-1}-\sum_{j=1}^{k-2}\frac{\partial\alpha_{k-2}}{\partial x_j}\varphi_j \quad qz_{k-1}(1+E_{k-1}(\tilde{x}_{k-1},\hat{\Theta}))\right] \tag{32}$$

$$\Psi_{(k-1)2}(\tilde{x}_{k-1}) = [\,0 \quad q\sum_{j=1}^{k-2}z_j^2\,] \tag{33}$$

where

$$E_{k-1}=\tilde{e}_{(k-1)(k-1)}^2+\sum_{s=1}^{k-2}\left(\tilde{e}_{(k-2)s}+\sum_{j=s+1}^{k-1}\tilde{e}_{(k-1)j}\eta_{(j-1)s}-\sum_{i=s}^{k-2}\frac{\partial\alpha_{k-2}}{\partial x_i}\tilde{e}_{is}-\sum_{i=s+1}^{k-2}\sum_{j=s+1}^{i}\frac{\partial\alpha_{k-2}}{\partial x_i}\tilde{e}_{ij}\eta_{(j-1)s}\right)^2$$

$\eta_{ij}(i=1,\cdots,k-1,\ j=1,\cdots,i)$ are functions satisfying the decompositions

$$\alpha_{i-1}(\tilde{x}_{i-1},\hat{\Theta}) = \sum_{j=1}^{i-1}x_j\tilde{\alpha}_{(i-1)j}(\tilde{x}_{i-1},\hat{\Theta}) = \sum_{j=1}^{i-1}z_j\eta_{(i-1)j}(\tilde{x}_{i-1},\hat{\Theta}) \tag{34}$$

since $\alpha_{i-1}(0,\cdot)=0$ and the definitions (8).

We now consider the k-th subsystem of (1) and will show in the following that for this subsystem if we construct a candidate of Lyapunov-Razumikhin function as

$$V_k(\tilde{z}_k,\hat{\Theta}) = V_{k-1}(\tilde{z}_{k-1},\hat{\Theta}) + \frac{1}{2}z_k^2 \tag{35}$$

then the time derivative of V_k satisfies the inequality similar to (30) and β_k also satisfies the formulation similar to (31).

Indeed, there is the following calculation for the time derivative of V_k along trajectories of the k-th subsystem and under the condition (2).

$$\dot{V}_k \le \dot{V}_{k-1}+z_k\left[x_{k+1}-\sum_{i=1}^{k-1}\frac{\partial\alpha_{k-1}}{\partial x_i}x_{i+1}-\frac{\partial\alpha_{k-1}}{\partial\hat{\Theta}}\dot{\hat{\Theta}}\right]+z_k(\varphi_k-\sum_{i=1}^{k-1}\frac{\partial\alpha_{k-1}}{\partial x_i}\varphi_i)\theta+N_k \tag{36}$$

where $N_k = |z_k|\left|e_k-\sum_{i=1}^{k-1}\frac{\partial\alpha_{k-1}}{\partial x_i}e_i\right||x_{1t}|.$

When the Razumikhin condition holds, the following inequality is satisfied by using (8) and the decompositions (21) and (34).

$$\begin{aligned} N_k &\le |z_k| \left| e_k - \sum_{i=1}^{k-1} \frac{\partial \alpha_{k-1}}{\partial x_i} e_i \right| q(\|z\| + \|\Theta\| + \|\hat{\Theta}\|) \\ &\le \frac{1}{2} q^2 z_k^2 (e_k - \sum_{i=1}^{k-1} \frac{\partial \alpha_{k-1}}{\partial x_i} e_i)^2 + \frac{1}{2}\|z\|^2 + \frac{1}{2} q^2 z_k^2 E_k \|\hat{\Theta}\|^2 + \frac{1}{2} \sum_{i=1}^{k} z_i^2 \\ &\quad + \frac{1}{2} q z_k^2 E_k \|\Theta\| + \frac{1}{2} q \sum_{i=1}^{k} z_i^2 \|\Theta\| \end{aligned} \tag{37}$$

where

$$E_k = \tilde{e}_{kk}^2 + \sum_{s=1}^{k-1} \left(\tilde{e}_{ks} + \sum_{j=s+1}^{k} \tilde{e}_{kj} \eta_{(j-1)s} - \sum_{i=s}^{k-1} \frac{\partial \alpha_{k-1}}{\partial x_i} \tilde{e}_{is} - \sum_{i=s+1}^{k-1} \sum_{j=s+1}^{i} \frac{\partial \alpha_{k-1}}{\partial x_i} \tilde{e}_{ij} \eta_{(j-1)s} \right)^2.$$

Substituting (37) into (36), we obtain

$$\begin{aligned} \dot{V}_k &\le z_k \left[x_{k+1} + z_{k-1} - \sum_{i=1}^{k-1} \frac{\partial \alpha_{k-1}}{\partial x_i} x_{i+1} - \frac{\partial \alpha_{k-1}}{\partial \hat{\Theta}} \dot{\hat{\Theta}} + \frac{1}{2} q^2 z_k (e_k - \sum_{i=1}^{k-1} \frac{\partial \alpha_{k-1}}{\partial x_i} e_i)^2 \right] \\ &\quad + \frac{1}{2} q^2 z_k^2 E_k \|\hat{\Theta}\|^2 + z_k (\varphi_k - \sum_{i=1}^{k-1} \frac{\partial \alpha_{k-1}}{\partial x_i} \varphi_i) \theta + q z_k^2 E_k \rho + q \sum_{i=1}^{k} z_i^2 \rho \\ &\quad + \frac{k}{2} \|z\|^2 + \frac{1}{2} \sum_{i=1}^{k} z_i^2 - \frac{2n-k+2}{2} \sum_{i=1}^{k-1} z_i^2 - (n-k+1) q \sum_{i=1}^{k-1} z_i^2 \hat{\rho} \\ &\quad + (\Theta - \hat{\Theta})^T \Gamma^{-1} (\beta_{k-1}(\tilde{x}_{k-1}, \hat{\Theta}) - \dot{\hat{\Theta}}) \end{aligned} \tag{38}$$

It is easy to see that the virtual control law

$$\begin{aligned} \alpha_k(\tilde{x}_k, \hat{\Theta}) &= -z_{k-1} + \sum_{i=1}^{k-1} \frac{\partial \alpha_{k-1}}{\partial x_i} x_{i+1} + \frac{\partial \alpha_{k-1}}{\partial \hat{\Theta}} \dot{\hat{\Theta}} - \frac{1}{2} q^2 z_k (e_k - \sum_{i=1}^{k-1} \frac{\partial \alpha_{k-1}}{\partial x_i} e_i)^2 \\ &\quad - \frac{1}{2} q^2 z_k E_k \|\hat{\Theta}\|^2 - \Psi_{k1} \hat{\Theta} - \frac{2n-k+2}{2} z_k - (n-k) q z_k \hat{\rho} \end{aligned} \tag{39}$$

renders

$$\begin{aligned} \dot{V}_k(\tilde{x}_k, \hat{\Theta}) &\le z_k z_{k+1} - \frac{2n-k+1}{2} \sum_{i=1}^{k} z_i^2 - (n-k) q \sum_{i=1}^{k} z_i^2 \hat{\rho} + \frac{k}{2} \|z\|^2 + \frac{1}{2} \sum_{i=1}^{k} z_i^2 \\ &\quad + (\Theta - \hat{\Theta})^T \Gamma^{-1} (\beta_k(\tilde{x}_k, \hat{\Theta}) - \dot{\hat{\Theta}}) \end{aligned} \tag{40}$$

whenever the Razumikhin condition holds, and

$$\beta_k(\tilde{x}_k, \hat{\Theta}) = \beta_{k-1} + \Gamma^{-1}\Psi_{k1}^T(\tilde{x}_k, \hat{\Theta})z_k + \Gamma^{-1}\Psi_{k2}^T(\tilde{x}_k) \tag{41}$$

where

$$\Psi_{k1}(\tilde{x}_k, \hat{\Theta}) = \left[(\varphi_k - \sum_{i=1}^{k-1} \frac{\partial \alpha_{k-1}}{\partial x_i}\varphi_i) \quad qz_k(1 + E_k(\tilde{x}_k, \hat{\Theta})) \right]$$

$$\Psi_{k2}(\tilde{x}_k) = \left[0 \quad q\sum_{i=1}^{k-1} z_i^2 \right]$$

The aforementioned inductive argument shows that (40) holds for $k = n$. In fact, at the n-th step, the Lyapunov-Razumikhin function is constructed as

$$V(z, \hat{\Theta}) = V_{n-1}(\tilde{z}_{n-1}, \hat{\Theta}) + \frac{1}{2}z_n^2 \tag{42}$$

then, the time derivative of V along the trajectories of the whole system (1) is obtained as follows:

$$\begin{aligned}\dot{V}(z, \hat{\Theta}) \le\ & z_n \left[u + z_{n-1} - \sum_{i=1}^{n-1} \frac{\partial \alpha_{n-1}}{\partial x_i} x_{i+1} - \frac{\partial \alpha_{n-1}}{\partial \hat{\Theta}}\dot{\hat{\Theta}} + \frac{1}{2}q^2 z_n (e_n - \sum_{i=1}^{n-1}\frac{\partial \alpha_{n-1}}{\partial x_i} e_i)^2 \right] \\ & + \frac{1}{2}q^2 z_n^2 E_n \|\hat{\Theta}\|^2 + z_n \Psi_{n1}\hat{\Theta} + \frac{n+1}{2}\|z\|^2 - \frac{n+2}{2}\sum_{i=1}^{n-1} z_i^2 \\ & + (\Theta - \hat{\Theta})^T \Gamma^{-1}(\beta_{n-1} + \Gamma^{-1}\Psi_{n1}^T z_n + \Gamma^{-1}\Psi_{n2}^T - \dot{\hat{\Theta}}) \end{aligned} \tag{43}$$

Thus, the adaptive robust stabilizing controller (7) can be obtained, where

$$\Psi_{n1}(x, \hat{\Theta}) = \left[(\varphi_n - \sum_{i=1}^{n-1} \frac{\partial \alpha_{n-1}}{\partial x_i}\varphi_i) \quad qz_n(1 + E_n(x, \hat{\Theta})) \right]$$

$$\Psi_{n2}(x) = \left[0 \quad q\sum_{i=1}^{n-1} z_i^2 \right]$$

$$E_n(x, \hat{\Theta}) = \tilde{e}_{nn}^2 + \sum_{s=1}^{n-1} \left(\tilde{e}_{ns} + \sum_{j=s+1}^{n} \tilde{e}_{nj}\eta_{(j-1)s} - \sum_{i=s}^{n-1} \frac{\partial \alpha_{n-1}}{\partial x_i}\tilde{e}_{is} - \sum_{i=s+1}^{n-1}\sum_{j=s+1}^{i} \frac{\partial \alpha_{n-1}}{\partial x_i}\tilde{e}_{ij}\eta_{(j-1)s} \right)^2$$

such that the time derivative of V satisfies

$$\dot{V}(z, \hat{\Theta}) \le -\frac{1}{2}\|z\|^2 \tag{44}$$

whenever the Razumikhin condition $\|\tilde{z}_t\| < q\|\tilde{z}\|$ holds.

Therefore, from (42) and (44), it is easy to show that the conditions (5) and (6) are satisfied. In view of Lemma 1, we conclude that the closed-loop system is stable in the sense of Lyapunov, and $z(t) \to 0$ as $t \to \infty$. Consequently, from (8) with $\alpha_i(0, \cdot) = 0$, it follows that $x(t) \to 0$ as $t \to \infty$ for any given $\phi(\cdot)$ and $\hat{\Theta}(0)$. □

4 Extension to a Broader Class of Systems

In the following, we further exploit the potential of this design approach to extend the result obtained in Theorem 1 to a broader class of uncertain nonlinear time-delay systems. The system under consideration is described as

$$\begin{cases} \dot{\xi} = f_0(\xi, x_1) + \varphi_0(\xi)\theta + e_0(\xi)d_0(\xi_t, \tilde{x}_{kt}) \\ \dot{x}_1 = x_2 + \varphi_1(\xi, x_1)\theta + e_1(\xi, x_1)d_1(\xi_t, \tilde{x}_{kt}) \\ \dot{x}_2 = x_3 + \varphi_2(\xi, \tilde{x}_2)\theta + e_2(\xi, \tilde{x}_2)d_2(\xi_t, \tilde{x}_{kt}) \\ \quad \vdots \\ \dot{x}_k = u + \varphi_k(\xi, \tilde{x}_k)\theta + e_k(\xi, \tilde{x}_k)d_k(\xi_t, \tilde{x}_{kt}) \end{cases} \tag{45}$$

where $\varphi_0(\cdot)$ and $e_0(\cdot)$ are known vector functions and $d_0(\cdot)$ is unknown vector function, $\varphi_0(0) = 0$, $e_0(0) = 0$. Another notations represent the same as those in the system (1).

Suppose that the delay related terms $d_i(\cdot)$ satisfy the following condition

$$\|d_i(\xi_t, \tilde{x}_{kt})\| \le \|\xi_t\|, \quad i = 0, 1, \cdots, k \tag{46}$$

Moreover, for the ξ-subsystem of (45), suppose that there exist a positive definite function $W(\xi)$, a smooth function $x_1^*(\xi, \hat{\Theta})$ with $x_1^*(0, \cdot) = 0$ and positive constants k_1, k_2 such that *A*1-*A*3 below hold, where $\hat{\Theta}$ is the estimate of $\Theta = [\theta^T \ \rho]^T$, $\rho = \|\theta\|$.

A1:

$$k_1\|\xi\|^2 \le W(\xi) \le k_2\|\xi\|^2 \tag{47}$$

A2: When the Razumikhin condition $\max_{-r\le\tau\le 0} \|\tilde{x}(t+\tau)\| < q\|\tilde{x}(t)\|$ holds, the following inequality is satisfied

$$\begin{aligned} \frac{\partial W}{\partial \xi}[f_0(\xi, x_1^*) + \varphi_0(\xi)\theta] \le & -\frac{2k+4}{2}\|\xi\|^2 - q^2\left\|\frac{\partial W}{\partial \xi}e_0\right\|^2 - \frac{1}{2}q^2\left\|\frac{\partial W}{\partial \xi}\tilde{e}_0\right\|^2\|\hat{\Theta}\|^2 \\ & -q\left\|\frac{\partial W}{\partial \xi}\tilde{e}_0\right\|^2\hat{\rho} - (k+1)q\|\xi\|^2\hat{\rho} + M(\xi)(\Theta - \hat{\Theta}) \end{aligned} \tag{48}$$

where $\tilde{e}_0(\cdot)$ is any function satisfying the decomposition $e_0(\xi) = \tilde{e}_0(\xi)\xi$, $M(\cdot)$ is a known real-valued function, $\tilde{x} = [\xi^T \ \tilde{x}_k^T]^T$, $q = \sqrt{p\lambda_2/\lambda_1}$ with $\lambda_2 = \max\{k_2, \frac{1}{2}\}$ and $\min\{k_1, \frac{1}{2}\}$.

A3: There exists a real-valued smooth function $\sigma_0(\xi, x_1) \ge 0$ such that

$$\left\|\frac{\partial W}{\partial \xi}\frac{\partial f_0(\xi, x_1)}{\partial x_1}\right\| \le (\|\xi\| + \|x_1\|)\sigma_0(\xi, x_1) \tag{49}$$

Theorem 2 *For the system (45), if the assumptions A1-A3 hold, then there exists an adaptive controller similar to the form (7) which renders the closed-loop system stable in the sense of Lyapunov and*

$$\lim_{t\to\infty} \xi(t) = 0, \quad \lim_{t\to\infty} x_i(t) = 0 \ (i = 1, \cdots, k)$$

for any given $\xi_0(\cdot)$, $x_{i0}(\cdot)$ $(i = 1, \cdots, k)$, *and* $\hat{\Theta}(0)$.

Proof The proof is similar to that of Theorem 1. The only difference is the first step that is the start point of the k steps recursive procedure, so we here give the first step of the proof only.

Define the nonlinear coordinate transformation (8) with $z_1 = x_1 - \alpha_0 = x_1 - x_1^*$ and the Razumikhin condition is (9) with $\tilde{z} = [\xi^T \ \tilde{z}_k^T \ (\Theta - \hat{\Theta})^T]^T$.

First, we consider the ξ-subsystem with a candidate of Lyapunov-Razumikhin function as

$$U(\xi, \hat{\Theta}) = W(\xi) + \frac{1}{2}(\Theta - \hat{\Theta})^T \Gamma (\Theta - \hat{\Theta}) \tag{50}$$

where $\Gamma \in R^{(p+1)\times(p+1)}$ is a given positive definite constant matrix.

Then, by using (46), the time derivative of U along any trajectories of the ξ-subsystem is derived as follows:

$$\begin{aligned}\dot{U}(\xi, \hat{\Theta}) &= \frac{\partial W}{\partial \xi}[f_0(\xi, x_1) + \varphi_0(\xi)\theta + e_0(\xi)d_0(\xi_t, \tilde{x}_{kt})] - (\Theta - \hat{\Theta})^T \Gamma \dot{\hat{\Theta}} \\ &\le \frac{\partial W}{\partial \xi}[f_0(\xi, x_1^*) + \varphi_0(\xi)\theta] + \left\|\frac{\partial W}{\partial \xi}\right\| \|e_0\| \|\xi_t\| \\ &\quad + \left\| \int_0^1 \frac{\partial W}{\partial \xi} \frac{\partial f_0(\xi, x_1)}{\partial x_1}\bigg|_{x_1 = x_1^* + \lambda z_1} d\lambda z_1 \right\| - (\Theta - \hat{\Theta})^T \Gamma \dot{\hat{\Theta}}\end{aligned} \tag{51}$$

Noting that the function decomposition $x_1^*(\xi, \hat{\Theta}) = \xi^T \eta_0(\xi, \hat{\Theta})$, and using (49), we obtain the following inequality

$$\begin{aligned}\left\| \int_0^1 \frac{\partial W}{\partial \xi} \frac{\partial f_0(\xi, x_1)}{\partial x_1}\bigg|_{x_1 = x_1^* + \lambda z_1} d\lambda z_1 \right\| &\le |z_1| \int_0^1 (\|\xi\| + |x_1^* + \lambda z_1|)\sigma_0(\xi, x_1^* + \lambda z_1) d\lambda \\ &\le |z_1| \int_0^1 (\|\xi\| + \|\xi\|\|\eta_0\| + |\lambda z_1|)\sigma_0(\xi, x_1^* + \lambda z_1) d\lambda \\ &\le |z_1| \int_0^1 (\|\xi\| + |\lambda z_1|)(1 + \|\eta_0\|)\sigma_0(\xi, x_1^* + \lambda z_1) d\lambda \\ &\le |z_1|(\|\xi\| + |z_1|) \int_0^1 (1 + \|\eta_0\|)\sigma_0(\xi, x_1^* + \lambda z_1) d\lambda \\ &\le \frac{1}{2}\|\xi\|^2 + z_1^2 \sigma_1(\xi, x_1, \hat{\Theta})\end{aligned} \tag{52}$$

where $\sigma_1(\xi, x_1, \hat{\Theta}) = (\frac{1}{2}\int_0^1 (1 + \|\eta_0\|)\sigma_0 d\lambda + 1) \int_0^1 (1 + \|\eta_0\|)\sigma_0 d\lambda$.

When the Razumikhin condition holds, we have

$$\left\|\frac{\partial W}{\partial \xi}\right\| \|e_0\|\|\xi_t\| \le \left\|\frac{\partial W}{\partial \xi}\right\| \|e_0\| q(\|\xi\| + \|\tilde{z}_k\| + \|\Theta\| + \|\hat{\Theta}\|)$$

$$\le q^2 \left\|\frac{\partial W}{\partial \xi} e_0\right\|^2 + \|\xi\|^2 + \frac{1}{2}\|\tilde{z}_k\|^2 + \frac{1}{2}q^2 \left\|\frac{\partial W}{\partial \xi}\tilde{e}_0\right\|^2 \|\hat{\Theta}\|^2$$

$$+\frac{1}{2}q\|\xi\|^2\|\Theta\| + \frac{1}{2}q\left\|\frac{\partial W}{\partial \xi}\tilde{e}_0\right\|^2 \|\Theta\| \tag{53}$$

Noting that $\|\Theta\| \le \|\theta\| + \rho = 2\rho$, then substituting (48), (52) and (53) into (51), we obtain

$$\dot{U}(\xi, \hat{\Theta}) \le -\frac{2k+1}{2}\|\xi\|^2 - kq\|\xi\|^2\rho + \frac{1}{2}\|\tilde{z}_k\|^2 + z_1^2\sigma_1$$

$$+(\Theta - \hat{\Theta})^T \Gamma\{\Gamma^{-1}M^T(\xi) + \Gamma^{-1}\Psi_0^T(\xi) - \dot{\hat{\Theta}}\} \tag{54}$$

whenever the Razumikhin condition holds, where

$$\Psi_0(\xi) = \left[0 \;\; q\left\|\frac{\partial W}{\partial \xi}\tilde{e}_0\right\|^2 + (k+1)q\|\xi\|^2\right]$$

We now consider the subsystem with the coordinate (ξ, x_1) and define a candidate of Lyapunov-Razumikhin function as

$$V_1(\xi, x_1, \hat{\Theta}) = U(\xi, \hat{\Theta}) + \frac{1}{2}z_1^2 \tag{55}$$

then, taking (54) and (46) into account, the time derivative of V_1 is calculated as follows:

$$\dot{V}_1(\xi, x_1, \hat{\Theta}) \le z_1\left\{x_2 + z_1\sigma_1 - \frac{\partial x_1^*}{\partial \xi}f_0 - \frac{\partial x_1^*}{\partial \hat{\Theta}}\dot{\hat{\Theta}} + (\varphi_1 - \frac{\partial x_1^*}{\partial \xi}\varphi_0)\theta\right\} + N_1$$

$$-\frac{2k+1}{2}\|\xi\|^2 - kq\|\xi\|^2\rho + \frac{1}{2}\|\tilde{z}_k\|^2$$

$$+(\Theta - \hat{\Theta})^T \Gamma\{\Gamma^{-1}M^T(\xi) + \Gamma^{-1}\Psi_0^T(\xi) - \dot{\hat{\Theta}}\} \tag{56}$$

where $N_1 = |z_1|\left|e_1 - \frac{\partial x_1^*}{\partial \xi}e_0\right| \|\xi_t\|$.

When the Razumikhin condition holds, N_1 satisfies the following inequality under $x_1 = z_1 + x_1^*$ and the decompositions $x_1^* = \xi^T\eta_0(\xi, \hat{\Theta})$, $e_0(\xi) = \tilde{e}_0(\xi)\xi$ and $e_1(\xi, x_1) = \tilde{e}_{10}(\xi, x_1)\xi + \tilde{e}_{11}(\xi, x_1)x_1$.

$$N_1 \le |z_1|\left|e_1 - \frac{\partial x_1^*}{\partial \xi}e_0\right| q(\|\xi\| + \|\tilde{z}_k\| + \|\Theta\| + \|\hat{\Theta}\|)$$

$$\le q^2 z_1^2 (e_1 - \frac{\partial x_1^*}{\partial \xi}e_0)^2 + \|\xi\|^2 + \frac{1}{2}\|\tilde{z}_k\|^2 + \frac{1}{2}q^2 z_1^2 E_1\|\hat{\Theta}\|^2 + \frac{1}{2}z_1^2$$

$$+\frac{1}{2}qz_1^2 E_1\|\Theta\| + \frac{1}{2}q\|\xi\|^2\|\Theta\| + \frac{1}{2}qz_1^2\|\Theta\| \tag{57}$$

where $E_1 = \tilde{e}_{11}^2 + (\tilde{e}_{10} + \tilde{e}_{11}\eta_0 - \frac{\partial x_1^*}{\partial \xi}\tilde{e}_0)^2$.

Substituting (57) into (56), we obtain the following inequality under $\|\Theta\| \leq \|\theta\| + \rho = 2\rho$.

$$\begin{aligned}\dot{V}_1(\xi, x_1, \hat{\Theta}) \leq\ & z_1 \left\{ x_2 + z_1\sigma_1 - \frac{\partial x_1^*}{\partial \xi} f_0 - \frac{\partial x_1^*}{\partial \hat{\Theta}}\dot{\hat{\Theta}} + q^2 z_1 (e_1 - \frac{\partial x_1^*}{\partial \xi} e_0)^2 + \frac{1}{2} z_1 \right\} \\ & + \frac{1}{2} q^2 z_1^2 E_1 \|\hat{\Theta}\|^2 + z_1(\varphi_1 - \frac{\partial x_1^*}{\partial \xi}\varphi_0)\theta + q z_1^2 E_1 \rho + q z_1^2 \rho \\ & - \frac{2k-1}{2}\|\xi\|^2 - (k-1)q\|\xi\|^2\rho + \|\tilde{z}_k\|^2 \\ & + (\theta - \hat{\theta})^T \Gamma \{\Gamma^{-1} M^T(\xi) + \Gamma^{-1}\Psi_0^T(\xi) - \dot{\hat{\Theta}}\} \end{aligned} \tag{58}$$

Therefore, the virtual controller

$$\begin{aligned}\alpha_1(\xi, x_1, \hat{\Theta}) =\ & -\frac{2k+2}{2} z_1 - z_1\sigma_1 + \frac{\partial x_1^*}{\partial \xi} f_0 + \frac{\partial x_1^*}{\partial \hat{\Theta}}\dot{\hat{\Theta}} - q^2 z_1 (e_1 - \frac{\partial x_1^*}{\partial \xi} e_0)^2 \\ & - \frac{1}{2} q^2 z_1 E_1 \|\hat{\Theta}\|^2 - \Psi_1(\xi, x_1, \hat{\Theta})\hat{\Theta} - (k-1) q z_1 \hat{\rho} \end{aligned} \tag{59}$$

renders

$$\begin{aligned}\dot{V}_1(\xi, x_1, \hat{\Theta}) \leq\ & z_1 z_2 - \frac{2k-1}{2}\|\xi\|^2 - \frac{2k+1}{2} z_1^2 - (k-1)q\|\xi\|^2\rho \\ & - (k-1) q z_1 \hat{\rho} + \|\tilde{z}_k\|^2 + (\Theta - \hat{\Theta})^T \Gamma (\beta_1 - \dot{\hat{\Theta}}) \end{aligned} \tag{60}$$

where

$$\begin{aligned}\beta_1(\xi, x_1, \hat{\Theta}) &= \Gamma^{-1} M^T(\xi) + \Gamma^{-1}\Psi_0^T(\xi) + \Gamma^{-1}\Psi_1^T z_1 \\ \Psi_1(\xi, x_1, \hat{\Theta}) &= [(\varphi_1 - \frac{\partial x_1^*}{\partial \xi}\varphi_0) \quad q z_1 (1 + E_1)] \end{aligned}$$

The remaining $k-1$ steps of the proof with V_1 and x_1^* is analogous to those of Theorem 1 and is, therefore, omitted. □

Remark 2. The assumption $A2$ means that the time derivative of $W(\xi)$ along the ξ-subsystem with virtual control law x_1^* contains not only the terms representing sufficient stability margin, but also the additional terms for dominating the delay-related uncertainties. However, it should be noted that $A2$ is required only at the point where the Razumikhin condition holds. Moreover, the term of the error between Θ and $\hat{\Theta}$ means that Θ is linearly parameterized.

Remark 3. Note that $\partial W/\partial\xi|_{x_1=0} = 0$, thus, there exists a smooth function $H(\xi, x_1)$ for smooth function $(\partial W/\partial\xi)(\partial f_0/\partial x_1)$ such that

$$\frac{\partial W}{\partial \xi}\frac{\partial f_0}{\partial x_1} = H(\xi, x_1) x_1$$

which means $A3$ is always true.

5 Conclusions

In this paper, we have investigated the problem of adaptive stabilization for uncertain nonlinear systems with time-delay. A similar backstepping recursive design approach to adaptive stabilizing controller has been proposed for a class of nonlinear time-delay systems with triangular structure, which is based on recursively constructing the Lyapunov-Razumikhin function that guarantees the convergence of the system state.

References

1. Krstic, M. Kabellakopoulos, I. and Kokotovic, P.V. (1995), *Nonlinear And Adaptive Control Design*, Wiley, New York.
2. Marino, R. and Tomei, P.(1995), *Nonlinear Control Design Geometric, Adaptive and Robust*, Prentice Hall.
3. Seto, D. Annaswamy, A.M. and Bailieul, J. (1994), Adaptive Control of Nonlinear Systems with a Triangular Structure, *IEEE Trans. Automat. Contr.* Vol.39 No.7, pp.1411-1427.
4. Lin, W. and Qian, C. (2002), Adaptive Control of Nonlinearly Parameterized Systems: the Smooth Feedback Case, *IEEE Trans. Automat. Contr.*, Vol.47, No.8, pp.1249-1266.
5. Hale, J.K. and Lunel, S.M. (1993), *Introduction to functional differential equations*, Springer-Verlag, New York.
6. Haddock, J.R. (1983) Lyapunov-Razumikhin functions and an invariance principle for functional differential equations, *Journal of Differential Equations* Vol.48, pp.95-122.
7. Xu, B. and Liu, Y. (1994), An improved Razumikhin-type theorem and its applications, *IEEE Trans. Automat. Contr.* Vol.39 No.4 pp.839-841.
8. Mao, X. (1997), Comments on an improved Razumikhin-type theorem and its applications, *IEEE Trans. Automat. Contr.* Vol.42 No.3 pp.429-430.
9. Nguang, S.K. (2000), Robust stabilization of a class of time-delay nonlinear systems, *IEEE Trans. Automat. Contr.* Vol.45 No.4, pp.756-762.
10. Zhou, S. Feng, G. and Nguang, S.K. (2002), Comments on "Robust stabilization of a class of time-delay nonlinear systems", *IEEE Trans. Automat. Contr.* Vol.47 No.9, pp.1586.
11. Jiao, X. Shen, T. and Tamura, K. (2005), Adaptive Robust Stabilization of Cascaded Nonlinear Systems with Uncertain Time-Delay, *Transaction of IEE of Japan*, Vol.125-C, No.2, February, (appear).

Robust Input-to-State Stability and Small Gain Theorem for Nonlinear Systems Containing Time-Varying Uncertainty*

Zhiyong Chen and Jie Huang

Department of Automation and Computer-Aided Engineering
The Chinese University of Hong Kong, Hong Kong, P.R.China
zychen@acae.cuhk.edu.hk; jhuang@acae.cuhk.edu.hk

1 Introduction

Consider nonlinear systems containing time-varying uncertainty as follows

$$\dot{x} = f(x, u, d(t)) \tag{1}$$

$$y = h(x, u, d(t)), \quad t \geq t_0 \geq 0 \tag{2}$$

where $x \in R^n$ is the plant state, $u \in R^m$ the input, $y \in R^p$ the output, t_0 the initial time, $f(x, u, d) : R^n \times R^m \times R^{n_d} \mapsto R^n$ is piecewise continuous in d, and locally Lipschitz in $\text{col}(x, u)$, and $h(x, u, d) : R^n \times R^m \times R^{n_d} \mapsto R^p$ is continuous in all its arguments. The so-called time-varying uncertainty $d(t) : [t_0, \infty) \mapsto R^{n_d}$ is *a family of* piecewise continuous function of t, representing the external disturbance and/or the internal unknown parameter variation. And suppose $f(0, 0, d(t)) = 0$ and $h(0, 0, d(t)) = 0$. System (1) and (2) arises naturally from the robust output regulation problem where $d(t)$ models the external disturbance and unknown plant parameters (see [2] - [4], etc.).

Sontag introduced the notion of input-to-state stability (ISS) for time-invariant nonlinear systems in [11], that is, for systems of the form (1) where f does not rely on $d(t)$. Jiang, *et al*, further extended the ISS concept in several aspects [6], [7]. First, they generalized the ISS concept to time-varying systems, that is, to systems of the form (1) with $d(t) = t$. Second, they introduced such concepts as input-to-output stability (IOS) and unbounded observability (UO) for systems described by both (1) and (2) with $d(t) = t$. Moreover, in the context of IOS and UO concepts, a result known as the small gain theorem was established for time-invariant systems in [7] and for time-varying systems in [6]. The result gives a condition (known as the small gain condition now)

* The work described in this paper was partially supported by the Hong Kong Research Grant Council under Grant CUHK4168/03E, and the National Natural Science Foundation of China under Grant 60374038.

under which certain inter-connection of two IOS subsystems is still IOS. Since then, the result has played an important role in studying the stabilization and output regulation of nonlinear systems [2] - [4].

As will be seen in the next section, the small gain condition given in [6], [7] involves two somewhat complex inequalities. Recently, by using some equivalent characterization of IOS concept, Isidori showed that, for the class of time-invariant systems, the small gain condition can be reduced to one simple inequality [5]. In [1], Chen and Huang introduced the concept of the robust input-to-state stability (RISS) for systems of the form (1) and showed that, for certain simple inter-connection of two RISS subsystems, the small gain condition can be simplified to one simple inequality. In this paper, we will further introduce the concepts of robust input-to-output stability (RIOS) and robust unbounded observability (RUO) for systems of the form (1) and (2) and then show that, for the same inter-connection of two RIOS subsystems of the form (1) and (2) as that given in [5], the small gain condition can be reduced to the same simple inequality as the one given by Isidori. It should be noted that our proof is completely different from that of Isidori because the equivalent characterization of IOS concept does not exist for time-varying systems.

2 Preliminaries

In this section, we will first extend ISS, IOS and UO concepts to system (1) and (2), thus leading to such concepts as robust ISS, robust IOS and robust UO. At the beginning, let us first clarify some notations. Let L_∞^m be the set of all piecewise continuous bounded functions $u : [t_0, \infty) \mapsto R^m$ with a finite supremum norm $\|u_{[t_0,\infty)}\| \stackrel{\text{def}}{=} \sup_{t\geq t_0} \|u(t)\|$. And denote the supremum norm of the truncation of $u(t)$ in $[t_1, t_2]$ with $t_0 \leq t_1 \leq t_2$ by $\|u_{[t_1,t_2]}\| \stackrel{\text{def}}{=} \sup_{t_1\leq t\leq t_2} \|u(t)\|$, and the identity function by Id.

Definition 1. *System (1) is said to be robustly (with respect to $d(t)$) input-to-state stable (RISS) if there exist a class $\mathcal{KL}$ function β, and a class $\mathcal{K}$ function γ, independent of $d(t)$, such that for any initial time $t_0 \geq 0$, any initial state $x(t_0) \in R^n$, any piecewise continuous $d(t)$, and any input function $u(t) \in L_\infty^m$, the solution $x(t)$ exists and satisfies*

$$\|x(t)\| \leq \beta(\|x(t_0)\|, t - t_0) + \gamma\left(\|u_{[t_0,t]}\|\right), \; t \geq t_0. \tag{3}$$

□

Remark 1. The RISS concept was first given by [10] for a special case of system (1) where $d(t)$ is an unknown constant. □

Remark 2. Since $\max\{\beta,\gamma\} \leq \beta+\gamma \leq \max\{2\beta, 2\gamma\}$ for any pair $\beta \geq 0$, $\gamma \geq 0$, an equivalent way to characterize the RISS of (1) is that there exist a class $\mathcal{KL}$

function β, and a class $\mathcal{K}$ function γ such that for any initial state $x(t_0) \in R^n$ and any input function $u(t) \in L_\infty^m$, the solution $x(t)$ exists and satisfies

$$\|x(t)\| \leq \max\left\{\beta(\|x(t_0)\|, t - t_0), \gamma\left(\|u_{[t_0,t]}\|\right)\right\}, \ t \geq t_0. \tag{4}$$

□

While the RISS concept connects the bound of the state to the bounds of the initial state and the input, the RIOS concept to be introduced below further connects the bound of the output to the bounds of the initial state and the input.

Definition 2. *System (1) and (2) is said to be robustly (with respect to $d(t)$) input-to-output stable (RIOS) if there exist a class $\mathcal{KL}$ function β, and a class $\mathcal{K}$ function γ, independent of $d(t)$, such that for any initial time $t_0 \geq 0$, any initial state $x(t_0) \in R^n$, any piecewise continuous $d(t)$, and any input function $u(t) \in L_\infty^m$, the output $y(t)$ satisfies*

$$\|y(t)\| \leq \max\left\{\beta(\|x(t_0)\|, t - t_0), \gamma\left(\|u_{[t_0,t]}\|\right)\right\}, \ t \geq t_0. \tag{5}$$

□

Definition 3. *System (1) and (2) is said to have the robust (with respect to $d(t)$) unbounded observability (RUO) property if there exist class $\mathcal{K}$ functions α_x, α_u and α_y, independent of $d(t)$, such that for any initial time $t_0 \geq 0$, any initial state $x(t_0) \in R^n$, any piecewise continuous $d(t)$, and any input function $u(t) \in L_\infty^m$, the solution $x(t)$ exists and satisfies*

$$\|x(t)\| \leq \max\left\{\alpha_x(\|x(t_0)\|), \alpha_u\left(\|u_{[t_0,t]}\|\right), \alpha_y\left(\|y_{[t_0,t]}\|\right)\right\}, \ t \geq t_0. \tag{6}$$

Moreover, it is said to have the robust strong unbounded observability (RSUO) property if (6) holds with $\alpha_x(\|x(t_0)\|)$ replaced by $\beta(\|x(t_0)\|, t - t_0)$ for some class $\mathcal{KL}$ function β. □

Remark 3. The concepts RIOS, RUO and RSUO follow straightforwardly from the concepts IOS, UO, and SUO given in [6]. Therefore, some relations among IOS, UO, and SUO established in [6] can be directly extended to RIOS, RUO and RSUO. For example, it can be verified that

$$\text{RSUO} \Rightarrow \text{RUO},$$

and

$$\text{RISS} \Leftrightarrow \text{RIOS} + \text{RSUO}$$

if $h(x, u, d(t))$ is bounded for all $t \geq t_0$ when x and u are in every compact subsets. □

To conclude this section, let us rephrase the small gain theorem from [6] for the purpose of the comparison.

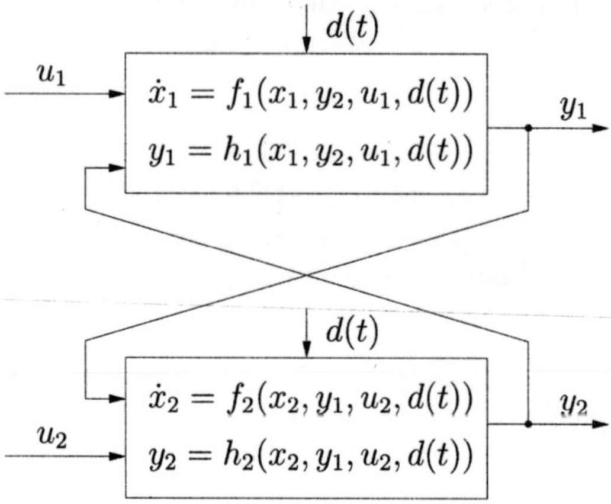

Fig. 1. Inter-connection of two subsystems

Consider the inter-connection of the following two systems as depicted in Fig. 1,

$$\begin{aligned} \dot{x}_1 &= f_1(x_1, y_2, u_1, d(t)) \\ y_1 &= h_1(x_1, y_2, u_1, d(t)), \;\; t \geq t_0 \geq 0 \end{aligned} \tag{7}$$

$$\begin{aligned} \dot{x}_2 &= f_2(x_2, y_1, u_2, d(t)) \\ y_2 &= h_2(x_2, y_1, u_2, d(t)), \;\; t \geq t_0 \geq 0 \end{aligned} \tag{8}$$

where, for $i = 1, 2$, $x_i \in R^{n_i}$, $u_i \in R^{m_i}$, $y_i \in R^{p_i}$, the functions $f_1(x_1, y_2, u_1, d)$ and $f_2(x_2, y_1, u_2, d)$ are piecewise continuous in d and locally Lipschitz in $\mathrm{col}(x_1, y_2, u_1)$ and $\mathrm{col}(x_2, y_1, u_2)$, respectively, and $d : [t_0, \infty) \mapsto R^{n_d}$ is piecewise continuous. And suppose the following assumption holds.

Assumption 1: There exists a C^1 function h such that

$$\mathrm{col}(y_1, y_2) = h(x_1, x_2, u_1, u_2, d(t))$$

is the unique solution of the equations

$$\begin{aligned} y_1 &= h_1(x_1, y_2, u_1, d(t)) \\ y_2 &= h_2(x_2, y_1, u_2, d(t)). \end{aligned}$$

Theorem 1. [**6**] *Consider the system composed of (7) and (8) with $d(t) = t$. Assume that subsystem (7) is IOS viewing y_1 as output and col(y_2, u_1) as input and (8) is IOS viewing y_2 as output and col(y_1, u_2) as input, i.e., there exist class $\mathcal{KL}$ functions β_1, β_2, class $\mathcal{K}$ functions γ_1^y, γ_2^y, γ_1^u, γ_2^u, independent of $d(t)$, such that, for any $x_1(t_0) \in R^{n_1}$, col$(y_2, u_1) \in L_\infty^{p_2+m_1}$, the solution of (7) exists and satisfies, for all $t \geq t_0$,*

$$\|y_1(t)\| \le \beta_1(\|x_1(t_0)\|, t-t_0) + \gamma_1^y\left(\|y_{2[t_0,t]}\|\right) + \gamma_1^u\left(\|u_{1[t_0,t]}\|\right), \tag{9}$$

and for any $x_2(t_0) \in R^{n_2}$, $col(y_1, u_2) \in L_\infty^{p_1+m_2}$, the solution of (8) exists and satisfies, for all $t \ge t_0$,

$$\|y_2(t)\| \le \beta_2(\|x_2(t_0)\|, t-t_0) + \gamma_2^y\left(\|y_{1[t_0,t]}\|\right) + \gamma_2^u\left(\|u_{2[t_0,t]}\|\right). \tag{10}$$

Further assume that systems (7) and (8) have the UO property. Then, under Assumption 1 and the condition that there exist two $\mathcal{K}_\infty$ functions ρ_1 and ρ_2 such that

$$(Id+\rho_1)\circ\gamma_1^y\circ(Id+\rho_2)\circ\gamma_2^y(r) \le r,\ \forall r \ge 0 \tag{11}$$
$$(Id+\rho_2)\circ\gamma_2^y\circ(Id+\rho_1)\circ\gamma_1^y(r) \le r,\ \forall r \ge 0, \tag{12}$$

the system composed of (7) and (8) is IOS viewing $x = col(x_1, x_2)$ as state, $y = col(y_1, y_2)$ as output and $u = col(u_1, u_2)$ as input, and has the UO property. In particular, for any $\mathcal{K}_\infty$ functions γ_3 and ρ_3, there exists a class $\mathcal{KL}$ function β such that

$$\|y(t)\| \le \beta(\|x(t_0)\|, t-t_0) + \gamma\left(\|u_{[t_0,t]}\|\right),\ \forall t \ge t_0, \tag{13}$$

where $\gamma = \gamma_1 + \gamma_2 + \gamma_3$ with

$$\gamma_1(r) = (Id+\rho_1^{-1})\circ(Id+\rho_3)^2\circ[\gamma_1^u + \gamma_1^y\circ(Id+\rho_2^{-1})\circ(Id+\rho_3)^2\circ\gamma_2^u](r)$$
$$\gamma_2(r) = (Id+\rho_2^{-1})\circ(Id+\rho_3)^2\circ[\gamma_2^u + \gamma_2^y\circ(Id+\rho_1^{-1})\circ(Id+\rho_3)^2\circ\gamma_1^u](r).$$

□

3 Small Gain Theorem

In this section, we will give a more clear-cut version of the small gain theorem for the same inter-connection as shown in Fig. 1. Our proof will based on the following technical lemma which is established in [1].

Lemma 1. *Let β be a class $\mathcal{KL}$ function, γ a class $\mathcal{K}$ function such that $\gamma(r) < r$, $\forall r > 0$, and $\mu \in (0,1]$ a real number. For any nonnegative real numbers s and M, and any nonnegative real function $z(t) \in L_\infty^1$ satisfying*

$$z(t) \le \max\left\{\beta(s,t), \gamma\left(\|z_{[\mu t,t]}\|\right), M\right\},\ \forall t \ge 0, \tag{14}$$

there exists a class $\mathcal{K}_\infty$ function $\hat{\beta}$ such that

$$z(t) \le \max\left\{\hat{\beta}(s,t), M\right\},\ \forall t \ge 0. \tag{15}$$

□

Remark 4. As pointed out in [1], Lemma 1 plays the same role in the proof of Theorem 2 as Lemma A.1 of [7] does in the proof of Theorem 1. However, instead of assuming the condition $r - \gamma(r) > 0, \forall r > 0$ of Lemma 1, Lemma A.1 of [7] has assumed a more complex condition as follows

$$r - \gamma(r) > \epsilon(r), \forall r > 0$$

where $\epsilon(\cdot)$ is some class $\mathcal{K}_\infty$ function. □

Theorem 2. *Assume that subsystem (7) is RIOS viewing y_1 as output and $col(y_2, u_1)$ as input and (8) is RIOS viewing y_2 as output and $col(y_1, u_2)$ as input, i.e., there exist class $\mathcal{KL}$ functions β_1, β_2, class $\mathcal{K}$ functions γ_1^y, γ_2^y, γ_1^u, γ_2^u, independent of $d(t)$, such that, for any $x_1(t_0) \in R^{n_1}$, $col(y_2, u_1) \in L_\infty^{p_2+m_1}$, the solution of (7) exists and satisfies, for all $t \geq t_0$,*

$$\|y_1(t)\| \leq \max\{\beta_1(\|x_1(t_0)\|, t - t_0), \gamma_1^y\left(\|y_{2[t_0,t]}\|\right), \gamma_1^u\left(\|u_{1[t_0,t]}\|\right)\}, \quad (16)$$

and for any $x_2(t_0) \in R^{n_2}$, $col(y_1, u_2) \in L_\infty^{p_1+m_2}$, the solution of (8) exists and satisfies, for all $t \geq t_0$,

$$\|y_2(t)\| \leq \max\{\beta_2(\|x_2(t_0)\|, t - t_0), \gamma_2^y\left(\|y_{1[t_0,t]}\|\right), \gamma_2^u\left(\|u_{2[t_0,t]}\|\right)\}. \quad (17)$$

Further assume that systems (7) and (8) have the RUO property. Then, under Assumption 1 and the condition

$$\gamma_1^y \circ \gamma_2^y(r) < r, \ \forall r > 0, \quad (18)$$

the system composed of (7) and (8) is RIOS viewing $x = col(x_1, x_2)$ as state, $y = col(y_1, y_2)$ as output and $u = col(u_1, u_2)$ as input, and has the RUO property. In particular,

$$\|y(t)\| \leq \max\left\{\beta(\|x(t_0)\|, t - t_0), \gamma\left(\|u_{[t_0,t]}\|\right)\right\}, \ \forall t \geq t_0, \quad (19)$$

for some class $\mathcal{KL}$ function β, and any class $\mathcal{K}_\infty$ function γ satisfying

$$\gamma(r) \geq \max\{2\gamma_1^y \circ \gamma_2^u(r), 2\gamma_1^u(r), 2\gamma_2^y \circ \gamma_1^u(r), 2\gamma_2^u(r)\}, \ \forall r \geq 0. \quad (20)$$

Proof: First, we note that the inequality (18) and the following one,

$$\gamma_2^y \circ \gamma_1^y(r) < r, \ \forall r > 0, \quad (21)$$

imply each other [5].

Now we will prove that the solution of the system composed of (7) and (8) is bounded for all $t \geq t_0$. To this end, we will first prove that the outputs y_1 and y_2 are bounded in a way similar to the proof of Theorem 10.6.1 of [5]. Suppose this is not the case, for every number $R > 0$, there exists a time $T > t_0$, such that $\|y_1(T)\| > R$ or $\|y_2(T)\| > R$. Without lose of generality, we only consider the case where $\|y_1(T)\| > R$. Choose R such that

$$R > \max\left\{\beta_1(\|x_1(t_0)\|,0), \gamma_1^y \circ \beta_2(\|x_2(t_0)\|,0), \gamma_1^y \circ \gamma_2^u\left(\left\|u_{2[t_0,\infty)}\right\|\right),\right.$$
$$\left.\gamma_1^u\left(\left\|u_{1[t_0,\infty)}\right\|\right)\right\}.$$

¿From (16) and (17), we have

$$\left\|y_{1[t_0,T]}\right\| \le \max\{\beta_1(\|x_1(t_0)\|,0), \gamma_1^y\left(\left\|y_{2[t_0,T]}\right\|\right), \gamma_1^u\left(\left\|u_{1[t_0,T]}\right\|\right)\} \quad (22)$$
$$\left\|y_{2[t_0,T]}\right\| \le \max\{\beta_2(\|x_2(t_0)\|,0), \gamma_2^y\left(\left\|y_{1[t_0,T])}\right\|\right), \gamma_2^u\left(\left\|u_{2[t_0,T]}\right\|\right)\}. \quad (23)$$

Substituting (23) into (22) gives

$$\left\|y_{1[t_0,T]}\right\| \le \max\left\{\beta_1(\|x_1(t_0)\|,0), \gamma_1^y \circ \beta_2(\|x_2(t_0)\|,0), \gamma_1^y \circ \gamma_2^y\left(\left\|y_{1[t_0,T]}\right\|\right),\right.$$
$$\left.\gamma_1^y \circ \gamma_2^u\left(\left\|u_{2[t_0,T]}\right\|\right), \gamma_1^u\left(\left\|u_{1[t_0,T]}\right\|\right)\right\}.$$

Since

$$\left\|y_{1[t_0,T]}\right\| > \gamma_1^y \circ \gamma_2^y\left(\left\|y_{1[t_0,T]}\right\|\right),$$

we have

$$\left\|y_{1[t_0,T]}\right\| \le \max\left\{\beta_1(\|x_1(t_0)\|,0), \gamma_1^y \circ \beta_2(\|x_2(t_0)\|,0), \gamma_1^y \circ \gamma_2^u\left(\left\|u_{2[t_0,\infty)}\right\|\right),\right.$$
$$\left.\gamma_1^u\left(\left\|u_{1[t_0,\infty)}\right\|\right)\right\} < R, \quad (24)$$

which contradicts $\|y_1(T)\| > R$. Therefore, the output of the system composed of (7) and (8) is bounded for all $t \ge t_0$. And by the RUO assumptions, the solution of the inter-connected system is bounded.

The RUO property of the inter-connected system follows straightforwardly from the RUO properties of the two subsystems. The remaining part is to prove the RIOS of the inter-connected system, which can be given in a way parallel to the proof of Theorem 2.1 of [7]. Let

$$M_1 = \max\left\{\gamma_1^y \circ \gamma_2^u(\left\|u_{2[t_0,\infty)}\right\|), \gamma_1^u(\left\|u_{1[t_0,\infty)}\right\|)\right\}$$
$$M_2 = \max\left\{\gamma_2^y \circ \gamma_1^u(\left\|u_{1[t_0,\infty)}\right\|), \gamma_2^u(\left\|u_{2[t_0,\infty)}\right\|)\right\}.$$

Using (24), and the symmetry of y_1 and y_2, we have, for all $t \ge t_0 \ge 0$,

$$\begin{aligned}\|y_1(t)\| &\le \max\left\{\beta_1(\|x_1(t_0)\|,0), \gamma_1^y \circ \beta_2(\|x_2(t_0)\|,0), M_1\right\}\\ &\le \max\left\{\delta_1(\|x(t_0)\|), M_1\right\}\\ \|y_2(t)\| &\le \max\left\{\beta_2(\|x_2(t_0)\|,0), \gamma_2^y \circ \beta_1(\|x_1(t_0)\|,0), M_2\right\}\\ &\le \max\left\{\delta_2(\|x(t_0)\|), M_2\right\}\end{aligned}$$

with $\delta_1(s) = \max\{\beta_1(s,0), \gamma_1^y \circ \beta_2(s,0)\}$ and $\delta_2(s) = \max\{\beta_2(s,0), \gamma_2^y \circ \beta_1(s,0)\}$. Therefore,

$$\begin{aligned}\|y(t)\| &\le \|y_1(t)\| + \|y_2(t)\|\\ &\le \max\left\{2\delta_1(\|x(t_0)\|), 2\delta_2(\|x(t_0)\|), 2M_1, 2M_2\right\}\\ &= \max\left\{\delta_3(\|x(t_0)\|), M_3\right\} \stackrel{\text{def}}{=} y_\infty\end{aligned} \quad (25)$$

with $\delta_3(s) = \max\{2\delta_1(s), 2\delta_2(s)\}$ and $M_3 = \gamma(\|u_{[t_0,\infty]}\|)$. Hence, $\|y_{[t_0,\infty)}\| \le y_\infty$.

Since the inter-connected system has the RUO property, there exist class $\mathcal{K}$ functions $\alpha_x, \alpha_u, \alpha_y$ such that, for $t \ge t_0$,

$$\begin{aligned}\|x(t)\| &\le \max\left\{\alpha_x(\|x(t_0)\|), \alpha_u\left(\|u_{[t_0,t]}\|\right), \alpha_y\left(\|y_{[t_0,t]}\|\right)\right\} \\ &\le \max\left\{\alpha_x(\|x(t_0)\|), \alpha_u\left(\|u_{[t_0,\infty]}\|\right), \alpha_y(y_\infty)\right\} \stackrel{\text{def}}{=} x_\infty.\end{aligned}$$

Next, for any time $t_1 \ge 0$, we have

$$\begin{aligned}&\|y_1(t_1+t_0)\| \\ &< \max\{\beta_1(\|x_1(t_1/2+t_0)\|, t_1/2), \gamma_1^y\left(\|y_{2[t_1/2+t_0,t_1+t_0]}\|\right), \gamma_1^u\left(\|u_{1[t_0,\infty)}\|\right)\} \\ &\le \max\left\{\beta_1(x_\infty, t_1/2), \gamma_1^y\left(\|y_{2[t_1/2+t_0,t_1+t_0]}\|\right), \gamma_1^u\left(\|u_{1[t_0,\infty)}\|\right)\right\} \qquad (26)\end{aligned}$$

and for $\tau \in [t_1/2, t_1]$,

$$\begin{aligned}&\|y_2(\tau+t_0)\| \\ &\le \max\{\beta_2(\|x_2(t_1/4+t_0)\|, \tau - t_1/4), \gamma_2^y\left(\|y_{1[t_1/4+t_0,\tau+t_0]}\|\right), \gamma_2^u\left(\|u_{2[t_0,\infty)}\|\right)\} \\ &\le \max\left\{\beta_2(x_\infty, t_1/4), \gamma_2^y\left(\|y_{1[t_1/4+t_0,t_1+t_0]}\|\right), \gamma_2^u\left(\|u_{2[t_0,\infty)}\|\right)\right\}. \qquad (27)\end{aligned}$$

Substituting (27) into (26) gives

$$\begin{aligned}&\|y_1(t_1+t_0)\| \\ &\le \max\{\beta_1(x_\infty, t_1/2), \gamma_1^y \circ \beta_2(x_\infty, t_1/4), \gamma_1^y \circ \gamma_2^y\left(\|y_{1[t_1/4+t_0,t_1+t_0]}\|\right), \\ &\quad \gamma_1^y \circ \gamma_2^u\left(\|u_{2[t_0,\infty)}\|\right), \gamma_1^u\left(\|u_{1[t_0,\infty)}\|\right)\} \\ &\le \max\left\{\bar{\beta}_1(x_\infty, t_1), \gamma_1^y \circ \gamma_2^y\left(\|y_{1[t_1/4+t_0,t_1+t_0]}\|\right), M_1\right\}\end{aligned}$$

for a class $\mathcal{KL}$ function $\bar{\beta}_1$ satisfying

$$\bar{\beta}_1(s,t) \ge \max\{\beta_1(s,t/2), \gamma_1^y \circ \beta_2(s,t/4)\}.$$

Next, denoting $z_1(t_1) = \|y_1(t_1+t_0)\|$ gives

$$z_1(t_1) \le \max\left\{\bar{\beta}_1(x_\infty, t_1), \gamma_1^y \circ \gamma_2^y\left(\|z_{1[t_1/4,t_1]}\|\right), M_1\right\}.$$

By Lemma 1, there exists a class $\mathcal{KL}$ function $\hat{\beta}_1$ such that $z_1(t_1) \le \max\{\hat{\beta}_1(x_\infty, t_1), M_1\}$. Hence,

$$\|y_1(t)\| \le \max\{\hat{\beta}_1(x_\infty, t - t_0), M_1\}, \ \forall t \ge t_0.$$

By symmetry of y_1 and y_2, we have,

$$\|y_2(t)\| \le \max\{\hat{\beta}_2(x_\infty, t - t_0), M_2\}, \ \forall t \ge t_0,$$

for some class $\mathcal{KL}$ function $\hat{\beta}_2$. As a result,

$$\begin{aligned}\|y(t)\| &\le \|y_1(t)\| + \|y_2(t)\| \\ &\le \max\{2\hat{\beta}_1(x_\infty, t-t_0), 2\hat{\beta}_2(x_\infty, t-t_0), 2M_1, 2M_2\} \\ &\le \max\{\beta_3(x_\infty, t-t_0), M_3\},\ \forall t \ge t_0 \end{aligned} \quad (28)$$

for $\beta_3(s,t) = \max\{2\hat{\beta}_1(s,t), 2\hat{\beta}_2(s,t)\}$. To complete the proof of the theorem, we need to show that inequality (28) implies inequality (19). To this end, consider the following two cases.
(i) $\delta_3(\|x(t_0)\|) \ge M_3$: We have $y_\infty = \delta_3(\|x(t_0)\|)$, and $\gamma^{-1} \circ \delta_3(\|x(t_0)\|) \ge \gamma^{-1}(M_3) = \left\|u_{[t_0,\infty]}\right\|$. Hence,

$$x_\infty = \max\left\{\alpha_x(\|x(t_0)\|), \alpha_u\left(\left\|u_{[t_0,\infty]}\right\|\right), \alpha_y \circ \delta_3(\|x(t_0)\|)\right\} \le \delta_4(\|x(t_0)\|)$$

for some class $\mathcal{K}$ function δ_4 satisfying

$$\delta_4(r) \ge \max\left\{\alpha_x(r), \alpha_u \circ \gamma^{-1} \circ \delta_3(r), \alpha_y \circ \delta_3(r)\right\}.$$

As a result, (28) gives

$$\|y(t)\| \le \max\{\beta_3(\delta_4(\|x(t_0)\|), t-t_0), M_3\},\ \forall t \ge t_0$$

(ii) $\delta_3(\|x(t_0)\|) \le M_3$: From (25), we have $\|y(t)\| \le M_3$ directly.
In both cases, we have obtained the following inequality

$$\|y(t)\| \le \max\left\{\beta(\|x(t_0)\|, t-t_0), \gamma\left(\left\|u_{[t_0,\infty)}\right\|\right)\right\},\ \forall t \ge t_0 \quad (29)$$

for some class $\mathcal{KL}$ function $\beta(s,t) = \beta_3(\delta_4(s),t)$. Since the solution $y(t)$ depends only on $u(\tau)$ for $t_0 \le \tau \le t$, the supremum on the right-hand side of (29) can be taken over $[t_0, t]$ which yields (19). As a result, the proof is completed □

Remark 5. For the class of time-invariant systems, Teel showed that a system is ISS if and only if the system has the *globally stable (GS)* property and *asymptotic gain (AG)* property, respectively, that is, there exist class $\mathcal{K}$ functions γ_0 and γ such that for any initial state $x(0) \in R^n$ and any input function $u(t) \in L_\infty^m$, the solution $x(t)$ exists and satisfies

$$\left\|x_{[0,\infty)}\right\| \le \max\left\{\gamma_0(\|x(0)\|), \gamma\left(\left\|u_{[0,\infty)}\right\|\right)\right\}, \quad (30)$$

and, respectively,

$$\lim_{t\to\infty}\left\|x_{[t,\infty)}\right\| \le \gamma\left(\lim_{t\to\infty}\left\|u_{[t,\infty]}\right\|\right). \quad (31)$$

Using this equivalent relation, Isidori has simplified the small gain condition for time-invariant systems to exactly the same inequality (18). Nevertheless, Isidori's proof cannot be carried over to the proof of the small gain theorem for time-varying systems because the above equivalent characterization of ISS property does not hold for time-varying systems as shown in [1].
□

Remark 6. In the special case where $h_1(x_1, y_2, u_1, d(t)) = x_1$, $h_2(x_2, y_1, u_2, d(t)) = x_2$, and $u_1 = u_2$, systems (7) and (8) become the same as systems (17) and (18) in [1] and Theorem 2 reduces to Theorem 2.1 of [1]. □

4 Conclusion

In this paper, we have given a simplified version of the small gain theorem for a general inter-connection of two RIOS and RUO subsystems. The result generalized our previous result in [1]. This result has closed the gap between the small gain theorem for time-varying systems and the small gain theorem for time-invariant systems. It is also noted that the simplified small gain theorem can find applications in some global robust stabilization and global robust output regulation problems [2] - [4].

References

1. Z. Chen and J. Huang (2004), A variation of the small gain theorem, *Proceedings of the 8th International Conference on Control, Automation, Robotics and Vision,* pp.338-343.
2. Z. Chen and J. Huang (2004), Dissipativity, stabilization, and regulation of cascade-connected systems, *IEEE Transactions on Automatic Control*, vol. 49, pp. 635-650.
3. Z. Chen and J. Huang (2004), Global robust servomechanism of lower triangular systems in the general case, *Systems and Control Letters,* vol. 52, pp. 209-220.
4. J. Huang and Z. Chen (2004), A general framework for tackling the output regulation problem, *IEEE Transactions on Automatic Control*, vol. 49, pp.2203-2218.
5. A. Isidori (1999), *Nonlinear Control Systems, Volume II*, Springer Verlag, New York.
6. Z.P. Jiang and I. Mareels (1997), A small-gain control method for nonlinear cascaded systems with dynamic uncertainties, *IEEE Transactions on Automatic Control*, vol. 42, pp. 292-308.
7. Z.P. Jiang, A. Teel, and L. Praly (1994), Small-gain theorem for ISS systems and applications, *Mathematics of Control, Signals and Systems*, vol. 7, pp. 95-120.
8. Y. Lin (1992), Lyapunov function techniques for stabilization, Ph.D. thesis, Rutgers University.
9. Y. Lin, E.D. Sontag, and Y. Wang (1993), Recent results on Lyapunov theoretic techniques for nonlniear stability, *Report SYCON-93-09.*
10. Y. Lin, E.D. Sontag, and Y. Wang (1995), Input to state stability for parameterized families of systems, *International Journal of Robust and Nonlinear Control*, vol. 5, pp.187-205.
11. E.D. Sontag (1989), Smooth stabilization implies coprime factorization, *IEEE Transactions on Automatic Control*, vol. 34, pp. 435 - 443.
12. E.D. Sontag and Y. Wang (1996), New characterizations of input-to-state stability, *IEEE Transactions on Automatic Control*, vol. 41, pp. 1283-1294.

A Survey on the Stabilization Control for Stochastic Nonlinear Systems*

Yun-Gang Liu[1] and Ji-Feng Zhang[2]

[1] School of Control Science and Engineering, Shandong University, Ji'nan 250061, P.R. China lygfr@263.net
[2] Institute of Systems Science, Chinese Academy of Sciences, Beijing 100080, P.R.China jif@iss.ac.cn

1 Introduction

The stabilization control of stochastic nonlinear systems is an important problem in control theory. Since its emergence in the 1960's, much progress has been obtained from stabilization to optimization, including the risk-sensitive control problem and other problems. The research of stochastic stabilization was renewed by Florchinger in 1990's in [11, 12], where the *control Lyapunov method* in control design of stochastic affine nonlinear systems was studied. Then the design of the controls for strict-feedback stochastic nonlinear systems was considered by means of state-feedback ([8], [12], [24]) and output-feedback ([1], [6], [7], [18], [19], [21], [22], [27]). Under the assumption (A): "the disturbance vector field vanishes at the origin", [7], [8] and [12] studied the problem of designing a control to asymptotically stabilize the closed-loop systems in the large. While [1], [6], [18], [19], [21], [22], [24] and [27] considered the control design to achieve the boundedness in probability of the closed-loop system without using the assumption (A). Specifically, [8] considered the disturbance attenuation problem; [6] and [27] considered the stabilization problem of systems with stable zero-dynamics; [1], [18], [21] and [24] considered the design of satisfaction control under a quadratic, a quartic regulation and quadratic tracking risk-sensitive cost criterion, respectively. [1] used the assumption (B): "the gain functions of stochastic noise are uniformly bounded", while [18], [21] and [24] did not; [19] and [22] considered the reduced-order observer-based stabilization control design of the single-input multi-output stochastic nonlinear systems.

Here we would like to give a brief survey on the stability theory of stochastic systems, and present some recent results on control design of stochastic nonlinear systems.

*This work was supported by the National Natural Science Foundation of China.
All correspondences should be addressed to the second author: Ji-Feng Zhang.

The remainder of this paper is organized as follows. Section 2 recalls the stability theory of stochastic systems first. Section 3 gives control Lyapunov method and gives the stochastic version of Sontag's formula. Section 4 discusses the stabilization control design by backstepping approach. Section 5 discusses risk-sensitive control. Section 6 includes some concluding remarks.

2 Stochastic Stability Theory

In this section, some results on stability theory of stochastic nonlinear systems will be summarized, which have widely been used in the stability analysis of closed-loop stochastic nonlinear control systems. For more details, the readers are referred to the classical books [9, 17].

Consider the system with stochastic noise

$$dx = f(t, x)dt + h(t, x)dw, \tag{1}$$

where $x \in \mathbb{R}^n$ is system state with initial condition $x(0) = x_0$, $w(t) \in \mathbb{R}^s$ is a standard Brownian motion, defined on a probability space $(\Omega, \mathcal{F}, \mathcal{P})$; and $f(\cdot, \cdot) : \mathbb{R}_+ \times \mathbb{R}^n \to \mathbb{R}^n$ and $h(\cdot, \cdot) : \mathbb{R}_+ \times \mathbb{R}^n \to \mathbb{R}^{n\times s}$ are Lipschitz continuous and denote the nonlinearity and disturbance coefficient of the system, respectively.

For convenience of citation, in the sequel, for a neighborhood $\mathbb{U}$ of the origin of $\mathbb{R}^n$ we will use $\mathcal{C}^{1,2}(\mathbb{R}_+ \times \mathbb{U})$ to denote the class of such functions of the form $V(t, x)$ defined in the domain $\mathbb{R}_+ \times \mathbb{U}$, once continuously differentiable with respect to t, and twice continuously differentiable with respect to x. For any given $V(t, x) \in \mathcal{C}^{1,2}(\mathbb{R}_+ \times \mathbb{R}^n)$, we define a differential operator $\mathcal{L}$ associated with system (1) as follows

$$\mathcal{L}V = \frac{\partial V}{\partial t} + \frac{\partial V}{\partial x} f(t, x) + \frac{1}{2}\mathrm{tr}\left(\frac{\partial^2 V}{\partial x^2} h(t, x)h^\tau(t, x)\right). \tag{2}$$

Before presenting the stability theory for stochastic system (1), we would like to review some auxiliary results about the properties of martingales and supermartingales first.

Let $(\Omega, \mathcal{F}, \mathcal{P})$ be a probability space, $\mathcal{F}_t \subset \mathcal{F}$ be a nondecreasing σ-algebra family of event in Ω, that is $\mathcal{F}_s \subset \mathcal{F}_t$ for $s < t$. Suppose that $\{\xi(t), \mathcal{F}_t\}$ is an adapted stochastic process with finite expectation. If $\mathbf{E}(\xi(t)|\mathcal{F}_s) \leq \xi(s)$ a.s. for any $s < t$, then $\{\xi(t), \mathcal{F}_t\}$ is called a supermartingale. If $\mathbf{E}(\xi(t)|\mathcal{F}_s) = \xi(s)$, then $\{\xi(t), \mathcal{F}_t\}$ is called a martingale.

Lemma 1. *Let $\mathbb{U}$ be a neighborhood of the origin of $\mathbb{R}^n$, and $\tau(t) = \min(\tau_{\mathbb{U}}, t)$, where $\tau_{\mathbb{U}}$ is the first exit time from $\mathbb{U}$ of the solution process $\{x(t), t \geq 0\}$ of system (1). Suppose there is a function $V(t, x) \in \mathcal{C}^{1,2}(\mathbb{R}_+ \times \mathbb{U})$ such that in $\mathbb{R}_+ \times \mathbb{U}$, $V(t, x)$ is bounded and $\mathcal{L}V(t, x) \leq 0$. Then for any initial state $x(0) \in \mathbb{U}$, the process $V(\tau_{\mathbb{U}}(t), x(\tau_{\mathbb{U}}(t)))$ is a supermartingale, and hence,*

$$\mathbf{E}(V(\tau_{\mathbb{U}}(t), x(\tau_{\mathbb{U}}(t)))) \leq \mathbf{E}(V(0, x(0))).$$

This lemma is a corollary of Lemma 2.4 in Chapter 5 of [17]. The applications of martingale theory to stability analysis of stochastic nonlinear systems are based on the following theorems which are listed here without proof.

Theorem 1. *If* $\{\xi(t),\, \mathcal{F}_t,\, t \geq 0\}$ *is a positive supermartingale, then the limit* $\xi_\infty = \lim_{t\to\infty} \xi(t)$ *exists almost surely and is finite. Moreover,* $\mathbf{E}(\xi_\infty) = \lim_{t\to\infty} \mathbf{E}(\xi(t))$.

Theorem 2. *Let* $w(t)$ *be* m*-dimensional vector-valued standard Wiener process on the time interval* $[0,\, \infty)$ *adapted to a nondecreasing* σ*-algebra family* $\mathcal{F}_t$ *of a probability space* $(\Omega,\, \mathcal{F},\, \mathcal{P})$, $g : [0,\, t_f] \times \Omega \to \mathbb{R}^m$ *be a vector-valued stochastic process adapted to* $\mathcal{F}_t$. *Set*

$$\zeta_0^t(g) = \exp\left(\int_0^t g^\tau(s)dw - \frac{1}{2}\int_0^t \|g(s)\|^2 ds\right).$$

Then, $\{\zeta_0^t(g),\, \mathcal{F}_t\}$ *is a supermartingale, and satisfies* $\mathbf{E}(\zeta_0^t(g)) \leq 1, \quad \forall t \in [0,\, \infty)$.

In the following subsections, various definitions on stochastic stability distributed in the literature are presented, and for each of them, at least one sufficiency theorem is given.

2.1 Stability in probability

Definition 1. *For the system (1), with* $f(t,\, 0) = 0$ *and* $h(t,\, 0) = 0$, *its solution* $x(t) = 0$ *is said to be stable in probability if for any initial state* x_0 *and some constant* $\varepsilon > 0$, *there exists*

$$\lim_{x_0\to 0} \mathbf{P}\left\{\sup_{t\geq 0} \|x(t)\| > \varepsilon\right\} = 0.$$

In order to state the main result of this subsection, we fist introduce the definitions of positively definite and radial unboundedness.

Definition 2. *A scalar function* $V : \mathbb{R}^n \to \mathbb{R}$ *is positively definite (semi-positively definite) if* $V(0) = 0$ *and* $V(x) > 0(\geq 0)$, $\forall x \in \mathbb{R}^n \setminus \{0\}$. V *is radially unbounded if* $V(x) \to \infty$ *as* $\|x\| \to \infty$.

For *stability in probability*, we have the following theorem which is analogous to the well-known theorem of Lyapunov for deterministic systems.

Theorem 3. *Assume that there is a function* $V(t, x) \in C^{1,2}(\mathbb{R}_+ \times \mathbb{R}^n)$ *satisfying* $W_1(x) \leq V(t,\, x) \leq W_2(x)$ *with* $W_1(x)$ *and* $W_2(x)$ *being positively definite. If* $f(t,\, 0) = 0$, $h(t,\, 0) = 0$, *and along system (1),*

$$\mathcal{L}V(t,\, x) \leq 0,$$

then the solution $x(t) = 0$ *of system (1) is stable in probability.*

Proof. Let ε be a number such that the ε-neighborhood $\mathbb{U}_\varepsilon$ of the origin is contained in $\mathbb{R}^n$. Set $V_\varepsilon = \inf_{t\geq 0,\, x\in\mathbb{R}^n\setminus\mathbb{U}_\varepsilon} V(t,\, x)$. By Lemma 1, for $\|x(0)\| < \varepsilon$ we have

$$\mathbf{E}(V(\tau_{\mathbb{U}_\varepsilon}(t),\, x(\tau_{\mathbb{U}_\varepsilon}(t)))) \leq \mathbf{E}(V(0,\, x(0))).$$

This together with *Chebysev's Inequality* leads to

$$\mathbf{P}\left\{\sup_{0\leq s\leq t} \|x(s)\| > \varepsilon\right\} = \mathbf{P}\left\{\tau_{\mathbb{U}_\varepsilon}(t) < t\right\}$$
$$\leq \frac{\mathbf{E}(V(\tau_{\mathbb{U}_\varepsilon}(t),\, x(\tau_{\mathbb{U}_\varepsilon}(t))))}{V_\varepsilon} \leq \frac{\mathbf{E}(V(0,\, x(0)))}{V_\varepsilon}.$$

Letting $t \to \infty$, we finally get

$$\mathbf{P}\left\{\sup_{0\leq s} \|x(s)\| > \varepsilon\right\} \leq \frac{\mathbf{E}(V(0,\, x(0)))}{V_\varepsilon}.$$

This implies the desired assertion, since $V(0,\, 0) = 0$ and the function $V(0,\, x)$ is continuous with respect to x. □

2.2 Asymptotic stability in probability

Definition 3. *For the system (1), with $f(t,\, 0) = 0$ and $h(t,\, 0) = 0$, its solution $x(t) = 0$ is said to be asymptotically stable in probability if it is stable in probability and satisfies*

$$\lim_{x_0\to 0} \mathbf{P}\left\{\lim_{t\to\infty} x(t) = 0\right\} = 1,$$

for any initial state x_0.

Definition 4. *For the system (1), with $f(t,\, 0) = 0$ and $h(t,\, 0) = 0$, its solution $x(t) = 0$ is said to be asymptotically stable in the large if it is asymptotically stable in probability and also for any initial state x_0,*

$$\mathbf{P}\left\{\lim_{t\to\infty} x(t) = 0\right\} = 1.$$

The following theorem gives a sufficient condition for *asymptotical stability.*

Theorem 4. *Let $\mathbb{U} \subset \mathbb{R}^n$ be a neighborhood of the origin, and $V(t,\, x)$ be a function in $\mathcal{C}^{1,2}(\mathbb{R}_+ \times \mathbb{U})$ satisfying $W_1(x) \leq V(t,\, x) \leq W_2(x)$ with $W_1(x)$ and $W_2(x)$ being positively definite. Suppose that for any sufficiently small $r > \varepsilon > 0$, the solution of system (1) with $f(t,\, 0) = 0$ and $h(t,\, 0) = 0$, beginning in the set $\{x :\ \varepsilon < \|x(t)\| < r\}$, reaches the boundary of this set in finite time almost surely. If along system (1),*

$$\mathcal{L}V(t,\, x) \leq 0,$$

then the solution $x(t) = 0$ of system (1) is asymptotically stable in probability.

If W_1 and W_2 are radially unbounded, and the set $\mathbb{U}$ is replaced by $\mathbb{R}^n$, then we can get the following theorem about asymptotic stability in the large.

Theorem 5. *Assume that there exists a function* $V(t, x) \in \mathcal{C}^{1,2}(\mathbb{R}_+ \times \mathbb{R}^n)$ *satisfying* $W_1(x) \leq V(t, x) \leq W_2(x)$ *with* $W_1(x)$ *and* $W_2(x)$ *being positively definite and radially unbounded. If along systems (1) with* $f(t, 0) = 0$ *and* $h(t, 0) = 0$, *there exists positively definite and radially unbounded function* $W(x)$ *such that*

$$\mathcal{L}V(t, x) \leq -W(x),$$

then the solution $x(t) = 0$ *of system (1) is asymptotically stable in the large.*

2.3 Exponential stability in mean square

Definition 5. *The solution* $x(t) = 0$ *of system (1) is said to be exponentially stable in mean square if there exist positive constants* c_1 *and* c_2 *such that for any initial state* $x_0 \in \mathbb{R}^n$,

$$\mathbf{E}(\|x(t)\|^2) \leq c_1 \mathbf{E}(\|x_0\|^2) e^{-c_2 t}.$$

The following theorem presents a sufficient condition to *exponential stability in mean square.*

Theorem 6. *Suppose that there exist a function* $V(t, x) \in \mathcal{C}^{1,2}(\mathbb{R}_+ \times \mathbb{R}^n)$ *and positive constants* c_1, c_2 *and* c_2 *such that for all* $x \in \mathbb{R}^n$,

$$c_1\|x\|^2 \leq V(x) \leq c_2\|x\|^2, \qquad \mathcal{L}V(x) \leq -c_3\|x\|^2. \tag{3}$$

Then, the solution $x(t) = 0$ *of system (1) is exponentially stable in mean square.*

Proof. By (3) and stochastic differential equation (1), we have

$$\mathbf{E}(V(x(t))) \leq \mathbf{E}(V(x_0)) - c_3 \int_0^t \mathbf{E}(\|x(s)\|^2) ds,$$

and furthermore,

$$c_1 \mathbf{E}(\|x(t)\|^2) \leq \mathbf{E}(V(x_0)) - c_3 \int_0^t \mathbf{E}(\|x(s)\|^2) ds.$$

Using *Comparison Lemma* [16], we have

$$\mathbf{E}(\|x(t)\|^2) \leq g(t),$$

where $g(t)$ with initial condition $g(0) = \mathbf{E}(V(x_0))$ satisfies

$$g(t) = \mathbf{E}(V(x_0)) - c_3 \int_0^t g(s) ds.$$

From this equation it follows

$$g(t) = \frac{\mathbf{E}(V(x_0))}{c_1} e^{-c_1^{-1} c_3 t}.$$

Thus,

$$\mathbf{E}(\|x(t)\|^2) \leq \frac{\mathbf{E}(V(x_0))}{c_1} e^{-c_1^{-1} c_3 t} \leq \frac{c_2}{c_1} \mathbf{E}(\|x_0\|^2) e^{-c_1^{-1} c_3 t}.$$

This implies the exponential stability in mean square. □

2.4 Boundedness in probability

Definition 6. *The solution process $\{x(t),\, t \geq 0\}$ of system (1) is said to be bounded in probability if*

$$\lim_{c \to \infty} \sup_{t \in [0,\infty)} \mathbf{P}\{\|x(t)\| > c\} = 0.$$

In order to study the control design problem in the sense of *boundedness in probability*, we state the following theorem which gives a sufficient condition of *boundedness in probability* for the solution $x(t) = 0$ of the system (1) with $f(t,0) = 0$ and $h(t,0) = 0$.

Theorem 7. *Assume that there exists a function $V(t,\, x) \in \mathcal{C}^{1,2}(\mathbb{R}_+ \times \mathbb{R}^n)$ satisfying $W_1(x) \leq V(t,\, x) \leq W_2(x)$ with $W_1(x)$ and $W_2(x)$ being positively definite and radially unbounded. If along system (1), with $f(0) = 0$ and $h(0) = 0$, there exists positive definite and radially unbounded function $W(x)$ and a constant $\delta > 0$, such that*

$$\mathcal{L}V(t,\, x) \leq -W(x) + \delta,$$

then the solution process $\{x(t),\, t \geq 0\}$ of system (1) is bounded in probability.

In addition, if $\delta = 0$, then the solution $x(t) = 0$ is asymptotically stable in the large.

Proof. We decompose the sample space into two mutually exclusive events $A_1 = \{\omega : W(x) \geq \delta\}$ and $A_2 = \{\omega : W(x) < \delta\}$.

Because of the radial unboundedness of $W(x)$ and the finite boundedness of δ, it is clear that

$$\lim_{\|x\| \to \infty} \mathbf{P}\{A_2\} = \lim_{c \to \infty} \int_{A_2} \mathbf{I}_{\{\|x\| \geq c\}}(\omega) \mathcal{P}(d\omega) = 0.$$

Setting $\alpha(c) = \inf_{\|x\| \geq c} W_1(x)$, we have

$$\begin{aligned}
\lim_{c\to\infty} \mathbf{P}\{\|x\| \geq c\} &= \lim_{c\to\infty} \int_{A_1} \mathbf{I}_{\{\|x\|\geq c\}}(\omega)\mathcal{P}(d\omega) + \lim_{c\to\infty} \int_{A_2} \mathbf{I}_{\{\|x\|\geq c\}}(\omega)\mathcal{P}(d\omega) \\
&= \lim_{c\to\infty} \int_{A_1} \mathbf{I}_{\{\|x\|\geq c\}}(\omega)\mathcal{P}(d\omega) \\
&= \lim_{c\to\infty} \int_{A_1} \frac{\mathbf{I}_{\{\|x\|\geq c\}} V(t,\, x)}{V(t,\, x)} \mathcal{P}(d\omega) \\
&\leq \lim_{c\to\infty} \int_{A_1} \frac{\mathbf{I}_{\{\|x\|\geq c\}} V(t,\, x)}{W_1(x)} \mathcal{P}(d\omega) \leq \lim_{c\to\infty} \frac{\mathbf{E}(V(t,\, x))}{\alpha(c)} \\
&\leq \lim_{c\to\infty} \frac{\mathbf{E}(V(0,\, x_0)) + \sup_{W(x)\leq\delta} W_2(x)}{\alpha(c)} = 0.
\end{aligned}$$

Thus, the stochastic boundedness follows. □

This theorem will play an important role in the tracking or stabilization control designs of the stochastic nonlinear systems in the case where the noise coefficient functions do not vanish at the origin 0.

As a special case of Theorem 7, we have the following theorem.

Theorem 8. *Under the condition of Theorem 7, if in addition, there are $c_1 > 0$ and $c_2 \geq 0$ such that*

$$\mathcal{L}V(t,\, x) \leq -c_1 V(t,\, x) + c_2, \tag{4}$$

then the solution process $\{x(t),\, t \geq 0\}$ of systems (1) is bounded in probability.

When $c_2 = 0$, the solution $x(t) = 0$ is asymptotically stable in the large.

Proof. From (4) and stochastic differential equation (1), it follows that

$$\mathbf{E}(V(t,\, x)) \leq \mathbf{E}(V(0,\, x_0)) + c_2 t - c_1 \int_0^t \mathbf{E}(V(s,\, x))ds.$$

By *Comparison Lemma* [16] we have

$$\mathbf{E}(V(t,\, x)) \leq g(t),$$

where $g(t)$ with the initial value $\mathbf{E}(V(0,\, x_0))$ satisfies the following equation

$$g(t) = \mathbf{E}(V(0,\, x_0)) + c_2 t - c_1 \int_0^t g(s)ds,$$

which gives

$$g(t) = \mathbf{E}(V(0,\, x_0))e^{-c_1 t} + \frac{c_2}{c_1}(1 - e^{-c_1 t}).$$

Thus,

$$\mathbf{E}(V(t,\, x)) \leq \mathbf{E}(V(0,\, x_0))e^{-c_1 t} + \frac{c_2}{c_1}(1 - e^{-c_1 t}).$$

Let $\alpha(c) = \inf_{\|x\|\geq c} W_1(x)$. Then we have

$$\begin{aligned}\mathbf{P}\{\|x\| \geq c\} &= \int_{\Omega} \mathbf{I}_{\{\|x\|\geq c\}}(\omega)\mathcal{P}(d\omega) = \int_{\Omega} \frac{\mathbf{I}_{\{\|x\|\geq c\}}V(t,\, x)}{V(t,\, x)}\mathcal{P}(d\omega) \\ &\leq \frac{\mathbf{E}(V(t,\, x))}{\alpha(c)} \leq \frac{\mathbf{E}(V(0,\, x_0))e^{-c_1 t} + \frac{c_2}{c_1}(1 - e^{-c_1 t})}{\alpha(c)} \\ &\leq \frac{\mathbf{E}(V(0, x_0)) + c_2/c_1}{\alpha(c)}.\end{aligned}$$

Clearly, $\alpha(c) \to \infty$ as $c \to \infty$. Thus, the stochastic boundedness follows. □

The following theorem is the revised version of Theorem 8, which is often used in the adaptive control design for stochastic nonlinear systems.

Theorem 9. *Assume that there exist C^2 functions $V_1(x) : \mathbb{R}^n \to \mathbb{R}_+$ and $V_2(\theta) : \mathbb{R}^m \to \mathbb{R}_+$, and positive constants $c_1 > 0$, $c_2 \geq 0$ such that*

$$\mathcal{L}V(X) \leq -c_1 V_1(x) + c_2, \tag{5}$$

where $X = [x^\tau,\, \theta^\tau]^\tau$ and $V(X) = V_1(x) + V_2(\theta)$. If θ is uniformly bounded, then the solution process $\{x(t),\, t \geq 0\}$ of the stochastic differential equation (1) is bounded in probability and satisfies

$$\lim_{t\to\infty} \frac{1}{t}\mathbf{E}\left(\int_0^t V_1(x(s))ds\right) \leq \frac{c_2}{c_1}.$$

In addition, when $c_2 = 0$, the solution process is asymptotically stable in the large.

Proof. Set $\alpha(c) = \inf_{\|X\|\geq c} V(X)$. Then we have

$$\begin{aligned}\mathbf{P}\{\|X\| \geq c\} &= \int_{\Omega} \mathbf{I}_{\{\|X\|\geq c\}}(\omega)\mathcal{P}(d\omega) = \int_{\Omega} \frac{\mathbf{I}_{\{\|X\|\geq c\}}V(X)}{V(X)}\mathcal{P}(d\omega) \\ &\leq \frac{\mathbf{E}(V(X))}{\alpha(c)} \leq \frac{\mathbf{E}(V(X(0))) + \sup_{s\in[0,\,\infty)} V_2(\theta(s)) + c_2/c_1}{\alpha(c)}.\end{aligned}$$

Clearly, $\alpha(c) \to \infty$ as $c \to \infty$. Thus, the stochastic boundedness follows.

Secondly, we have

$$\mathbf{E}(V(X)) - \mathbf{E}(V(X(0))) \leq -c_1\mathbf{E}\left(\int_0^t V_1(x(s))ds\right) + c_2 t.$$

From this it follows that

$$\frac{1}{t}\mathbf{E}\left(\int_0^t V_1(x(s))ds\right) \leq \frac{c_2}{c_1} + \frac{1}{t} \cdot \frac{\mathbf{E}(V(X(0))) - \mathbf{E}(V(X))}{c_1},$$

which gives

$$\lim_{t\to\infty} \frac{1}{t}\mathbf{E}\left(\int_0^t V_1(x(s))ds\right) \leq \frac{c_2}{c_1}.$$

In addition, if $c_2 = 0$, then by Theorem 7, the solution process is asymptotically stable in the large. □

2.5 Noise-to-state stability

The concept of a noise-to-state Lyapunov function was introduced in [8], which is a stochastic extension of Songtag's ISS (Input-to-State Stability) Lyapunov function, and developed an adaptive stabilization scheme (based on tuning functions) that requires no *a priori* knowledge of a bound on the covariance of the stochastic noise. The purpose of this subsection is to present some concepts and conditions on noise-to-state stability of the system (1) for the case where the infinitesimal covariance $\Sigma(t) \triangleq \mathbf{E}(dw(t)(dw(t))^{\tau})/dt$ of the Brownian motion $w(t)$ is not identity.

Definition 7. $\gamma : \mathbb{R}_+ \to \mathbb{R}_+$ *is said to be a* $\mathcal{K}$ *function if it is continuous, strictly increasing and* $\gamma(0) = 0$. *It is said to be a* $\mathcal{K}_\infty$ *function if* $\gamma \in \mathcal{K}$ *and* $\gamma(r) \to \infty$ *as* $r \to \infty$.

Definition 8. $\gamma(\cdot, \cdot) : \mathbb{R}_+ \times \mathbb{R}_+ \to \mathbb{R}_+$ *is said to be a* $\mathcal{KL}$ *function if, for each fixed* s, $\gamma(r, s)$ *is a* $\mathcal{K}$ *function with respect to* r, *and for each fixed* r, $\gamma(r, s)$ *decreases to zero as* s *increases to* ∞.

Definition 9. *The solution* $x(t) = 0$ *of system (1) is said to be* noise-to-state stable in probability *if* $\forall \varepsilon \in (0, 1)$ *there exists a* $\mathcal{K}$ *function* $\gamma_1(\cdot)$ *and a* $\mathcal{KL}$ *function* $\gamma_2(\cdot, \cdot)$ *such that*

$$\mathbf{P}\left\{\|x(t)\| < \gamma_1\left(\sup_{0\le s\le t}\|\Sigma(s)\|\right) + \gamma_2\left(\mathbf{E}(\|x(0)\|), t\right)\right\} \ge 1 - \varepsilon.$$

The following theorem gives a sufficient condition for noise-to-state stability in probability.

Theorem 10. *Assume that there exists a* C^2 *function* $V : \mathbb{R}^n \to \mathbb{R}_+$, *a constant* $c > 0$, *and* $\mathcal{K}_\infty$ *functions* $\alpha_1, \alpha_2, \alpha_3$, *such that*

$$\alpha_1(\|x\|) \le V(x) \le \alpha_2(\|x\|), \quad \mathcal{L}V(x) \le -cV(x) + \alpha_3(\|\Sigma\|).$$

Then the solution $x(t) = 0$ *of system is* noise-to-state stable in probability.

Proof. By Theorem 8, we have

$$\mathbf{E}(V(x(t))) \le e^{-ct}\mathbf{E}(V(x(0))) + \frac{1}{c}\gamma\left(\sup_{0\le s\le t}\|\Sigma(s)\|\right)(1 - e^{-ct}),\ \gamma \in \mathcal{K},\ \forall t \ge 0.$$

From this we immediately get the assertion. □

3 Lyapunov-Like Technique

Lyapunov function method was mainly used to analyze the stability performance of dynamical systems before 1980. With the discovery of the well-known

Sontag's formula [29], the method became an effectual tool for control design. It was Florchinger who promoted the research on the stabilization control of the stochastic nonlinear systems in 1990's. He introduced and extended the concept of control Lyapunov functions from deterministic system to stochastic systems, and gave the stochastic version of Sontag's stabilization formula [11, 12].

Consider the stochastic differential equation:

$$dx = f(x)dt + g(x)udt + h(x)dw, \tag{6}$$

where $x \in \mathbb{R}^n$ is system state; $u \in \mathbb{R}^m$ is control law; $w \in \mathbb{R}^s$ is standard Brownian motion, defined on a probability space $(\Omega, \mathcal{F}, \mathcal{P})$; f, g and h are Lipschitz functions mapping $\mathbb{R}^n$ into $\mathbb{R}^n$, $\mathbb{R}^{n\times m}$ and $\mathbb{R}^{n\times s}$, respectively, vanishing in the origin and such that

$$\|f(x)\| + \|g(x)\| + \|h(x)\| \le K(1 + \|x\|),$$

for some nonnegative constant K and all $x \in \mathbb{R}^n$.

Definition 10. *The system (6) satisfies a* stochastic Lyapunov *condition at the origin if there exist a neighborhood* $\mathbb{U}$ *of the origin in* $\mathbb{R}^n$ *and a Lyapunov function* V *defined on* $\mathbb{U}$ *such that for every* $x \in \mathbb{U}\backslash\{0\}$,

$$\frac{\partial V(x)}{\partial x} g(x) = 0 \implies \mathcal{L}V(x) \le 0.$$

A Lyapunov function V satisfying the above condition is called as a control Lyapunov function of the system (6).

The following theorem, established in [12], is an extension of Artstein's theorem [2] to the feedback stabilization of stochastic nonlinear systems.

Theorem 11. *The stochastic nonlinear system (6) satisfies a stochastic Lyapunov condition at the origin if and only if it is asymptotically stabilizable by means of a control law* $u(x)$ *smooth in the neighborhood of the origin.*

By extending the Sontag's formula from deterministic system to stochastic system, we can design a stabilization control law for system (6) as follows

$$u = \begin{cases} \left(\dfrac{-\dfrac{\partial V}{\partial x} f - \dfrac{1}{2}\mathrm{tr}\left\{ \dfrac{\partial^2 V}{\partial x^2} h h^{\tau} \right\}}{\left\| \dfrac{\partial V}{\partial x} g \right\|^2} - 1 \right) \dfrac{\partial V}{\partial x} g, & \dfrac{\partial V}{\partial x} g(x) \ne 0, \\ \text{any real number}, & \dfrac{\partial V}{\partial x} g(x) = 0, \end{cases}$$

where $V(x)$ is a control Lyapunov function of system (6).

It can be seen that under above control law u, the solution $x(t) = 0$ of system (6) is asymptotically stable in the large.

4 Integrator Backstepping Approach

The introduction of the integrator backstepping design approach is a great breakthrough to global stabilization control design. It provides a general constructive tool for designing global stabilization controls for nonlinear systems in or feedback equivalent to strict-feedback form. Recently, a series of research results on stochastic nonlinear systems have been obtained by using this method. Among others, [6, 7, 8, 19, 22, 25, 26, 27] considered the stabilization control design problem based on state-feedback or output-feedback, and [18, 20, 21, 24] studied the satisfactory control design problem under risk-sensitive index.

The systems studied by using backstepping approach in the literature are in strict-feedback form:

$$\begin{aligned} dx_i &= x_{i+1}dt + f_i(x_1, \ldots, x_i)dt + h_i(x_1, \ldots, x_i)dw, \\ &\quad i = 1, \ldots, n-1, \\ dx_n &= udt + f_n(x)dt + h_n(x)dw, \end{aligned} \tag{7}$$

when state-feedback control problems are considered [24]; or

$$\begin{aligned} dx_i &= x_{i+1}dt + f_i(y)dt + h_i(y)dw, \\ &\quad i = 1, \ldots, n-1, \\ dx_n &= udt + f_n(y)dt + h_n(y)dw, \\ y &= x_1, \end{aligned} \tag{8}$$

when output-feedback control problems are considered [6, 7].

The following strict-feedback nonlinear systems in observer canonical form with stable zero-dynamics were studied in [20, 27]:

$$\begin{aligned} dx_i &= x_{i+1}dt + f_i(y)dt + h_i(y)dw, \\ &\quad i = 1, \ldots, \rho-1, \\ dx_j &= x_{j+1}dt + f_j(y) + g(y)b_{j-\rho}udt + b_j(y)dw, \\ &\quad j = \rho, \ldots, n-1, \\ dx_n &= udt + f_n(y)dt + g(y)b_{n-\rho}udt + h_n(y)dw, \\ y &= x_1. \end{aligned} \tag{9}$$

By using backstepping approach, Pan and Basar were the first to solve the control problem for the class of strict-feedback systems of the form (7) in [24]. Deng and Kristić gvae the first design scheme by means of output-feedback for system (8) in [6, 7]. Pan, Liu and Shi presented a stochastic stabilization control design procedure for system (9) in [27]. All the results can be summarized in the following theorem.

Theorem 12. *For the systems (7), (8) and (9), suppose that*

A1). *All functions* f_i, h_i, $i = 1, ..., n$, *are smooth with respect to their arguments;*

A2). *All the roots of the polynomial* $b_m s^m + ... + b_1 s + b_0$, $b_m \neq 0$ $(m = n - \rho)$, *have negative real parts.*

Then, there exists a control law based on state-feedback for systems (7), and a control law based on output-feedback for the systems (8) and (9) such that the solution $x(t) = 0$ *of the corresponding closed-loop systems is asymptotically stable in the large when the functions* f_i *and* h_i $(i = 1, ..., n)$ *vanish at the origin 0, and otherwise, the solution process* $\{x(t), t \geq 0\}$ *is bounded in probability. In addition, the control laws can be designed iteratively by using the backstepping approach.*

5 Risk-Sensitive Control Design

The research on stochastic risk-sensitive control can be traced back to the early 1970s, when Jacobson introduced the linear exponential quadratic Gaussian (LEQG) problem [13]. Then, Whittle put a risk-sensitivity parameter into the cost function, and solved the problem for linear discrete-time systems [30]. Bensoussan and van Schuppen considered the continuous-time case in [5]. But the significance of the risk-sensitive control was not fully realized until 1990s. It is known that stochastic risk-sensitive control is more general than H_∞ control and H_2 control, and closely related to differential game problems [10, 14, 23, 28, 31]. For example, when the noise vanishes, the large deviation limit of the stochastic risk-sensitive control is nothing but a deterministic differential game problem. These connections initialled and accelerated the research on stochastic risk-sensitive controls over the last decade.

Pan and Basar considered the suboptimal risk-sensitive control design in [24] via state-feedback by integrator backstepping approach. Liu *et al* studied the output-feedback risk-sensitive control realization in [18, 20]. In [24], for the stochastic nonlinear system (7), the control law was designed iteratively to ensure the closed-loop system to be bounded in probability and the output to satisfy a pre-given long-term average risk-sensitive cost value. The output feedback case was studied in [18] for systems of the form (8).

The problem of output feedback control design for a class of stochastic nonlinear systems in observer canonical form with stable zero-dynamics (9) was investigated in [20] under a quadratic tracking risk-sensitive cost criterion. In general, the design of output-feedback control is more difficult and challenging than that of full state-feedback control. Since the early 1990s, a general framework to study output-feedback control problem has been developed. The key thought is first to introduce the so-called *information state*, which is a generalization of observer or filter, and then, by a measure transformation, to change the output-feedback control design problem into a full state-feedback one of an augmented system [4, 3, 14, 15]. However, generally speaking, the equality (or inequality) of the information state satisfied is infinite-dimensional, to which an explicit finite-dimensional solution exists only for linear or special nonlinear systems [4]. The method presented in [20] is completely different from the information state one, and can be used to deal with more general inherently nonlinear systems. The objective of [20] is practical: to search for a *satisfaction* control rather than an optimal one. This makes it possible to avoid the measure transformation. For strict-feedback nonlinear systems, explicit formula of the control is given. It is found that for any given risk-sensitivity parameter and desired tracking risk-sensitive cost value, a dynamic output-feedback control can be constructively designed so that the closed-loop system is bounded in probability and the long-time average risk-sensitive cost is upper bounded by the desired value. In summary, we have the following theorem.

Theorem 13. *For the system (9), suppose that*

A1) *The nonlinear functions $f_i(\cdot)$ and $h_i(\cdot)$ $(i = 1, \ldots, n)$ are smooth, i.e., $f_i(\cdot) \in C^\infty(\mathbb{R}^n)$ and $h_i(\cdot) \in C^\infty(\mathbb{R}^n)$; the nonlinear function $g(\cdot)$ is continuous, and for any $y \in \mathbb{R}$, $g(y) \neq 0$;*

A2) *All the roots of the polynomial $b_m s^m + \cdots + b_1 s + b_0$, $b_m \neq 0$ $(m = n - \rho)$, have negative real parts;*

A3) *Desired system output y_d is deterministic, it and its derivatives $\dot{y}_d, \ldots, y_d^{(\rho)}$ are known, and uniformly bounded.*

Then, for any pre-given constant $c > 0$, one can design iteratively by backstepping approach an output-feedback tracking control law such that the corresponding closed-loop system is bounded in probability and the tracking error satisfies

$$J_\theta(y) = \limsup_{T\to\infty} \frac{1}{T} \cdot \frac{2}{\theta} \ln\left(\mathbf{E}\left(\exp\left(\frac{\theta}{2}\int_0^T (y - y_d)^2\, dt\right)\right)\right) \leq c.$$

6 Concluding Remarks

In this paper, we present several important concepts and theorems on stability theory of stochastic processes, and accordingly give sufficient conditions. These concepts, conditions and theorems are often encountered in the related literature and have been widely used for control design and stability analysis of stochastic nonlinear control systems. We also introduce concisely the main results obtained recently on control design of stochastic nonlinear control systems, including state-feedback stabilization control, output-feedback stabilization control and satisfactory risk-sensitive control etc.

References

1. Arslan G. and Basar T. (2002), Risk-sensitive adaptive trackers for strict-feedback systems with output measurements, *IEEE Transactions Automatic Control*, Vol. 47, pp. 1754-1758.
2. Artstein Z. (1983), Stabilizations with relaxed controls, *Nonlinear Analysis*, Vol. 7, pp. 1163-1173.
3. Bensoussan A., Elliott R. J. (1996), General finite-dimensional risk-sensitive problems and small noise limits, *IEEE Transactions on Automatic Control*, Vol. 41, pp. 210-215.
4. Bensoussan A. (1992), *Stochastic control of partially observable systems*, Cambridge: Cambridge University Press.
5. Bensoussan A., Van Schuppen J. H. (1985), Optimal control of partially observable stochastic systems with an exponential-of-integral performance index, *SIAM Journal on Control and Optimization*, Vol. 23, pp. 599-613.
6. Deng H., Krstić M. (2000), Output-feedback stabilization of stochastic nonlinear systems driven by noise of unknown covariance, *Systems & Control Letters*, Vol. 39, pp. 173-182.
7. Deng H., Krstić M. (1999), Output-feedback stochastic nonlinear stabilization, *IEEE Transactions on Automatic Control*, Vol. 44, pp. 328-333.
8. Deng H., Krstić M., Williams R. (2001), Stabilization of stochastic nonlinear systems driven by noise of unknown covariance, *IEEE Transactions on Automatic Control*, Vol. 46, pp. 1237-1253.
9. Doob J. L. (1953), *Stochastic Processes*, New York: John Wiley & Sons.
10. Fleming W. H., McEneaney W. M. (1995), Risk-sensitive control on an infinite time horizon, *SIAM Journal on Control and Optimization*, Vol. 33, pp. 1881-1915.
11. Florchinger P. (1993), A universal formula for the stabilization of control stochastic differential equations, *Stochastic Analysis and Applications*, pp. 155-162.
12. Florchinger P. (1995), Lyapunov-like techniques for stochastic stability, *SIAM Journal on Control and Optimization*, Vol. 33, pp. 1151-1169.
13. Jacobson D. H. (1973), Optimal stochastic linear systems with exponential performance criteria and their relation to deterministic differential games, *IEEE Transactions on Automatic Control*, Vol. 18, pp. 167-172.
14. James M. R., Baras J., Elliott R. J. (1994), Risk-sensitive control and dynamic games for partially observed discrete-time nonlinear systems, *IEEE Transactions on Automatic Control*, Vol. 39, pp. 780-792.
15. James M. R.,Baras J. (1996), Partially observed differential games, infinite dimensional HJI equations, and nonlinear H_∞ control, *SIAM J. Control and Optim.*, Vol. 34, pp. 1342-1364.
16. H. Khalil, *Nonlinear Systems*, 3rd Edt., New York: Macmillan, 2002.
17. Khas'minskii R. Z. (1980), *Stochastic Stability of Differential Equations*, Rockville: S and N International Publishers.
18. Liu Y., Pan Z., Shi S. (2003), Output feedback control design for strict-feedback stochastic nonlinear systems under a risk-sensitive cost, *IEEE Transactions on Automatic Control*, Vol. 47, 509-513.
19. Liu Y., Zhang J. (2004), Minimal-order observer and output-feedback stabilization control design of stochastic nonlinear systems, *Science in China*, Series F, Vol. 47, pp. 527-544.

20. Liu Y., Zhang J. (2004), Practical oput-feedback risk-sensitive control for stochastic nonlinear systems with zero-dynamics, submitted to *SIAM Journal on Control and Optimization* and revised recently.
21. Liu Y., Zhang J., Pan Z. (2003), Design of satisfaction output feedback controls for stochastic nonlinear systems under quadratic tracking risk-sensitive index, *Science in China*, Series F, Vol.46, pp. 126-144.
22. Liu Y., Zhang J. (2004), Reduced-order observer-based control design for stochastic nonlinear systems, *Systems & Control Letters*, Vol. 52, pp. 123-135.
23. Nagai H. (1996), Bellman equations of risk-sensitive control, *SIAM Journal of Control and Optimization*, Vol. 34, pp. 74-1016.
24. Pan Z., Basar T. (1999), Backstepping controller design for nonlinear stochastic systems under a risk-sensitive cost criterion, *SIAM Journal on Control and Optimization*, Vol. 37, pp. 957-995.
25. Pan Z. (2002), Canonical forms for stochastic nonlinear systems, *Automatica*, Vol. 38, pp. 1163-1170.
26. Pan Z. (2001), Differential geometric condition for feedback complete linearization of stochastic nonlinear, *Automatica*, Vol. 37, pp. 145-149, 2001.
27. Pan Z., Liu Y., Shi S. (2001), Output-feedback stabilization for stochastic nonlinear systems in observer canonical form with stable zero-dynamics, *Science in China*, Series F, Vol. 44, pp. 292-308.
28. Runolfsson T. (1994), The equivalence between infinite horizon control of stochastic systems with exponential-of-integral performance index and stochastic differential games, *IEEE Transactions on Automatic Control*, Vol. 39, pp. 1551-1563.
29. Sontag E. D., Sussmann H. (1989), Further comments on stability of the angular velocity of a rigid body, *Systems* & Control Letters, Vol. 12, pp. 457-473.
30. Whittle, P. (1981), Risk-sensitive linear-quadratic-gaussian control, *Advances in Applied Probability*, Vol. 13, pp. 764-777.
31. Whittle P. (1990), *Risk-Sensitive Optimal Control*, Chichester: John Wiley and Sons.

Nonlinear Control of A Class of Differential-Algebraic Equation Systems*

Shengwei Mei[1], Feng Liu[2] and Qiang Lu[1]

[1] Electrical Engineering Dept. of Tsinghua University, Beijing, 100084, P.R.China
meishengwei@mail.tsinghua.edu.cn

[2] Institute of Systems Science, Chinese Academy of Sciences, Beijing 100080, P.R.China liufeng@mx.amss.ac.cn

1 Introduction

The last two decades witnessed considerable research efforts aimed at the development of systematic design methodologies for nonlinear feedback systems, one of which most important contributions is the geometric theory of nonlinear control systems [1]. However, this methodology is mainly applied to deal with the control problems based on Ordinary Differential Equations (ODE). As well known, many physical systems can be modelled as Differential-Algebraic Equations (DAE), such as power systems [2], chemical processes [3] and electrical circuits [4]. Since the conventional geometric theory of control systems described by ODE can not be directly applied to DAE systems, it is necessary to develop appropriate tools for that.

So far, lots of remarkable efforts have been done on DAE control systems. The design of a feedback control law for a class of nonlinear DAE systems was considered by Krishnan and Mcclamroch [5], which was applied to the tracking control of constrained robot systems. Further works, including the problems of input-output decoupling, exact linearization and asymptotic output tracking, were done by Liu Xiaoping [6]. However, their results require affine structure of algebraic variables. Some other works focus on index one, semi-explicit DAE systems. S. Kaprielian [7] studied the input-output linearization of MIMO semi-explicit DAE systems and its application to the control of AC/DC power systems, whose essence is the implicit function theorem and the chain rule of differentiation. In the work of Wang and Chen [8], some new definitions (M derivative, M bracket) of DAE systems are given, which are similar to the definitions and theorems in classical differential geometric theory. With this tool, the geometric control of index 1, semi-explicit DAE systems can be systematically formulated and solved with certain assumptions. Nevertheless,

*This work was supported by NSFC (50377018, 50595411) and by 973 Project (G2004CB217902).

such a tool cannot provide more profit for the control design of DAE systems since we can obtain the same results from the work of S. Kaprielian without such a tool. Noticing the definition of M derivative is given by the link principle of differentiation, the conclusion is straightforward. However, it is obvious that the proposed tool results in a much simpler and beautiful form of description.

In order to extend the standard geometric control theory to DAE, the relationship between vector field and DAE should be studied. Reich [9] introduced the notion of a regular DAE as a DAE to which a vector field uniquely corresponds. From the geometric point of view, the essence of such a notion is to guarantee the existence and uniqueness of the solution of DAE system [10]. Motivated by that, we introduce some assumptions for the studied control systems described by DAE to avoid the complicated discuss on the existence and uniqueness of solutions and treat the studied DAE systems as differential equations on manifold. With certain assumptions of differential index, the index of studied DAE systems can be reduced. Sequentially we can obtain the equivalent constrained systems. Then, there exists a local state-space realization of DAE systems on the so-called configuration space [9]. Based on the state-space realization, standard geometric control theory can be applied to solve the control problems of DAE system such as the stabilization or output tracking.

The aim of this paper is to establish a geometric interpretation for the nature of control systems described by a more general class of DAE with high index, consequently provide a theoretical justification and corresponding systematic approach for the geometric control of interested DAE systems.

2 DAEs as differential equations on manifolds

Consider a class of control systems described by a set of nonlinear DAEs in the form of

$$\begin{cases} \dot{x} = f(x,y) + g(x,y)\rho(x,y,u) \\ 0 = k(x,y) \\ z = h(x,y) \end{cases} \tag{1}$$

where $(x, y) \in M$ with M an $n + r$ dimensional manifold; x is the dynamic state variable vector with dimension n; y the instantaneous state variable vector with dimension r; $z \in R^m$ the output of the system and $u \in R^m$ the control input.

Let:

$$g(x,y) = \left[g_1(x,y), \cdots g_m(x,y) \right];$$

$$\rho(x,y,u) = \left[\rho_1(x,y,u), \cdots \rho_m(x,y,u) \right]^T;$$

$$k(x,y) = \left[\, k_1(x,y), \cdots k_r(x,y) \,\right]^T;$$

$$h(x,y) = \left[\, h_1(x,y), \cdots h_m(x,y) \,\right]^T;$$

where $f(x,y)$, $g(x,y)$, $\rho(x,u)$, $k(x,y)$ and $h(x,y)$ are all smooth.

For DAE system (1), there ordinarily does not exist a state realization. Hence, for the sake of guaranteeing the existence and uniqueness of solutions of DAE system (1), some assumptions are necessary.

A1 : $\rho_i(\cdot,\cdot,\cdot)$ *is continuous with* $|\partial\rho(x,y,u)/\partial u| \neq 0$ *and* $\rho(x_0,y_0,0)=0$;

A2: the DAE system has a constant vector differential index:

$$\kappa = \left[\, \kappa_1, \cdots \kappa_r \,\right]$$

with $n > r + \sum_{i=1}^{r} \kappa_i$. *It means that, for each* $\kappa_i \geq 2$, *there are*

(a). $\frac{\partial(L_f^j k_i(x,y))}{\partial y} = 0 \quad j = 1,2,\cdots,\kappa_i - 2; i = 1,2,\cdots,r$

(b). $rank S_1(x,y) = r$ *on the neighborhood of* (x_0,y_0), *where*

$$S_1(x,y) = \left[\frac{\partial}{\partial y} \left(\begin{bmatrix} L_f^{\kappa_1 - 1} k_1(x,y) \\ \vdots \\ L_f^{\kappa_r - 1} k_r(x,y) \end{bmatrix} \right) \right]_{r \times r}$$

A3: for each $k_i(x,y) = 0$, *there are* $L_{g_l} L_f^j k_i(x,y) = 0$, *for* $j = 0,1,\cdots,(\kappa_i - 2); i = 1,\cdots,r; l = 1,\ldots m$

In order to process the stability analysis and controller design, the information about the existence and uniqueness of the solutions is necessary to be provided. Considering DAE systems as differential equations on a manifold is an intuitive approach that allows for deep insight into the properties of the existence and uniqueness theory of DAE system and its geometric structure [10]. Consequentially, we could define a class of well-posed systems to avoid some unexpected analytical difficulties during the study.

As in the reference [5], we define

$$N^* = \{(x,y) \in M | L_f^j k_i(x,y) = 0; \quad j = 0,\cdots,\kappa_i - 1; i = 1,\cdots,r\}$$

Lemma 1. N^* *is an* $(n+r-\kappa_1-\kappa_2-\cdots\kappa_r)$ *dimensional smooth submanifold of* M.

Proof. In the light of the classical literature of Isidori [1], it is easy to prove the linear independence of $dL_f^j k_i(x,y)$ under the assumptions A2 and A3, where, dH denotes the row vector

$$dH = \left[\frac{\partial H(x,y)}{\partial x_1} \cdots \frac{\partial H(x,y)}{\partial x_n} \frac{\partial H(x,y)}{\partial y_1} \cdots \frac{\partial H(x,y)}{\partial y_r} \right]$$

Then Lemma 1 is a direct consequence. □

The submanifold N^* plays an important role in the state-space description and geometric analysis of the DAE system, which will be discussed in detail later.

To guarantee the studied DAE system is well posed, we introduce the following theorem.

Theorem 1. *Suppose the input function* $u : [0,\infty) \to R^r$ *is an integrable function. If the initial condition satisfies* $(x_0, y_0) \in N^*$, *then there exists a unique solution* $(x(t), y(t))$ *(at least locally defined) of the initial value problem corresponding to DAE system (1) satisfying assumptions A1~A3. Moreover, the solution satisfies* $(x(t), y(t)) \in N^*$ *for each* $t \geq 0$.

Proof. Firstly let

$$v = [v_1 \cdots v_m]^T = \rho(x, y, u) \tag{2}$$

Then it is obvious that, according to A1, v is also integrable.

In order to prove the theorem, we need to construct an equivalent system of the original one. In the light of the definition of differential index of DAE systems, and with assumption A2, the following equations can be obtained by calculating the derivatives of the algebraic constraints:

$$\begin{array}{l} k_1(x,y)=0 \\ L_f^1 k_1(x,y)=0 \\ \cdots \\ L_f^{\kappa_1-1} k_1(x,y)=0 \\ \cdots \\ k_r(x,y)=0 \\ \cdots \\ L_f^{\kappa_r-1} k_r(x,y)=0 \end{array} \tag{3}$$

and

$$\begin{array}{l} L_f^{\kappa_1} k_1(x,y) + \sum\limits_{l=1}^{m} L_{g_l} L_f^{\kappa_1-1} k_1(x,y) v_l + \frac{\partial L_f^{\kappa_1-1} k_1(x,y)}{\partial y} \dot{y} = 0 \\ \cdots \\ L_f^{\kappa_r} k_r(x,y) + \sum\limits_{l=1}^{m} L_{g_l} L_f^{\kappa_r-1} k_r(x,y) v_l + \frac{\partial L_f^{\kappa r-1} k_r(x,y)}{\partial y} \dot{y} = 0 \end{array} \tag{4}$$

They can be rewritten in compact forms as

$$L_f^j k_i(x,y) = 0 \quad (j = 1, \cdots \kappa_i - 1; i = 1, \cdots r) \tag{5}$$

and

$$S_1(x,y)\dot{y} + S_2(x,y)v + f_1(x,y) = 0 \tag{6}$$

where

$$f_1(x,y) = \begin{bmatrix} L_f^{\kappa_1} k_1(x,y) \\ \vdots \\ L_f^{\kappa_r} k_r(x,y) \end{bmatrix}$$

and

$$S_2(x,y) = \begin{bmatrix} L_{g_1} L_f^{\kappa_1 - 1} k_1(x,y) & \cdots & L_{g_m} L_f^{\kappa_1 - 1} k_1(x,y) \\ \vdots & \ddots & \vdots \\ L_{g_1} L_f^{\kappa_r - 1} k_r(x,y) & \cdots & L_{g_m} L_f^{\kappa_r - 1} k_r(x,y) \end{bmatrix}$$

With the assumptions A3, there is $|S_1(x,y)| \neq 0$. It immediately yields

$$\dot{y} = -S_1^{-1}(x,y)\left[f_1(x,y) + S_2(x,y)v\right] \tag{7}$$

Besides, the original dynamic equations of x still hold. Hence we obtain

$$\dot{x} = f(x,y) + g(x,y)v \tag{8}$$

Eqs. (5), (7) and (8) form a new constrained system that is equivalent to the original system (1) with the unchanged output $z = h(x,y)$.

For the new system, the input function v is integrable. Given an initial condition (x_0, y_0) satisfies

$$L_f^j k_i(x_0, y_0) = 0 \quad (j = 1, \cdots \kappa_i - 1, i = 1, \cdots r) \tag{9}$$

Then the initial value problem defined by (5), (7)and (8) with initial value (9) has a unique solution. It would be noted that the existence of constraints (5) maybe have the solution curves break down. However, at least locally, the unique solution is well defined. Since the constraints (5) hold for all time $t > 0$, the solution $(x(t), y(t))$ must satisfy $L_f^j k_i(x(t), y(t)) = 0$ $(j = 1, \cdots \kappa_i - 1, i = 1, \cdots r)$ for each $t \geq 0$.

Note that a smooth submanifold is defined by Eq. (5), which is just N^*. Hence, the theorem is proved. □

There are some remarks on this theorem as follows.

Remark 1. In fact, there exist various kinds of complex phenomenon in DAE systems, especially for the cases of high-index. Such complicated dynamic behaviors make the dynamic analysis control design extremely difficult. Much worse, there is even not a standard theorem about the existence and uniqueness of solution. Motivated by the notion of regular DAE systems [9], the conditions A2 and A3 are assumed to guarantee "nice" behavior of the control system and avoid analytical difficulties.

Remark 2. The equivalence between the two systems makes it is possible to study the constrained system defined by Eqs. (5), (7)and (8) instead of the original DAE system (1) with assumptions A1~A3. One advantage of such a procedure is that the obtained constrained system has much clearer geometric structure that is favorable to the profit of stability analysis and controller design. Actually the previous discuss reveals that the studied DAE system can in some sense be viewed as differential equations on manifolds. Or more specifically, they can be reduced to be a dynamic system on the submanifold N^*. In fact, with a proper local coordinate chat, the submanifold N^* imposes an $(n+r-\kappa_1-\kappa_2-\cdots\kappa_r)$ dimensional state-space for the DAE system under consideration. In the later section, the detail about the local expression of the reduced state-space realization will be presented.

Corollary 1. *For DAE system (1) with assumptions A1, if it is of index 1, then the system has an n dimensional state-space realization.*

This corollary is helpful to the control design of lots of realistic DAE systems that are of index 1, such as electric circuit systems and power systems. Actually, some important contributions have been done for that [8]. This corollary puts forward a united theoretical justification for those works.

3 Decoupling and linearization via feedback

For the sake of simplicity, we rewrite Eqs. (5), (7)and (8) in a compact form as

$$\begin{cases} \dot{\tilde{x}} = F(\tilde{x}) + G(\tilde{x})v \\ 0 = L_f^j k_i(\tilde{x}) \quad (j=1,\cdots\kappa_i-1, i=1,\cdots r) \\ z = h(\tilde{x}) \end{cases} \tag{10}$$

where

$$\tilde{x} = \begin{bmatrix} x \\ y \end{bmatrix}; \quad F(\tilde{x}) = \begin{bmatrix} f(\tilde{x}) \\ -S_1^{-1}(\tilde{x})f_1(\tilde{x}) \end{bmatrix} \text{;and } G(\tilde{x}) = \begin{bmatrix} g(\tilde{x}) \\ -S_1^{-1}(\tilde{x})S_2(\tilde{x}) \end{bmatrix};$$

For system (10), we introduce further assumptions:

A4: The equivalent constrained system (10) has a relative degree μ_i with respect to an output z_i, which satisfies $(n+r) \geq \sum_{i=1}^r \kappa_i + \sum_{i=1}^m \mu_i$. That is, the vector

$$L_G L_F^j h_i(\tilde{x}) \underline{\underline{\Delta}} [L_{G_1} L_F^j h_i(\tilde{x}) \cdots L_{G_m} L_F^j h_i(\tilde{x})] = \mathbf{0}$$

$$(j=0,\cdots\mu_i-2; i=1,\cdots m)$$

$$L_G L_F^{\mu_i-1} h_i(\tilde{x}) \underline{\underline{\Delta}} [L_{G_1} L_F^{\mu_i-1} h_i(\tilde{x}) \cdots L_{G_m} L_F^{\mu_i-1} h_i(\tilde{x})] \neq \mathbf{0}$$

$$(i=1,2,\cdots m)$$

A5: the decoupling matrix, $A(\tilde{x})$, is nonsingular, where

$$A(\tilde{x}) = \begin{bmatrix} L_{G_1}L_F^{\mu_1-1}h_1(\tilde{x}) & \cdots & L_{G_m}L_F^{\mu_1-1}h_1(\tilde{x}) \\ \vdots & \ddots & \vdots \\ L_{G_1}L_F^{\mu_m-1}h_1(\tilde{x}) & \cdots & L_{G_m}L_F^{\mu_m-1}h_1(\tilde{x}) \end{bmatrix}_{m\times m}$$

For system (10), a coordinate transformation is chosen as

$$\begin{aligned} &\phi_{1,1}(\tilde{x}) = k_1(\tilde{x}) = k_1(x,y) \\ &\phi_{1,2}(\tilde{x}) = L_F k_1(\tilde{x}) = L_f k_1(x,y) \\ &\vdots \\ &\phi_{1,\kappa_1}(\tilde{x}) = L_F^{\kappa_1-1}k_1(\tilde{x}) = L_f^{\kappa_1-1}k_1(x,y) \\ &\vdots \\ &\phi_{r,1}(\tilde{x}) = k_r(\tilde{x}) = k_r(x,y) \\ &\phi_{r,2}(\tilde{x}) = L_F k_r(\tilde{x}) = L_f k_r(x,y) \\ &\vdots \\ &\phi_{r,\kappa_r}(\tilde{x}) = L_F^{\kappa_1-1}k_r(\tilde{x}) = L_f^{\kappa_1-1}k_r(x,y) \end{aligned} \tag{11}$$

and

$$\begin{aligned} &\varphi_{1,1}(\tilde{x}) = h_1(\tilde{x}) \\ &\varphi_{1,2}(\tilde{x}) = L_F h_1(\tilde{x}) \\ &\cdots \\ &\varphi_{1,\mu_1}(\tilde{x}) = L_F^{\mu_1-1}h_1(\tilde{x}) \\ &\varphi_{m,1}(\tilde{x}) = h_m(\tilde{x}) \\ &\cdots \\ &\varphi_{m,\mu_m}(\tilde{x}) = L_F^{\mu_m-1}h_m(\tilde{x}) \end{aligned} \tag{12}$$

First of all, we need to introduce several lemmas as the following.

Lemma 2. *For system (1) with assumptions A1∼A5, the row vectors*

$$\begin{aligned} &d\phi_{1,1}, d\phi_{1,2} \cdots d\phi_{1,\kappa_1} \\ &\cdots \\ &d\phi_{r,1}, d\phi_{r,2} \cdots d\phi_{r,\kappa_r} \\ &d\varphi_{1,1}, d\varphi_{1,2} \cdots d\varphi_{1,\mu_1} \\ &\cdots \\ &d\varphi_{m,1}, d\varphi_{m,2} \cdots d\varphi_{m,\mu_m} \end{aligned}$$

are linearly independent.

Proof. Along the lines of the proof of reference [1] (Lemma 5.1.2), the lemma is easy to prove with some appropriate modifications. □

Lemma 3. *For system (1) with A1∼A5, if* $(n+r) > \sum_{i=1}^{r}\kappa_i + \sum_{i=1}^{m}\mu_i$ *is strictly satisfied, it is always possible to find* $s = (n+r) - \sum_{i=1}^{r}\kappa_i - \sum_{i=1}^{m}\mu_i$ *functions* $\xi_1(x,y), \cdots \xi_s(x,y)$ *such that the mapping*

$$T(x,y) = \begin{bmatrix} \phi(x,y) \\ \varphi(x,y) \\ \xi(x,y) \end{bmatrix} \tag{13}$$

has the Jacobian matrix nonsingular at (x_0, y_0) *and therefore qualifies as a local coordinates transformation in a neighborhood of* (x_0, y_0), *where*

$$\phi(x,y) = \begin{bmatrix} \phi_{1,1}(x,y) \\ \cdots \\ \phi_{1,\kappa_1}(x,y) \\ \cdots \\ \phi_{r,1}(x,y) \\ \cdots \\ \phi_{r,\kappa_r}(x,y) \end{bmatrix}; \ \varphi(x,y) = \begin{bmatrix} \varphi_{1,1}(x,y) \\ \cdots \\ \varphi_{1,\mu_1}(x,y) \\ \cdots \\ \varphi_{m,1}(x,y) \\ \cdots \\ \varphi_{m,\mu_m}(x,y) \end{bmatrix}; \ \xi(x,y) = \begin{bmatrix} \xi_1(x,y) \\ \cdots \\ \xi_s(x,y) \end{bmatrix}$$

This lemma is the direct consequence of Lemma 2.

Based on these two lemmas, we can deduce the theorem about the decoupling problem of DAE system (10). However, for the sake of convenience of description, we first prove the following lemma.

Lemma 4. *Consider an affine nonlinear DAE system described by*

$$\begin{cases} \dot{x} = f(x,y) + g(x,y)v \\ 0 = k(x,y) \\ z = h(x,y) \end{cases} \tag{14}$$

If it satisfies assumptions A2∼A5, then the decoupling problem in the sense of input-output is solvable by a state feedback $v = \alpha(x,y) + \beta(x,y)\bar{v}$. *One of the solutions is*

$$\alpha(x,y) = -A^{-1}(\tilde{x})b(\tilde{x}) = -A^{-1}(x,y)b(x,y)$$
$$\beta(x,y) = A^{-1}(\tilde{x}) = A^{-1}(x,y)$$

where

$$b(\tilde{x}) = b(x,y) = \begin{bmatrix} L_F^{\mu_1}h_1(x,y) \\ L_F^{\mu_2}h_2(x,y) \\ \cdots \\ L_F^{\mu_m}h_m(x,y) \end{bmatrix}$$

Proof. From the proof of theorem 1, it is easy to see that the system (14) can also be transferred to the constrained system (10). we can study the decoupling problem directly on system (10). Choose coordinates transformation as (13), which is local diffeomophism in the neighborhood of (x_0, y_0). Then the system (10) can be rewritten as

$$\begin{cases} \dot{\phi}_{1,1} = \phi_{1,2} \\ \cdots \\ \dot{\phi}_{1,\kappa_1-1} = \phi_{1,\kappa_1} \\ \dot{\phi}_{1,\kappa_1} = 0 \\ \cdots \\ \dot{\phi}_{r,1} = \phi_{r,2} \\ \cdots \\ \dot{\phi}_{r,\kappa_r-1} = \phi_{r,\kappa_r} \\ \dot{\phi}_{r,\kappa_r} = 0 \end{cases} ; \quad \begin{cases} \dot{\varphi}_{1,1} = \varphi_{1,2} \\ \cdots \\ \dot{\varphi}_{1,\mu_1-1} = \varphi_{1,\mu_1} \\ \dot{\varphi}_{1,\mu_1} = L_F^{\mu_1} h_1(x,y) + \sum_{i=1}^m L_{G_i} L_F^{\mu_1-1} h_1(x,y) v_i \\ \cdots \\ \dot{\varphi}_{m,1} = \varphi_{m,2} \\ \cdots \\ \dot{\varphi}_{m,\mu_m-1} = \varphi_{m,\mu_m} \\ \dot{\varphi}_{m,\mu_m} = L_F^{\mu_m} h_m(x,y) + \sum_{i=1}^m L_{G_i} L_F^{\mu_m-1} h_m(x,y) v_i \end{cases}$$

and

$$\dot{\xi} = \left[\frac{\partial \xi}{\partial x} \ \frac{\partial \xi}{\partial y} \right] [F(x,y) + G(x,y)v]$$

It should be noted that $\phi_{i,j}(x,y) \equiv 0$. On the other hand, by applying the coordinate transformation T, we have

$$\begin{aligned} \dot{\xi} &= \left[\frac{\partial \xi}{\partial x} \ \frac{\partial \xi}{\partial y} \right] [F(x,y) + G(x,y)v] \\ &= p(x,y) + q(x,y)v \\ &= p(T^{-1}(\phi,\varphi,\xi)) + q(T^{-1}(\phi,\varphi,\xi))v \\ &= p(T^{-1}(\varphi,\xi)) + q(T^{-1}(\varphi,\xi))v \\ &= \tilde{p}(\varphi,\xi) + q(\varphi,\xi)v \end{aligned}$$

and

$$h(x,y) = h(T^{-1}(\phi,\varphi,\xi)) = \tilde{h}(\varphi,\xi)$$

Hence the system is equivalent to the $(n+r-\kappa_1-\kappa_2-\cdots\kappa_r)$ dimensional state-space realization given by

$$\begin{aligned} &\dot{\varphi}_{i,j} = \varphi_{i,j+1} \quad (j = 1,2\cdots\mu_i - 1) \\ &\dot{\varphi}_{i,\mu_i} = L_F^{\mu_i} h_i(x,y) + \textstyle\sum_{l=1}^m L_{G_l} L_F^{\mu_i-1} h_i(x,y) v_l \\ &\qquad = L_F^{\mu_i} \tilde{h}_i(\varphi,\xi) + \textstyle\sum_{l=1}^m L_{G_l} L_F^{\mu_i-1} \tilde{h}_i(\varphi,\xi) v_l \quad (i = 1,2\cdots m) \\ &\dot{\xi} = \tilde{p}(\varphi,\xi) + q(\varphi,\xi)v \end{aligned} \tag{15}$$

Introduce the denotation $\bar{v}$ such that

$$\bar{v} = b(x,y) + A(x,y)v = \tilde{b}(\varphi,\xi) + \tilde{A}(\varphi,\xi)v \tag{16}$$

According to assumption A5, the system can be rewritten in the form of

$$\begin{aligned} &\dot{\varphi}_{i,j} = \varphi_{i,j+1} \quad (j = 1,2\cdots\mu_i - 1) \\ &\dot{\varphi}_{i,\mu_i} = \bar{v}_i \quad (i = 1,2\cdots m) \\ &\dot{\xi} = \tilde{p}(\varphi,\xi) - q(\varphi,\xi)\tilde{A}^{-1}(\varphi,\xi)\tilde{b}(\varphi,\xi) + q(\varphi,\xi)\tilde{A}^{-1}(\varphi,\xi)\bar{v} \end{aligned} \tag{17}$$

Obviously, the equations have a decoupled structure in the sense of input-output. It can be seen that Eq. (16) implies that

$$v = -A^{-1}(x,y)b(x,y) + A^{-1}(x,y)\bar{v} = \alpha(x,y) + \beta(x,y)\bar{v}$$

This completes the proof of this Lemma. □

Theorem 2. *For DAE system (1) with assumptions A1∼A5, the problem of decoupling in the sense of input-output is solvable by a state feedback.*

Proof. Firstly, let

$$v = [v_1 \cdots v_m]^T = \rho(x,y,u) \tag{18}$$

Then system (1) is the same as system (14). From Lemma 4, the decoupling problem can be solved by state feedback $v = \alpha(x,y) + \beta(x,y)\bar{v}$. According to assumption A1, in the view of implicit function theorem, there must exist $u = \psi(x,y,v)$ for each v. This completes the proof of this theorem. □

Noticing the linear structure in Eq. (17), we can easily obtain the theorem about the linearization problem in the sense of input-output.

Theorem 3. *For DAE system (1) with assumptions A1∼A5, the problem of linearization in the sense of input-output is solvable by a state feedback.*

Remark 3. The proofs of theorem 2 and theorem 3 follow the standard nonlinear control theory based on differential geometry. Hence, these two theorems can be regarded as extension of the standard geometric control theory. The proof is constructive so that it also provides a design approach for the decoupling problems of DAE systems.

Remark 4. System (15) provides a reduced state-space realization of the DAE control system (1) with the assumptions A1∼A5. The procedure is also constructive. It would be seen that the state-space is just the submanifold N^*. In the literature [9], such a submanifold is called construction space. From another viewpoint, the coordinates transformation divides the complete state-space ((x,y) space) into three subspaces spanned respectively by. $\phi(x,y)$, $\varphi(x,y)$ and $\xi(x,y)$. The latter two subspaces are tangent to the manifold N^* and the first one is perpendicular to N^*. Since $\phi(x,y)$ is strictly equal to zero, it can be viewed as "*absolute*" zero dynamics of the system. That is the essential reason why the reduced state-space realization can fully describe the original system. Consequentially, the feedback control can be regarded as a map: $u : N^* \to R^m$. Furthermore, $\varphi(x,y)$ and $\xi(x,y)$ divide the tangent space of manifold N^* at the neighborhood of (x,y). $\varphi(x,y)$ represents the linearized subspace in the sense of input-output, while $\xi(x,y)$ is tangent to it. Hence, the introduced coordinate transformation provides a geometric interpretation of the DAE system under consideration.

With Theorem 2 and Theorem 3, we can easily deduce the following two corollaries.

Corollary 2. *For DAE system (1) with assumptions A1∼A5, if* $(n+r) = \sum_{i=1}^{r} \kappa_i + \sum_{i=1}^{m} \mu_i$*, then the system has an exact linearization and decoupling realization in the sense of input-output.*

4 Stabilization and output tracking

In this section we consider the problems of stabilization and output tracking of DAE systems under consideration.

Theorem 4. *For DAE system (1) with assumptions A1∼A5, its state-space realization has the form of system (15). One of the sufficient conditions of the solvability of the stabilization problem is that the zero dynamics of the reduced system (15) are globally asymptotically stable at* $\xi = 0$.

Proof. The state-space realization (15) has the decoupling form of (17) with the feedback law, Eq. (16). For each decoupling block, we choose the i–th feedback control as

$$\bar{v}_i = -c_{i,0}\varphi_{i,1} - c_{i,1}\varphi_{i,2} \cdots - c_{i,\mu_i-1}\varphi_{i,\mu_i} \tag{19}$$

such that all the roots of the corresponding characteristic polynomials

$$\eta_i(s) = c_{i,0} + c_{i,1}s + \cdots + c_{i,\mu_i-1}s^{\mu_i-1} + s^{\mu_i}$$

are Hurwitz

In view of classical nonlinear control theory based on differential geometry [1], with assumption of the asymptotical stability of zeros dynamics, the system can be asymptotically stabilized by the feedback control at $\xi = 0$. Finally, from the equivalence of systems (1) and (15), we can obtain the conclusion of Theorem 4. □

Similarly, the theorem about the output tracking problem can also be obtained. Suppose the control outputs $z(t)$ are required to track the reference output $z_R(t)$, then we have the following theorem.

Theorem 5. *For DAE system (1) with assumptions A1 ∼A5, its state-space realization has the form of system (15). One of the sufficient conditions of the solvability of the output tracking problem is that the zero dynamics of the reduced system (15) are globally asymptotically stable at* $\xi = 0$ *and globally Lipschitz.*

Proof. The proof is also similar to that of Theorem 4. Firstly we define an "error" $e(t)$ as the difference between the real output $z(t)$ and the reference output $z_R(t)$, i.e. $e(t) = z(t) - z_R(t)$. Then we choose the i–th feedback control as

$$\bar{v}_i = z_R^{(\mu_i)} - c_{i,0}e - c_{i,1}e^{(1)} \cdots - c_{i,\mu_i-1}e^{(\mu_i-1)} \tag{20}$$

such that the corresponding characteristic polynomials

$$\eta_i(s) = c_{i,0} + c_{i,1}s + \cdots + c_{i,\mu_i-1}s^{\mu_i-1} + s^{\mu_i}$$

is Hurwitz. Then in the light of classical nonlinear control theory based on differential geometry [1], with assumption of the asymptotical stability of zeros dynamics, the tracking error for the output of closed-loop system (17) asymptotically approaches to zero. Finally, according to the equivalence of system (1) and (15), we can obtain the conclusion of theorem 5. □

Combining results given in previous sections, it is apparent that the asymptotical stability of the zero dynamics is a sufficient condition to achieve decoupling control or output tracking with an internally asymptotically stable closed-loop. Actually from the proof above, it is easy to obtain solutions of stabilization or output tracking control problems.

5 Conclusion

In this paper, the control problems of nonlinear systems described by a class of DAE have been considered, which includes the decoupling and linearization in the sense of input-output, the asymptotical stabilization and output tracking problems. With the assumptions A2 and A3, this work provides some sufficient conditions of guaranteeing the existence and uniqueness of solution, on which the sequential analysis and design are based. The constructive proof is given so an approach of reducing the index of DAE systems and establishing their equivalent constrained systems is obtained. By applying the standard input-output linearization scheme, the problems such as decoupling and linearization in the sense of input-output are solvable with some assumptions. In addition, this work shows that the decoupled and linearized form of the system is one of the state-space realizations of the original DAE system. Finally, it can be concluded that most of results in the geometric control theory based on vector field can be applied to solve the control problems of DAE systems.

References

1. Alberto Isidori, Nonlienar Control System, Spirng-Venag, 1989

2. D. J. Hill, and I. M. Y. Mareels, stability theory for differential-algebraic systems with application to power systems. IEEE Trans. On Circuits and Systems, 1990(37), pp1461-1423
3. A. Kumar and P. Daoutidis, Feedback control of nonlinear differential-algebraic equation systems. A. I. Ch. E. J.,1995(41),pp619-636
4. R.W. Newcomb, The semistate description of nonlinear time-variable circuits, IEEE Trans. On Circuits and Systems, 1981(28), pp62-71
5. Hariharan Krishnan and N. Harris McClamroch, Tracking in nonlinear differential-algebraic control systems with applications to constrained robot systems. Automatica, Vol. 30, No. 12, 1994, pp1885-1897
6. Xiaoping Liu, Asymptotic output tracking of nonlinear differential-algebraic control systems. Automatica, Vol. 34, No. 3, 1998, pp393-397
7. S. Kaprielian, J. Turi and L.R. Hunt, Vector input-output linearization for a class of descriptor systems. Proceedings of the 30 th Conference on D&C, Brighton, England, Dec. 1991, pp1949-1954
8. Wang Jie, Chen Chen, Wu Hua, etal, Theory and method for parametric adaptive control design of multi-machines power systems, Proceedings of the CSEE, 2002, 22(5), pp5-9
9. Rebastian Reich, On a geometrical interpretation of differential-algebraic equations, Circuits Systems Signal Process, Vol. 9, NO.4, 1990
10. Werner C. Rheinboldt, Differential-algebraic systems as differential equations on manifolds, Mathematics of Computation, Vol.43, No.168, Oct. 1984, pp473-482

Optimal Design of Grazing Behavior for Multi-Agent Robots

Xiaoming Hu[1], Yi Huang[2] and Daizhan Cheng[2]

[1] Optimization and Systems Theory, Royal Institute of Technology, Stockholm, Sweden
[2] Institute of Systems Science, Chinese Academy of Sciences, Beijing 100080, P.R. China

1 Introduction

Grazing behavior is one of the most fundamental social behaviors for animals such as cattle, where animals would search for and eat food. Many robotic applications such as cleaning, grass mowing and land-mines detection, can be deduced to a control problem that is similar to the grazing behavior. In order to cover a large field efficiently, the use of multiple robots is essential. For such applications with multi-agent robots [2], a basic problem is the complete coverage path planning (CCPP) or region filling problem. There is already a considerable deal of research regarding multiple robots in coverage path planning, see for example, [6, 1, 3, 8, 5, 7] and the references therein. However, many existing methods for robots with limited working area have overlooked the issue of optimality, and/or only work in environments with rather limited size.

In this paper we consider the grazing problem in large environments such as open fields. The grazing problem considered is similar to the complete coverage problem such as in [5]. However, a positive dwell time at each point is required to complete the work (thus the term "grazing"), which is necessary for many applications and would also make the path planning more complicated.

Our basic assumption is that the cost for a robot purely to move from one point to another is negligible in comparison to the cost of grazing. Based on this assumption, we will first show mathematically what the optimal path planning algorithms are in terms of maximizing the grazed area per unit time. Then we will show how the algorithms together with a formation control can be implemented online for nonholonomic robots.

2 Regular field

In this section we consider the grazing behavior of a team of robots on a regular field, namely a rectangular field.

First, we define our problem: Let the field $F \subset \mathbb{R}^2$ be a compact set. The grazing radius of a robot when standing still is r. That is, assume the position of the robot is x, then it can graze the disc

$$D_x = \{y \in F \mid \|y - x\| \le r\}.$$

The dwelling time for each point $p \in F$ to be served (grazed) is $T > 0$, namely p should stay inside the grazing disc for at least T.

In order to simplify the problem, we assume in the beginning that the field is large enough so that the boundary constraints can be ignored.

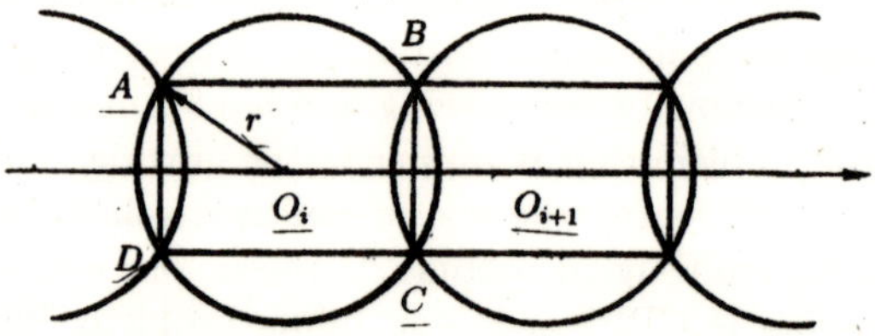

Fig. 1. Best Horizontal Distribution

Assume the robots are lined on a row and moving forward simultaneously. (refer to Fig. 1) Then a simple analysis shows that the effective region for each robot is the inscribed rectangle $ABCD$, if a positive dwelling time for each point is required. Denote $\overline{AD} = d$ and $\overline{DC} = l$, and N the number of robots. Now it is easy to see that the maximum speed for keeping enough dwelling time is

$$v = \frac{d}{T}.$$

Then the effective sweeping area of each robot per unit time is

$$S = lv = \frac{ld}{T} = \frac{2}{T} l \sqrt{r^2 - (\frac{l}{2})^2}. \tag{1}$$

Now assume our goal is to make grazing done as soon as possible, which is equivalent to

$$\max_l S(l).$$

An elementary calculus computation shows that the best solution is

$$\frac{dS(l)}{dl} = 0 \quad \Rightarrow \quad l^* = \sqrt{2}r. \tag{2}$$

That is, the rectangular inscribed in each circle is a square. Equivalently, the optimal length for all the robots is $L^* = \sqrt{2}Nr$.

Next, we consider the case when the field is a strip with width L satisfied $L^* < L < 2L^*$. The reason for considering this is the following: Assume the field F is a rectangle. Then it can be grazed column by column with the optimal width L^*. If the width of the field can be divided by L^* we have no problem at all. But if it is not like this, we have to consider the case of $L^* < L < 2L^*$.

In fact, it is not difficult to see that as $L \geq \sqrt{2}L^*$ we have to run twice. Now assume $L^* < L < \sqrt{2}L^*$. Then we have two options: (i) run once; (ii) run twice. It is easy to see that for case (i) the effective area per unit time is:

$$S_1 = \frac{2\sqrt{r^2 - (\frac{L}{2N})^2}}{T} L, \tag{3}$$

and for case (ii) the effective area per unit time is:

$$S_2 = \frac{2\sqrt{r^2 - (\frac{L}{4N})^2}}{T} \frac{L}{2}. \tag{4}$$

Comparing (3) with (4), one sees easily that $L_0 = \frac{4r}{\sqrt{5}}N$ is a critical point.

Remark 2.1 *1. It is obvious that the optimal strategy is independent of the dwelling time $T > 0$ (as long as the robots can go fast enough);*

2. The same argument can show that the uniform distribution of the robots is the best strategy. That is, any two adjacent two robots should keep the same distance. This strategy is used through the paper.

Summarizing the above argument we have the following result:

Proposition 2.1 *When a positive dwelling time is required and assume the term of N robots are on a row. If the width of the field is free, the best distance between two adjacent robots is $l^* = \sqrt{2}r$. For the remainder of the column-wise grazing, if the width L satisfies $L^* < L < 2L^*$, then the best strategy is one column strategy provided $L^* < L < \frac{4r}{\sqrt{5}}Nr$; if $\frac{4r}{\sqrt{5}}Nr < L < 2L^*$ then the best strategy is to go twice.*

3 Irregular strip and obstacles

Now we consider the case of irregular field. It still can be grazed column by column till the last column with an irregular boundary, which can also be considered as formed by obstacles in the field. So the problem comes down to grazing with an irregular strip. Moreover, we assume without loss of much generality that the strip is only with irregular boundary on one side.

We describe the irregular boundary of the strip by $f(x)$ with the regular side of boundary as x-axis (Refer to Fig. 2.). In Section 5 we will discuss how to implement this algorithm online based on only local information.

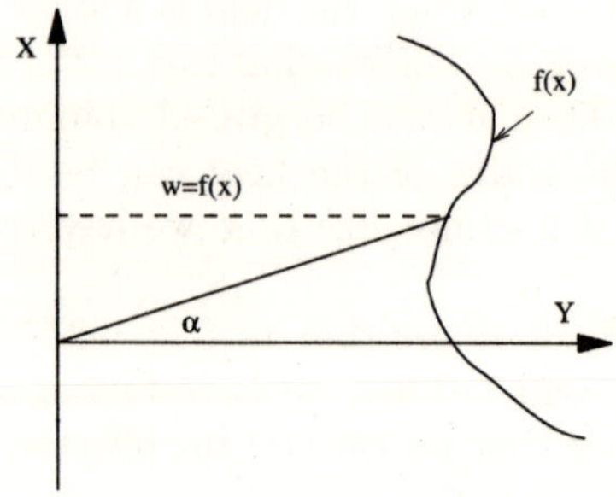

Fig. 2. Irregular Strip.

To avoid tedious but similar discussion, we assume

$$0 < f(x) < 2Nr.$$

According to the uniformly distributing law, the position of i-th robot should be

$$y_i = \frac{(2i-1)f(x)}{2N}, \quad i = 1, 2, \cdots, N. \tag{5}$$

According to the argument in the last session, we have the best speed as

$$\dot{x}_i = 2\sqrt{r^2 - \left(\frac{f(x)}{2N}\right)^2} \tag{6}$$

Using it to (5), we have

$$\dot{y}_i = \frac{2i-1}{2N} f'(x)\dot{x}_i. \tag{7}$$

Next, we consider the control problem. Through this paper we use the following model for a robot

$$\begin{cases} \dot{x}_i = v_i \cos\phi_i \\ \dot{y}_i = v_i \sin\phi_i \\ \dot{\phi}_i = w_i \end{cases} \tag{8}$$

Now we look for the corresponding optimal control. Since

$$\tan\phi_i = \frac{\dot{y}_i}{\dot{x}_i} = \frac{(2i-1)f'(x)}{2N},$$

then

$$v_i(x) = \frac{\dot{x}}{\sec(\phi)} = \frac{\dot{x}}{\left\{1+\left[\frac{(2i-1)f'(x)}{2N}\right]^2\right\}^{1/2}}$$
$$= 2\sqrt{\frac{4N^2r^2 - f^2(x)}{4N^2+[(2i-1)f'(x)]^2}}. \tag{9}$$

Note that

$$\phi_i = \tan^{-1}\left[\frac{(2i-1)f'(x)}{2N}\right].$$

Then we have

$$w_i(x) = \frac{1}{1+\left(\frac{[(2i-1)f'(x)]^2}{2N}\right)^2}(2i-1)f''(x)\dot{x}$$
$$= \frac{8N^2(2i-1)f''(x)\sqrt{4N^2r^2-f^2(x)}}{4N^2+[(2i-1)f'(x)]^2}. \tag{10}$$

4 Non-horizontal team's formation

In previous sections we discussed the grazing problem for team in a horizontal form. Precisely, the line of the team is perpendicular to the direction of advance. It has some advantages in organization. But this section will show that this distribution form is not "optimal".

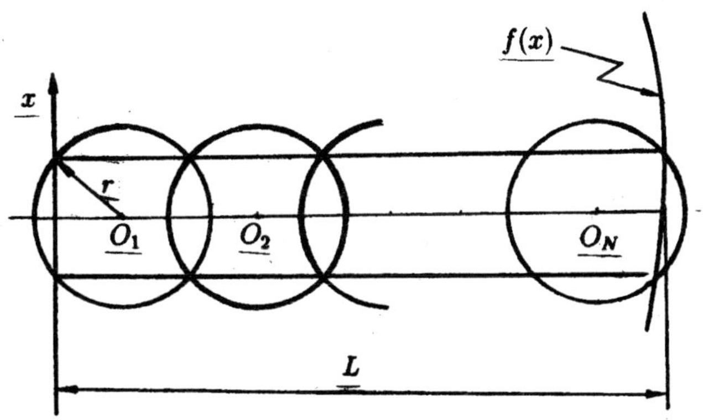

Fig. 3 Non-horizontal Team

Refer to Fig.3, it is clear that even though we still use squares inside each circle as the working area (later on, we can see that in this case square is not the optimal), we received an extra feasible working area – the parallelogram $BEHC$.

Now assume the horizontal angle of the team line is α and the horizontal angle of the vertical B of the circumscribed rectangle is β, then the effective working area between o_i and o_{i+1} is:

$$S_1 = \overline{o_i, p} \times \overline{B, C} = 2r \sin\beta \times 2r \cos(\beta - \alpha) \cos\alpha.$$

The whole effective working area is:

$$S(\alpha, \beta) = (N-1)S_1 + \overline{BC} \times \overline{AB} = 4r^2[(N-1)\cos(\beta-\alpha)\cos\alpha + \cos\beta]\sin\beta. \tag{11}$$

To get maximum value, we set

$$\begin{aligned} \tfrac{\partial S(\alpha,\beta)}{\partial \alpha} &= 4r^2 \sin(\beta)\sin(\beta - 2\alpha) := 0, \\ \tfrac{\partial S(\alpha,\beta)}{\partial \beta} &= 4r^2 \left[(-(N-1)\sin(\beta-\alpha)\cos\alpha - 2\sin\beta)\sin\beta \right. \\ &\quad \left. + ((N-1)\cos(\beta-\alpha)\cos\alpha + 2\cos\beta)\cos\beta\right] := 0. \end{aligned} \tag{12}$$

From the first equation we have

$$\alpha = \frac{\beta}{2}. \tag{13}$$

Plugging it into the second equation of (12) yields

$$\frac{N+3}{2}[\cos^2\beta - \sin^2\beta] + \frac{N-1}{2}\cos\beta = 0. \tag{14}$$

Solving it, we have the optimal solution

$$\cos\beta^* = \frac{-(N-1) + \sqrt{(N-1)^2 + 8(N+3)^2}}{4(N+3)}.$$

We cite the optimal angles for some small N. In the table the item W is the effective width and item S_u is the effective area per unit robot (assume $r = 1$).

Table 1. Optimal Angles and Area(in degree (^{o}))

N	2	3	4	5	6	7	8	9	10
β^*	48.7861	52.5524	51.0482	53.6248	54.4278	55.0514	55.5497	55.9569	56.2959
α^*	24.3930	25.5241	26.2762	26.8124	27.2139	27.5257	27.7748	27.9784	28.1480
W^*	2.9766	4.5147	6.0402	7.5584	9.0721	10.5827	12.0912	3.5982	15.1039
S_u	2.2392	2.3406	2.3977	2.4343	2.4597	2.4784	2.4927	2.5039	2.5130

It is clear from Table 1 that as N increasing, β increases and the effective area per unit robot, S_u, increases too.

In fact, no matter what β is, the optimal α always satisfies (13). Applying this fact to (11), we can obtain the per union effective area as ($r = 1$)

$$S_u = 2[\frac{N-1}{N} + \frac{N+1}{N}\cos\beta]\sin\beta. \tag{15}$$

As $N \to \infty$, we have

$$S_u|_{N=\infty} = 2(1+\cos\beta)\sin\beta.$$

Then it is easy to verify that the best β is

$$\beta|_\infty = \pi/3.$$

Then the area S_u, is

$$S_u|_\infty = \frac{3}{2}\sqrt{3} = 2.5981.$$

Next, we consider the case of a strip with fixed width W. First, it is easy to figure out that now the effective area per unit robot is

$$S_u = 2W\sin\beta/N. \tag{16}$$

It is obvious that the large the β (for $0 < \beta < \frac{\pi}{2}$) the large the S_u. Then we look for the largest feasible β. Note that (refer to Fig. 3) β is subject to the following restriction:

$$(N-1)\overline{O_iP} + 2\cos\alpha = W.$$

That is,

$$2(N-1)\cos(\beta-\alpha)\cos\alpha + 2\cos\beta = W.$$

Using trigonometric transformation yields

$$(N+1)\cos\beta + (N-1)\cos(\beta-2\alpha) = W. \tag{17}$$

Now

$$\max[\cos(\beta-2\alpha)] \Rightarrow \min\cos\beta \Rightarrow \sin\beta.$$

Hence (17) yields that

$$\alpha* = \frac{1}{2}\beta^*, \quad \cos\beta^* = \frac{W-(N-1)}{N+1}. \tag{18}$$

Note that a natural restriction is $W < 2N$. Denote $\Delta = 20\%(2N - W)$, the following two tables show the optimal α^* and the corresponding effective area per unit robot respectively.

Table 2. Optimal Team Angle α

$W\backslash N$	2	3	4	5	6	7	8	9	10
$W^*-4\Delta$	33.6480	35.3170	36.4373	37.2414	37.8467	38.3186	38.6970	39.0071	39.265
$W^*-3\Delta$	31.4934	33.0273	34.0541	34.7896	35.3424	35.7729	36.1177	36.4001	36.635
$W^*-2\Delta$	29.2526	30.6529	31.5881	32.2567	32.7586	33.1490	33.4615	33.7171	33.930
$W^*-\Delta$	26.8985	28.1653	29.0094	29.6120	30.0636	30.4148	30.6955	30.9251	31.116
W^*	24.3930	25.5241	26.2762	26.8124	27.2139	27.5257	27.7748	27.9784	28.148
$W^*+\Delta$	21.6782	22.6686	23.3260	23.7940	24.1441	24.4158	24.6328	24.8100	24.957
$W^*+2\Delta$	18.6571	19.4975	20.0545	20.4505	20.7465	20.9760	21.1592	21.3088	21.433
W^*+3D	15.1413	15.8142	16.2595	16.5757	16.8119	16.9949	17.1409	17.2601	17.359
W^*+4D	10.6434	11.1104	11.4189	11.6378	11.8012	11.9278	12.0286	12.1110	12.179

Table 3. Optimal Effective Area S_u

$W\backslash N$	2	3	4	5	6	7	8	9	10
$W^*-4\Delta$	3.9814	6.2764	8.5481	10.8017	13.0421	15.2728	17.4962	19.7140	21.9275
$W^*-3\Delta$	4.2097	6.6232	9.0270	11.4211	13.8072	16.1872	18.5625	20.9341	23.3028
$W^*-2\Delta$	4.3781	6.8781	9.3813	11.8826	14.3811	16.8769	19.3705	21.8623	24.3525
$W^*-\Delta$	4.4735	7.0202	9.5819	12.1489	14.7179	17.2876	19.8573	22.4270	24.9964
W^*	4.4783	7.0219	9.5907	12.1713	14.7582	17.3486	19.9413	22.5353	25.1304
$W^*+\Delta$	4.3681	6.8447	9.3549	11.8821	14.4190	16.9619	19.5088	22.0585	24.6101
$W^*+2\Delta$	4.1051	6.4295	8.7928	11.1767	13.5729	15.9769	18.3861	20.7991	23.2147
W^*+3D	3.6213	5.6698	7.7584	9.8689	11.9927	14.1250	16.2631	18.4055	20.5509
W^*+4D	2.7557	4.3134	5.9057	7.5172	9.1404	10.7713	12.4074	14.0474	15.6902

From the discussion in last section one sees that the the first key step in path planning is to determine the optimal α^*. Unlike the horizontal case, it is very difficult to obtain closed form solution for non-horizontal team distribution. In the following discussion we assume $f(x)$ is varying slowly. Precisely, $|f'(x)|$ is small enough. Then certain approximations are applicable. Refer to Fig. 4 we propose some approximations.

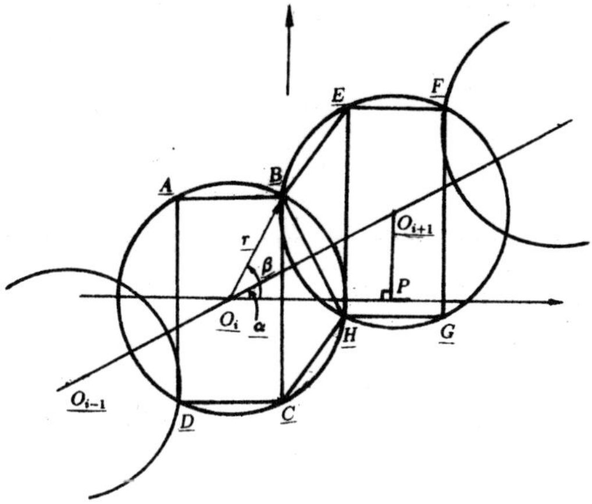

Fig. 4 Optimal Planning

Approximation 1.

$$W = f(X) \approx f(x).$$

Using Appr. 1, and (18), we have that

$$\alpha = \frac{1}{2}\cos^{-1}\left[\frac{f(x)-(N-1)}{N+1}\right]. \tag{19}$$

Approximation 2.

$$f(X) \approx f(x) + f'(x)(X-x).$$

Since

$$\frac{f(X)}{X-x} = \cot(\alpha),$$

we have

$$\frac{f(x)+f'(x)(X-x)}{X-x} \approx \cot(\alpha);$$

equivalently,

$$\frac{f(x)}{X-x} \approx \cot(\alpha) - f'(x),$$

$$X - x \approx \frac{f(x)}{\cot(\alpha)-f'(x)}.$$

Plugging it into (18) yields

$$\cos(2\alpha) \approx \frac{\frac{f(x)}{1-f'(x)\tan(\alpha)} - (N-1)}{N+1}. \tag{20}$$

Since $f'(x)$ is assumed to be small, so we may further assume that on the right hand side of (20)

$$\alpha \approx \alpha|_\infty = \frac{\pi}{6},$$

which leads to the following approximation

$$\alpha \approx \frac{1}{2}\cos^{-1}\left[\frac{\frac{f(x)}{1-\frac{\sqrt{3}}{3}f'(x)} - (N-1)}{N+1}\right]. \tag{21}$$

Next, we try to solve $\tan(\alpha)$ from (20). For notational ease, denote $t = \tan(\alpha)$. Then (20) can be expressed as

$$\frac{1-t^2}{1+t^2} = \frac{f(x) - (N-1)(1-f'(x)t)}{(N+1)(1-f'(x)t}.$$

An algebraic computation converts it to the following cubic equation.

$$2f'(x)t^3 - (f(x)+2)t^2 - 2Nf'(x)t + (2N - f(x)) = 0. \tag{22}$$

Setting

$$\begin{cases} p = -\left[\frac{(f(x)+2)^2}{12(f'(x))^2} + N\right], \\ q = -\frac{(f(x)+2)^3}{108(f'(x))^3} + \frac{2N-(N+3)f(x)}{6f'(x)}, \\ D = \left(\frac{p}{3}\right)^3 + \left(\frac{q}{2}\right)^2, \end{cases}$$

we have

$$\tan(\alpha) = \left[-\frac{q}{2} + \sqrt{D}\right]^{1/3} + \left[-\frac{q}{2} - \sqrt{D}\right]^{1/3}. \tag{23}$$

5 Online implementation and formation control

In this section we will focus on the case of horizontal teams. The state feedback controls (9)-(10) are theoretically correct. But in practical applications, we would have problems with robustness and feasibility. The main problems are first that x doesn't make sense if not all robots are on a line precisely and secondly that it is usually difficult or expensive to measure the global x. A possible solution would be that each robot uses its own x^i. Then the team will be broken, and no further cooperation could be realized.

We need an agent-based control such as in swarm behaviors. That a robot adjusts its velocity by observing the adjacent neighbors is an easy and reasonable way for a team work. Thus a typical formation control scheme is to use the nearest neighbor rules, see for example, [4].

We propose the following nearest neighbor rule: The first robot (as a leader) uses its own coordinates for control. That is:

$$v_1(x) = v_1(x^1); \quad w_1(x) = w_1(x^1).$$

and the follows use their nearest neighbors' information:

$$\begin{aligned} v_i(x) &= v_i(x^{i-1}, x^i, x^{i+1}); \\ w_i(x) &= w_i(x^{i-1}, x^i, x^{i+1}), \quad 1 < i < N. \end{aligned}$$

The last robot uses its only neighbor's information:

$$v_N(x) = v_N(x^{N-1}, x^N); \quad w_N(x) = w_N(x^{N-1,x^N}).$$

Remark 5.1: When there are obstacles in the path, the robot that initiate an obstacle avoidance will become the leader for the subteam.

For notational ease, we denote the optimal velocity as

$$V(x) = 2\sqrt{r^2 - \left(\frac{f(x)}{2N}\right)^2}.$$

A necessary condition for optimal control to satisfy is that each can have this velocity. So we propose the following speeds, where the first robot uses its own state as feedback variable and the second to the last second use their left and right neighbors coordinates, and the last uses only itself and the last second one. Finally the state equation is assumed as follows:

$$\begin{cases} \dot{x}_1 = V(x_1), \\ \dot{x}_2 = V(\frac{x_1+x_2+x_3}{3}) - \mu(x_2 - x_1), \\ \vdots \\ \dot{x}_{N-1} = V(\frac{x_{N-2}+x_{N-1}+x_N}{3}) - \mu(x_{N-1} - x_{N-2}), \\ \dot{x}_N = V(\frac{x_{N-1}+x_N}{2}) - \mu(x_N - x_{N-1}). \end{cases} \tag{24}$$

Note that in addition to the optimal velocity we add a linear term to assure the convergence of the difference with $\mu > 0$, which will be determined later.

The physical meaning of the linear term is clear: If a robot is behind, accelerate it, and if it is ahead, slow it down.

Now we show that under suitable $\mu > 0$ the motion of robots under model (24) will be lined up. Define

$$\delta_i = x_{i+1} - x_i, \quad i = 1, \cdots, N-1.$$

Denote by

$$D(x) = V'(x).$$

Then we have that

$$\begin{aligned} \dot{\delta}_1 &= V(\tfrac{x_1+x_2+x_3}{3}) - \mu(x_2 - x_1) - V(x_1) \\ &= V(x_1 + \tfrac{x_3-x_2+2(x_2-x_1)}{3}) \\ &\quad -V(x_1) - \mu(x_2 - x_1) \\ &= D(x_1)(\tfrac{2}{3}\delta_1 + \tfrac{1}{3}\delta_2) - \mu\delta_1 - O(\|\delta\|^2). \end{aligned}$$

Similarly, we have

$$\begin{aligned}\dot{\delta}_i &= D(x_{i-1})\frac{\delta_{i+1}+\delta_i+\delta_{i-1}}{3} - \mu(\delta_i - \delta_{i-1}) \\ &+O(\|\delta\|^2), \quad i = 2,\cdots,N-2; \\ \dot{\delta}_{N-1} &= D(x_{i-2})\frac{1}{3}\delta_{N-2} + \frac{1}{6}\delta_{N-1} - \mu(\delta_{N-1} - \delta_{N-2}) \\ &+O(\|\delta\|^2).\end{aligned}$$

Putting them together and denote by

$$\delta = (\delta_1, \delta_2, \cdots, \delta_{N-1})^T,$$

the equation about δ is obtained as

$$\dot{\delta} = A\delta + O(\|\delta\|^2), \tag{25}$$

where $A = \Psi Q + \mu M$ with

$$\Psi = \begin{pmatrix} 2/3 & 1/3 & 0 & & \\ 1/3 & 1/3 & 1/3 & & \\ 0 & 1/3 & 1/3 & 1/3 & \\ & & \ddots & \ddots & \ddots \\ & & & 1/3 & 1/6 \end{pmatrix}$$

$$Q = \begin{pmatrix} D(x_1) & 0 & 0 & \cdots & 0 \\ 0 & D(x_1) & 0 & \cdots & 0 \\ 0 & 0 & D(x_2) & \cdots & 0 \\ & & & \ddots & \\ 0 & 0 & 0 & \cdots & D(x_{N-2}) \end{pmatrix}$$

$$M = \begin{pmatrix} -1 & 0 & 0 & \cdots & 0 \\ 1 & -1 & 0 & \cdots & 0 \\ 0 & 1 & -1 & \cdots & 0 \\ & & \ddots & & \\ 0 & 0 & \cdots & 1 & -1 \end{pmatrix}.$$

It is obvious that M is a Hurwitz matrix, and Q is bounded, assuming $f'(x)$ is bounded. Hence as $\mu > 0$ large enough, (25) is asymptotically stable. Summarizing the above argument, we have the following result:

Proposition 5.1 *Consider the system (24). Assume $\mu > 0$ large enough, then*

$$\lim_{t\to\infty}[x_i(t) - x_j(t)] = 0, \quad \forall i, j$$

i.e, they are eventually lined up.

Finally, we have to find agent-based controls to realize (24). Note that it is reasonable to replace $f(x)$ in (5) by $f(x_i)$, then following the same argument as in the section 3 (10) becomes

$$w_i(x_i) = \frac{8N^2(2i-1)f''(x_i)\sqrt{4N^2r^2 - f^2(x_i)}}{4N^2 + [(2i-1)f'(x_i)]^2}, \quad i = 1, \cdots, N. \tag{26}$$

Combining it with (24) yields

$$v_i(x_i) = \begin{cases} \frac{V(x_1)}{\left\{1+\left[\frac{(2i-1)f'(x_i)}{2N}\right]^2\right\}^{1/2}}, & i = 1, \\ \frac{V(\frac{x_{i-1}+x_i+x_{i+1}}{3})-\mu(x_i-x_{i-1})}{\left\{1+\left[\frac{(2i-1)f'(x_i)}{2N}\right]^2\right\}^{1/2}}, & 1 < i < N, \\ \frac{V(\frac{x_{N-1}+x_N}{2})-\mu(x_N-x_{N-1})}{\left\{1+\left[\frac{(2i-1)f'(x_i)}{2N}\right]^2\right\}^{1/2}}, & i = N. \end{cases} \tag{27}$$

6 Conclusion

This paper considers the problem of razing behavior, which is one of the most fundamental social behaviors and very important in the formation control of robot teams.

Similar to the complete coverage problem but with dynamic moving behaviors, we investigated some optimal team control rules under the requirement of minimum dwell time, which is reasonable in completing a practical task.

The optimal path planning algorithms have been obtained in terms of maximizing the grazed area per unit time. Finally, the algorithms together with a formation control have been implemented online for nonholonomic robots.

References

1. T. Arai, D. Kurabayashi, J. Ota, and S. Ichikawa. Motion planning for cooperative sweeping with relocating obstacles. In *IEEE Int. Conf. Sys. Man Cyb.*, Beijing, 1996.
2. R.C. Arkin. *Behavior-based Robotics.* MIT Press, 1998.
3. Z. Butler, A. Rizzi, and R. Hollis. Cooperative coverage of rectilinear environments. In *IEEE International Conference on Robotics and Automation*, San Francisco, 2000.
4. Ali Jadbabaie, Jie Lin, and A. Stephen Morse. Coordination of groups of mobile autonomous agents using nearest neighbor rules. *IEEE Trans. Automat. Control*, 48(6):988–1001, 2003.

5. D. Kurabayashi, J. Ota, T. Arai, and E. Yoshida. Cooperative sweeping by multiple mobile robots. In *IEE International Conference on Robotics and Automation*, April 1996.
6. G. Luo and S.X. Yang. Real-time path planning with deadlock a voidance of multiple cleaning robots. In *IEEE International Conference on Robotics and Automation*, Taipei, 2003.
7. J. Ota, T. Arai, E. Yoshida, D. Kurabayashi, and J. Sasaki. Motion skills in multiple mobile robot system. In *Robotics and Autom. Sys.*, vol.19, 1996.
8. I. Rekleitis, G. Dudek, and E. Milios. Multi-robot collaboration for robust exploration. In *IEEE International Conference on Robotics and Automation*, San Francisco, 2000.

Part 2

Nonsmooth and Switched System Control

Stability Analysis and Adaptive Controller Design for a Class of Nonsmooth Systems

Takashi Nakakuki, Tielong Shen, Yasuhiko Mutoh and Katsutoshi Tamura

Department of Mechanical Engineering, Graduate Division of Science and Technology, Sophia University, 7-1, Kioicho, Chiyoda-ku, Tokyo, 1028554,
[t-nakaku, tetu-sin, y_mutou, k-tamura]@sophia.ac.jp

1 Introduction

Control and stability analysis of discontinuous systems have attracted considerable research interests in both mathematical issues and practical applications. In the mathematical issues, the definition of solutions of differential equations with discontinuous right hand side has been pioneered by Filippov [5], and the existence and uniqueness of the solution have been also explored [5, 2]. On the other hand, in practical applications, control problems of nonlinear systems with discontinuity such as stick-slip motion, backlash, switching control, etc. have been addressed in Filippov's framework [9, 10, 8, 11, 1].

As is well-known, the Lipschitz condition plays a crucial role in nonlinear system theory for assuring the existence and uniqueness of a solution of a differential equation which describes the system dynamics. However in nonsmooth systems the Lipschitz condition is often violated. Especially, the uniqueness of a solution of a system controlled by a switching or discontinuous controller should be evaluated carefully. Polycarpou [10] and Kim [6] addressed the problem of the existence and uniqueness of the solution. Polycarpou [10] derived a sufficient condition for the existence and uniqueness of the Filippov solution of general adaptive control systems including switching adaptive schemes while Kim [6] recently presented a sufficient condition for the existence of the Carathéodory solution in nonlinear systems with general discontinuous switching algorithms.

Meanwhile, various controller designs with discontinuous actions in practical applications have been developed [9, 1, 8, 11]. Paden *et al.* [9] hammered out a *VSS* design of robot manipulators by utilizing the concept of the Filippov solution. Alvarez *et al.* [1] proposed a switching control algorithm for a Coulomb friction oscillator. Although these papers theoretically challenged the control problems of nonsmooth systems, they did not rigorously address the basic problems such as the uniqueness of the solution of the closed loop system. Recently, Orlov *et al* [8] developed a chattering controller for an electrical servo-motors with discontinuity. In the paper, although the property of the solution on the discontinuous surface created by the controller is examined, the uniqueness of the solution on the other discontinuity is not discussed. Sekhavat *et al.* [11] showed not only a controller design for a hydraulic actuator with friction but also the existence and uniqueness of the solution of the closed loop system.

Adaptive control is one of the challenging tasks for nonsmooth systems. In fact, above-mentioned papers [9, 1, 8, 11] require the exact information of the system or preknowledge of the upper and lower bounds of uncertainty. For adaptive control of nonsmooth systems, there is a remarkable book by Tao *et al* [13]. In the book various mechanical systems with nonsmoothness are considered whereas the stability analysis for each system is individually made by each specialized methodology.

This paper is consisted of two main topics, that is, non-adaptive and adaptive approaches. Throughout the paper, a solid stability analysis with the concept of the Filippov solution and controller design employing rigorous nonsmooth analysis are comprehensively presented for an affine system with discontinuity. In case of non-adaptive approach, first, a sufficient condition for which any Lipschitz continuous controller can not achieve the asymptotic stability of the system is shown. Then, a switching control law with so-called L_gV-structure is proposed, and the controller is further modified to assure the uniqueness. In case of adaptive approach, uncertainty, which is linearly parameterized in an unknown vector and a known discontinuous function, is considered. More general and comprehensive system than [13] is addressed. For such a nonsmooth system, generally speaking, the classical stability analysis [7] can not be applied due to the presence of the discontinuous function which may create a sliding motion while our methodology assures the stability during the sliding motion as well as during the non-sliding motion. Although the uniqueness of the Filippov solution of the closed loop system is not necessarily guaranteed, global strong stability and attractiveness of the state of the controlled system to the origin are achieved.

The outline of this paper is as follows: In Section 2 non-adaptive approach for an affine nonsmooth system is discussed. In Section 3 adaptive approach for the system is addressed. In these two sections, some illustrated examples are prepared to make the material more understandable. Some conclusions are given in Section 4. In Appendix mathematical preliminaries are prepared.

2 L_gV Structure-Based Switching Control Approach

In this section we focus on a single input system to make clear the essential discussion. All results in this section can be straightforward extended to the multi-input case. If the preknowledge of uncertainty such as the upper and lower bounds is available, a L_gV structure-based switching controller is one of reasonable strategies to deal with the system.

2.1 Problem Statement

Consider the plant with nonlinear discontinuity $\mu(x)$

$$\dot{x} = Ax + b(\mu(x) + u) \tag{1}$$

where $x \in \mathbb{R}^n$ is the state vector and $u \in \mathbb{R}$ is the control input. The pair (A, b) is assumed to be controllable. The discontinuity $\mu(x)$ is given by

$$\mu(x) = \begin{cases} m_1 & \text{if } S(x) > 0 \\ -m_2 & \text{if } S(x) < 0 \end{cases} \tag{2}$$

where m_1, m_2 are positive constants. $S(x)$ is a smooth surface with $S(0) = 0$ which divides the state space into the domains S^+ and S^- as shown in Fig. ??. Without loss of generality we assume that A is an asymptotic stable matrix. The purpose of the controller design is to find a control law $u(x)$ which makes the closed loop system asymptotically stable and assures the uniqueness of the Filippov solution. Define $f(x)$ by

$$f(x) = Ax + b(\mu(x) + u(x)). \tag{3}$$

As shown in Proposition 1 stabilizing control law $u(x)$ should be a discontinuous function. Thus, the closed loop system should be described by the differential inclusion form

$$\dot{x} \in K[f](x) \tag{4}$$

where $K[\cdot]$ is the Filippov set [5] (See Appendix for definition). For the origin $x = 0$ being an equilibrium point, the following is required.

$$0 \in K[f](0). \tag{5}$$

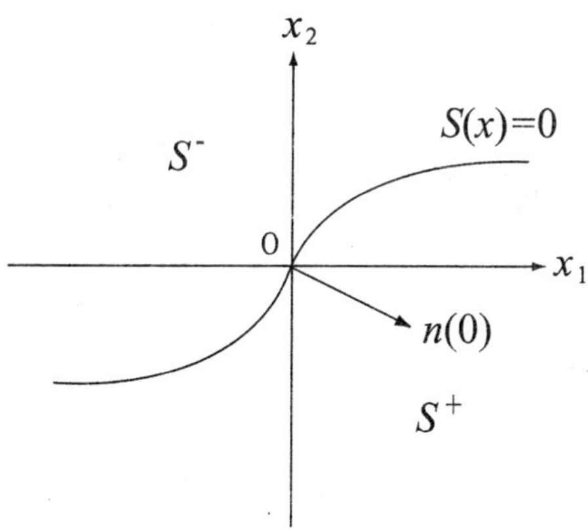

Fig. 1. $S(x)$, S^- and S^+

2.2 Controller Design and Stability Analysis

First, we discuss the stability property of the free system of (1)

$$\dot{x} = \bar{f}(x) = Ax + b\mu(x). \tag{6}$$

Since $S(0) = 0$,

$$K[\bar{f}](0) = \{\zeta \in \mathbb{R} \mid \zeta = (\alpha m_1 - (1-\alpha)m_2)b,\ 0 \le \alpha \le 1\}. \tag{7}$$

Thus, $0 \in K[\bar{f}](0)$, and $x = 0$ is a candidate of the equilibrium point. Let $n(x)$ be the normal vector to the surface $S(x)$ at a point x directed from S^- to S^+ as shown in Fig. 1. Let x be a point on the surface $S(x)$, and $\bar{f}^-(x)$ and $\bar{f}^+(x)$ be limiting values where x is approached from S^- and S^+ respectively. The projection of $\bar{f}^-$ and $\bar{f}^+$ to $n(x)$ is defined as

$$\begin{aligned} \bar{f}_n^+(x) &= \langle n(x), \bar{f}^+(x)\rangle \\ \bar{f}_n^-(x) &= \langle n(x), \bar{f}^-(x)\rangle. \end{aligned} \tag{8}$$

Proposition 1. *If $\bar{f}_n^+(0) > 0$ and $\bar{f}_n^-(0) < 0$, the origin $x = 0$ of system (6) is not stable. We call this d-unstable.*

Proof. Consider the trajectory starting from $x(0) = \delta_1 b \in B_\delta(0)$ (δ-neighborhood of the origin) for small $\delta_1 > 0$. For small $t_1 > 0$ the trajectory is given by

$$x(t) \simeq \delta_1 b + m_1 bt, \quad \forall t \in [0, t_1] \tag{9}$$

Since $\bar{f}_n^+(0) = m_1 n^T(0)b > 0$, $x(t) \in S^+$ for all $t \in [0, t_1]$. Simple $\epsilon - \delta$ argument implies that $x = 0$ is not a stable equilibrium point. □

When a control law $u(x)$ for system (1) is Lipschitz continuous, the same discussion as described in Proposition 1 is applied to obtain the following.

Proposition 2. *Let the origin of system (6) be d-unstable. Then any Lipschitz continuous controller $u(x)$ with $u(0) = 0$ can not achieve the asymptotic stability of the origin of system (1).* □

Now we move on to the controller design of system (1). Since Proposition 2 insists that a discontinuous control strategy should be considered, we introduce the following L_gV structure-based switching controller:

$$u(x) = -\gamma \operatorname{sgn}(b^T Px) \tag{10}$$

where $\gamma > \max(m_1, m_2) = m$ and P is a symmetric positive definite matrix which satisfies the following Lyapunov equation

$$PA + A^T P = -Q, \quad Q = Q^T > 0. \tag{11}$$

Then the closed loop system is given by

$$\dot{x} = Ax + b(\mu(x) - \gamma \mathrm{sgn}(b^T Px)). \tag{12}$$

Let $V(x)$ be a Lyapunov function candidate given by

$$V(x) = \frac{1}{2} x^T Px. \tag{13}$$

Calculating the set valued derivative (See Appendix for definition) along the trajectories of system (12),

$$\dot{\tilde{V}}(x) = \{v \in \mathbb{R} \mid v = -\frac{1}{2} x^T Qx + x^T PbK[\mu(x)] + x^T PbK[-\gamma \mathrm{sgn}(b^T Px)]\}. \tag{14}$$

Since $\max_{\zeta \in K[\mu](x)} |\zeta| = m$,

$$v \leq -\frac{1}{2} x^T Qx - (\gamma - m)|x^T Pb| < 0, \quad \forall x \neq 0, \ \forall v \in \dot{\tilde{V}}(x). \tag{15}$$

Note that the Lyapunov stability theory [12] for nonsmooth systems is given under the condition of the *existence and uniqueness* of the Filippov solution of the closed loop system. As a result, we should also examine the existence and uniqueness of the Filippov solution. The following example shows that the uniqueness of the Filippov solution of the closed loop system is not assured whereas the set valued derivative $\dot{\tilde{V}}(x)$ is negative definite.

Example 1. Consider the following system

$$\dot{x} = \begin{bmatrix} 0 & 1 \\ -2 & -3 \end{bmatrix} x + \begin{bmatrix} 0 \\ \mu(x) + u(x) \end{bmatrix} \tag{16}$$

where the discontinuous function is given by

$$\mu(x) = \begin{cases} 1 & \text{if } S(x) = 10x_1 + x_2 > 0 \\ -1 & \text{if } S(x) = 10x_1 + x_2 < 0. \end{cases} \tag{17}$$

From (10),

$$u = -\gamma \mathrm{sgn}(b^T Px) = -2\mathrm{sgn}(0.25x_1 + 0.25x_2). \tag{18}$$

Let $V(x)$ be a Lyapunov function candidate given by (13). Then, we can conclude that $\dot{\tilde{V}}(x) < 0$ for all $x \neq 0$. However we can also indicate that there exists a point failing to satisfy the sufficient condition for the uniqueness of the solution by Filippov [5].

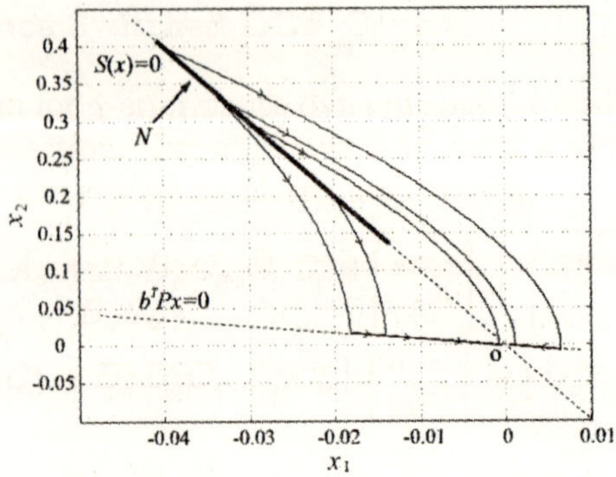

Fig. 2. Trajectories of system (16) with controller (18) (enlarged view around the set N)

In fact, consider a point $x \in N$ where

$$N = \{x \in \mathbb{R}^2 | \ S(x) = 0, \ 1/72 < |x_1| < 3/72\} \tag{19}$$

as shown by the solid line in Fig. 2. On the point $x^* \in \{x \in N | \ b^T Px > 0\}$ we have $n(x^*) = [10, \ 1]^T$ and

$$\begin{aligned} f^+(x^*) &= \begin{bmatrix} x_2 \\ -2x_1 - 3x_2 - 1 \end{bmatrix} \\ f^-(x^*) &= \begin{bmatrix} x_2 \\ -2x_1 - 3x_2 - 3 \end{bmatrix}. \end{aligned} \tag{20}$$

Calculating $f_n^+(x^*)$ and $f_n^-(x^*)$, we obtain

$$f_n^+(x^*) > 0, \quad f_n^-(x^*) < 0 \tag{21}$$

which implies that the condition for the uniqueness of the Filippov solution of the closed loop system is not satisfied at the point. As shown in Fig. 2 trajectories starting from $x(0) \in N$ have two different behaviors, that is, the non-uniqueness property. □

As might be expected, we can also conclude that the closed loop system with the L_gV controller (10) is *strongly* stable and attractive. However our purpose of this section is to design an asymptotic stabilizing controller with the *uniqueness* property of the solution. Example 1 gives us an idea of a modification of the controller, that is, the controller should modify the vector field on $S(x) = 0$. See the following theorem.

Theorem 1. *Consider system (1). Let the control law be given by*

$$u = -\gamma_1 \text{sgn}(b^T Px) - \gamma_2 \text{sgn}\left(\frac{\partial S}{\partial x} b\right) \text{sgn}(S(x)) \tag{22}$$

where $\gamma_1 > \max(m_1, m_2) + \gamma_2$ *and* $\gamma_2 > (m_1 + m_2)/2$. *Suppose that* $(\partial S/\partial x)b \neq 0$ *for all* $x \in \mathbb{R}^n$. *Then, the origin* $x = 0$ *of the closed loop system is globally asymptotically stable, and the uniqueness of the Filippov solution is assured.*

Proof. The closed loop system is given by

$$\dot{x} = Ax + b\{\mu(x) - \gamma_1 \text{sgn}(b^T Px) - \gamma_2 \text{sgn}\left(\frac{\partial S}{\partial x} b\right) \text{sgn}(S(x))\} \tag{23}$$

Replacing $\mu(x)$ by $\bar{\mu}(x) = \mu(x) - \gamma_2 \text{sgn}\left(\frac{\partial S}{\partial x} b\right) \text{sgn}(S(x))$, the similar derivation from (10) to (15) can be applied to obtain the negative definiteness of the set valued derivative $\dot{\tilde{V}}(x)$ of a Lyapunov function candidate $V(x)$ given by (13) because of $\gamma_1 > \max_{\zeta \in K[\bar{\mu}(x)]} |\zeta| = \max(m_1, m_2) + \gamma_2$.
[existence of the solution] On an arbitrary compact set $D \subset \mathbb{R}^n$ the right hand side of (23) satisfies

$$\|Ax + b(\bar{\mu}(x) - \gamma_1 \text{sgn}(b^T Px))\| \leq L \ \text{ a.e.} \tag{24}$$

where L is a constant. Thus the right hand side of (23) satisfies *Condition B* (See Appendix for definition) which is a sufficient condition for the existence of the Filippov solution.
[uniqueness of the solution] Let x be a point on $\Omega_o = \{x \in \mathbb{R}^n | \ S(x) = 0, \ x \neq 0\}$. Then we have

$$\begin{aligned} f^+(x) &= Ax + b\{m_1 - \gamma_1 \text{sgn}(b^T Px) - \gamma_2 \text{sgn}\left(\tfrac{\partial S}{\partial x} b\right)\} \\ f^-(x) &= Ax + b\{-m_2 - \gamma_1 \text{sgn}(b^T Px) + \gamma_2 \text{sgn}\left(\tfrac{\partial S}{\partial x} b\right)\} \end{aligned} \tag{25}$$

Since $\gamma_2 > (m_1 + m_2)/2$, we obtain

$$f_n^+(x) - f_n^-(x) = \frac{\partial S}{\partial x} b \left\{ m_1 + m_2 - 2\gamma_2 \text{sgn}\left(\frac{\partial S}{\partial x} b\right) \right\} < 0, \quad \forall x \in \Omega_o \tag{26}$$

which implies that the uniqueness of the solution on Ω_o is assured according to Theorem 14 in Filippov [5]. The uniqueness of the solution on $\Omega_c = \{x \in \mathbb{R}^n | \ b^T Px = 0, \ x \neq 0\}$ can be shown in a similar way.
Since for every Filippov solution starting from any initial condition the existence and uniqueness of the solution are established, from Theorem 3 with $D = \mathbb{R}^n$ and radially unboundedness of (13) we can conclude that $x = 0$ is globally asymptotically stable. □

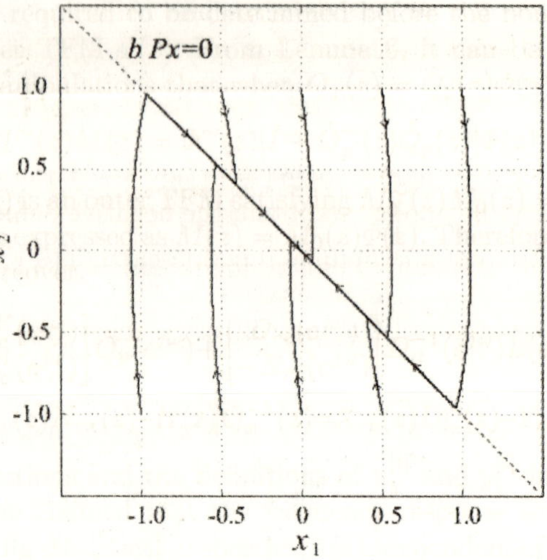

Fig. 3. Trajectories of system (16) with controller (27)

Example 2. Consider system (16) again. In order to assure the uniqueness of the solution in addition to the asymptotic stability of the closed loop system, we employ the following controller

$$u = -4\text{sgn}(0.25x_1 + 0.25x_2) - 2\text{sgn}(10x_1 + x_2) \tag{27}$$

according to (22) where $(\partial S/\partial x)b = 1$ for all x. Fig. 3 shows some trajectories of the closed loop system. In this case, the origin $x = 0$ is globally asymptotically stable and the uniqueness of the solution is assured unlike the result by the L_gV controller (18). □

3 Adaptive Control Approach

The L_gV controller in the previous section requires the upper and lower bounds of uncertainty. This section explores an adaptive controller design for nonsmooth systems with uncertainty. The controller design approach is based on the concept of the Filippov solution since a classical adaptive approach can not be applied to establish the stability of the adaptive control system. The adaptive control system in this section requires a strong stability concept since the uniqueness of the solution of the system is not necessarily guaranteed. It is shown that the adaptive control system is globally strongly stable and the state of the controlled system converges to the origin.

3.1 Problem Statement

Consider the plant

$$\dot{x} = Ax + b\big(\phi(x)\theta + u\big) \tag{28}$$

where $x \in \mathbb{R}^n$, $u \in \mathbb{R}^m$, and the pair (A, b) is assumed to be controllable. Without loss of generality we assume that A is a stable matrix. $\theta \in \mathbb{R}^p$ is an unknown vector, and $\phi : \mathbb{R}^n \to \mathbb{R}^{m\times p}$ is a known discontinuous function satisfying *Condition B* (See Appendix for definition) and is given by

$$\phi(x) = \begin{cases} \phi^+(x) & \text{if } S(x) > 0 \\ \phi^-(x) & \text{if } S(x) < 0 \end{cases} \tag{29}$$

where $\phi^+,\ \phi^- : \mathbb{R}^n \to \mathbb{R}^{m\times p}$ are locally Lipschitz, and $S(x)$ is a smooth function with $S(0) = 0$.

The objective of this section is to design an adaptive controller for system (28) in despite of the discontinuity of ϕ. The adaptive control problem for system (28) with *smooth* function ϕ has long been studied [7]. Notice that even if we employ a smooth control law $u(x)$ for system (28), the closed loop system should be described by the differential inclusion form

$$\dot{x} \in K[Ax + b(\phi\theta + u)](x) \tag{30}$$

due to the discontinuity of ϕ.

3.2 Adaptive Controller Design

In this section, we consider the adaptive controller design for system (28). Employing an estimate $\hat{\theta}(t)$ and an update law $\Phi(x)$, we can obtain the following certainty-equivalence form

$$\begin{aligned} u &= -\phi(x)\hat{\theta} \\ \dot{\hat{\theta}} &= \Phi(x). \end{aligned} \tag{31}$$

Then, we have the following closed loop differential inclusion [1]

$$\begin{bmatrix} \dot{x} \\ \dot{\tilde{\theta}} \end{bmatrix} \in K[f](x, \tilde{\theta}) \subset \begin{bmatrix} \{Ax\} + K[b\phi\theta] - K[b\phi\hat{\theta}] \\ -K[\Phi] \end{bmatrix} \tag{32}$$

where $\tilde{\theta} = \theta - \hat{\theta}$ is the parameter error.

It is very important to notice that in (32) we can not describe $K[b\phi\theta] - K[b\phi\hat{\theta}]$ as $K[b\phi\tilde{\theta}]$ since ϕ in (28) is different from ϕ in (32) in physical terms. The former ϕ is of a nature of the system while the latter ϕ is of an implementation of the actuator.

Thus, we can not proceed the adaptive controller design in a classical manner (e.g. [7]). In order to overcome the difficulty, we split the stability analysis into two regions $\Omega = \{(x, \tilde{\theta}) \in \mathbb{R}^{(n+p)} \mid S(x) = 0\}$ and $\overline{\Omega}$, which is the complementary set of Ω.

[1]Throughout the paper, the addition of two sets $A \subseteq \mathbb{R}^n$ and $B \subseteq \mathbb{R}^n$ is defined by $A + B = \{a + b \in \mathbb{R}^n \mid a \in A,\ b \in B\}$.

Theorem 2. *Consider system (28). Let the adaptive control law be given by*

$$\begin{aligned} u &= -\phi(x)\hat{\theta} \\ \dot{\hat{\theta}} &= \gamma\phi(x)^T b^T P x. \end{aligned} \tag{33}$$

Then, the closed loop system is globally strongly stable. Moreover for every solution $(x(t), \hat{\theta}(t)) \in \Xi$, *where* Ξ *is the set of solutions starting from* $(x(0), \hat{\theta}(0)) = (x_0, \hat{\theta}_0)$, $x(t)$ *converges to* 0 *as* t *tends to* ∞.

Proof. The closed loop system is given by

$$\begin{bmatrix} \dot{x} \\ \dot{\tilde{\theta}} \end{bmatrix} \in K[f](x, \tilde{\theta}) \subset \begin{bmatrix} \{Ax\} + K[b\phi\theta] - K[b\phi\hat{\theta}] \\ -K\left[\gamma\phi^T b^T P x\right] \end{bmatrix}. \tag{34}$$

Let $V : \mathbb{R}^{n+p} \to \mathbb{R}$ be a Lipschitz, regular and positive definite function given by

$$V(x, \tilde{\theta}) = x^T P x + \frac{1}{2\gamma}\tilde{\theta}^T\tilde{\theta}, \quad \gamma > 0. \tag{35}$$

where P is a symmetric positive definite matrix satisfying the Lyapunov equation

$$PA + A^T P = -Q, \quad Q = Q^T > 0. \tag{36}$$

Note that $V(x(t), \tilde{\theta}(t))$ is absolutely continuous in t by Theorem 2.2 in [12]. In what follows we will evaluate the behavior of V in two regions Ω and $\overline{\Omega}$, respectively.

(I) The evaluation of $\dot{V}$ with respect to $(x, \tilde{\theta}) \notin \Omega$

Since the discontinuity of the system occurs on Ω, in this case we can treat system (34) as the following usual differential equation

$$\begin{bmatrix} \dot{x} \\ \dot{\tilde{\theta}} \end{bmatrix} = f(x, \tilde{\theta}) = \begin{bmatrix} Ax + b\phi\tilde{\theta} \\ -\gamma\phi^T b^T P x \end{bmatrix}. \tag{37}$$

Thus, calculating the usual time derivative of V along the trajectories of system (37), we obtain

$$\begin{aligned} \dot{V} &= \begin{bmatrix} x^T P & \frac{1}{\gamma}\tilde{\theta}^T \end{bmatrix} \cdot \begin{bmatrix} Ax + b\phi\tilde{\theta} \\ -\gamma\phi^T b^T P x \end{bmatrix} \\ &\leq -\tfrac{1}{2}x^T Q x + x^T P b\phi\tilde{\theta} - \tilde{\theta}^T\phi^T b^T P x \\ &= -\tfrac{1}{2}x^T Q x, \quad \forall (x, \tilde{\theta}) \notin \Omega. \end{aligned} \tag{38}$$

(II) The evaluation of $\dot{V}$ with respect to $(x, \tilde{\theta}) \in \Omega$

In this case we should pay attention to two situations: one is that a trajectory $(x(t), \tilde{\theta}(t))$ intersects the surface Ω with Lebesgue measure 0 with respect to t, and another one is that a trajectory $(x(t), \tilde{\theta}(t))$ has a sliding motion on Ω. However, in the former case it is known that the time derivative of V on a point of a Lebesgue measure 0 set with respect to t can be ignored in evaluating the stability of the system. Thus, we focus on the latter case here.

$S(x) = 0$ is a smooth surface with $S(0) = 0$ which divides the state space into the domains $\Omega^+ = \{(x, \tilde{\theta}) \mid S(x) > 0\}$ and $\Omega^- = \{(x, \tilde{\theta}) \mid S(x) < 0\}$. Now let $N(x, \tilde{\theta})$ be the normal vector to the surface $S(x)$ at a point $(x, \tilde{\theta})$ directed from Ω^- to Ω^+. In this case, we have $N(x, \tilde{\theta}) = [n(x)^T \ 0]^T$. Let $(x, \tilde{\theta}) \in \Omega$, and $f^-(x, \tilde{\theta})$ and $f^+(x, \tilde{\theta})$ be limiting values where $(x, \tilde{\theta})$ is approached from Ω^- and Ω^+ respectively. The projection of $f^-(x, \tilde{\theta})$ and $f^+(x, \tilde{\theta})$ on $N(x, \tilde{\theta})$ is defined as

$$\begin{aligned} f_N^+(x, \tilde{\theta}) &= \langle N(x, \tilde{\theta}),\ f^+(x, \tilde{\theta})\rangle \\ f_N^-(x, \tilde{\theta}) &= \langle N(x, \tilde{\theta}),\ f^-(x, \tilde{\theta})\rangle. \end{aligned} \tag{39}$$

Then, $f^+(x, \tilde{\theta})$ and $f^-(x, \tilde{\theta})$ are calculated by

$$\begin{aligned} f^+(x, \tilde{\theta}) &= \begin{bmatrix} Ax + b\phi^+(x)\tilde{\theta} \\ -\gamma\phi^+(x)^T b^T P x \end{bmatrix} \\ f^-(x, \tilde{\theta}) &= \begin{bmatrix} Ax + b\phi^-(x)\tilde{\theta} \\ -\gamma\phi^-(x)^T b^T P x \end{bmatrix} \end{aligned} \tag{40}$$

which yield

$$\begin{aligned} f_N^+(x, \tilde{\theta}) &= n(x)^T Ax + n(x)^T b\phi^+(x)\tilde{\theta} \\ f_N^-(x, \tilde{\theta}) &= n(x)^T Ax + n(x)^T b\phi^-(x)\tilde{\theta}. \end{aligned} \tag{41}$$

According to Lemma 3 in [5], a trajectory during a sliding motion is dominated by the following differential equation

$$\begin{bmatrix} \dot{x} \\ \dot{\tilde{\theta}} \end{bmatrix} = \alpha f^+(x, \tilde{\theta}) + (1 - \alpha) f^-(x, \tilde{\theta}), \quad \alpha = \frac{f_N^-}{f_N^- - f_N^+} \tag{42}$$

subject to the constraint $f_N^+ \leq 0$, $f_N^- \geq 0$ and $f_N^- - f_N^+ > 0$. Thus, substituting (40) and (41) into (42),

$$\begin{aligned} \dot{x} = \frac{1}{n^T b(\phi^- - \phi^+)\tilde{\theta}} &\Big\{ n^T b(\phi^- - \phi^+)\tilde{\theta} Ax \\ &- n^T Axb(\phi^- - \phi^+)\tilde{\theta} + n^T b\phi^- \tilde{\theta} b\phi^+ \tilde{\theta} - n^T b\phi^+ \tilde{\theta} b\phi^- \tilde{\theta} \Big\} \end{aligned} \tag{43}$$

$$\begin{aligned} \dot{\tilde{\theta}} = \frac{-\gamma}{n^T b(\phi^- - \phi^+)\tilde{\theta}} &\Big\{ \left(n^T Ax + n^T b\phi^- \tilde{\theta}\right)(x^T P b\phi^+)^T \\ &- (n^T Ax + n^T b\phi^+ \tilde{\theta})(x^T P b\phi^-)^T \Big\}. \end{aligned} \tag{44}$$

Now we examine the time derivative of V along the trajectories of system (43) and (44) during the sliding motion on Ω. Calculating $\dot{V}$,

$$
\begin{aligned}
\dot{V} &= \left[x^T P \ \tfrac{1}{\gamma}\tilde{\theta}^T\right] \cdot \begin{bmatrix} \dot{x} \\ \dot{\tilde{\theta}} \end{bmatrix} \\
&= \tfrac{1}{n^T b(\phi^- - \phi^+)\tilde{\theta}} \cdot x^T P \left\{ n^T b(\phi^- - \phi^+)\tilde{\theta} A x \right. \\
&\quad \left. - n^T A x b(\phi^- - \phi^+)\tilde{\theta} + n^T b\phi^- \tilde{\theta} b \phi^+ \tilde{\theta} - n^T b \phi^+ \tilde{\theta} b \phi^- \tilde{\theta} \right\} \\
&\quad - \tfrac{1}{n^T b(\phi^- - \phi^+)\tilde{\theta}} \cdot \tilde{\theta}^T \left\{ \left(n^T A x + n^T b \phi^- \tilde{\theta}\right) (x^T P b \phi^+)^T \right. \\
&\quad \left. - \left(n^T A x + n^T b \phi^+ \tilde{\theta}\right) (x^T P b \phi^-)^T \right\} \\
&= \tfrac{n^T b(\phi^- - \phi^+)\tilde{\theta}}{n^T b(\phi^- - \phi^+)\tilde{\theta}} \cdot x^T P A x \\
&\leq -\tfrac{1}{2} x^T Q x, \quad \forall (x, \tilde{\theta}) \in \Omega \ .
\end{aligned} \tag{45}
$$

where $n^T b(\phi^- - \phi^+)\tilde{\theta} \neq 0$ for all t on Ω due to the constraint $f_N^- - f_N^+ > 0$.

(III) The stability of the adaptive system

From (38) and (45) we can conclude that for almost all t

$$\dot{V} \leq \frac{1}{2} x^T Q x < 0, \quad \text{a.e. } t, \quad \forall x \neq 0, \quad \forall \tilde{\theta}. \tag{46}$$

which implies that the origin of the closed loop system is globally strongly stable by directly applying Theorem 1 in [3]. In order to show the attractiveness of $x(t)$, consider a trajectory $(x(t), \hat{\theta}(t)) \in \Xi$. From (46), it follows that

$$\sup_{(x,\tilde{\theta}) \in \Xi} V(x(t), \tilde{\theta}(t)) \leq V(x_0, \tilde{\theta}_0) \leq M, \quad \forall t \geq 0 \tag{47}$$

where M is a sufficiently large positive constant. Thus, Ξ is uniformly bounded. Moreover, $\dot{x}(t)$ is also essentially uniformly bounded since $\dot{x} \in K[Ax + b\phi\theta - b\phi\hat{\theta}]$, and ϕ^+, ϕ^- are continuous functions. Thus, Ξ is equicontinuous. From (46) and (47),

$$\int_0^\infty x(t)^T Q x(t) dt \leq 2M. \tag{48}$$

Hence, Barbalat's lemma follows that $x(t)$ converges to 0. □

Remark 1. Unlike Theorem 1 in Section 2.2, we do not explore the uniqueness of the Filippov solution of the closed loop system. See the following example for the non-uniqueness of the adaptive system. □

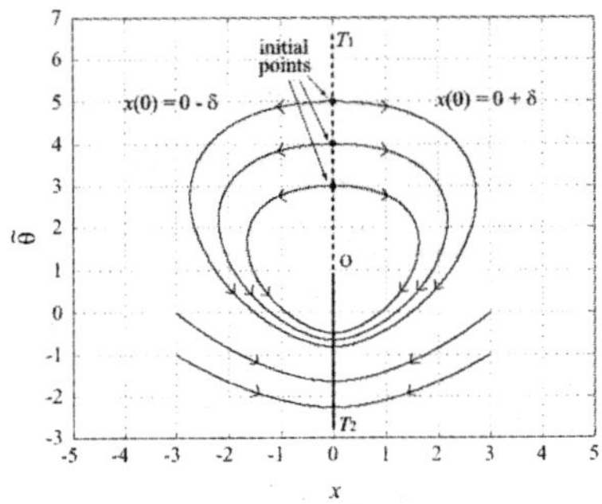

Fig. 4. Trajectories of system (49)

Example 3. Consider the first order affine system

$$\dot{x} = -x + \theta \mathrm{sgn}(x) + u \tag{49}$$

where $x \in \mathbb{R}$, $u \in \mathbb{R}$, θ is an unknown parameter and $S(x) = x$. Then, from (33) the adaptive control law is given by

$$\begin{aligned} u &= -\hat{\theta}\mathrm{sgn}(x) \\ \dot{\hat{\theta}} &= \gamma x \mathrm{sgn}(x), \quad \gamma > 0 \end{aligned} \tag{50}$$

which yields the following closed loop system

$$\begin{bmatrix} \dot{x} \\ \dot{\tilde{\theta}} \end{bmatrix} = f(x, \tilde{\theta}) = \begin{bmatrix} -x + \theta\mathrm{sgn}(x) - \hat{\theta}\mathrm{sgn}(x) \\ -\gamma x \mathrm{sgn}(x) \end{bmatrix} \tag{51}$$

where $\tilde{\theta} = \theta - \hat{\theta}$. Since the discontinuous surface is $\Omega = \left\{(x, \tilde{\theta}) \in \mathbb{R}^2 \mid S(x) = x = 0\right\}$, utilizing Theorem 14 in [5], we can show that the uniqueness and continuous dependence of the solution starting from an initial condition satisfying $x(0) = 0$ and $\tilde{\theta}(0) > 0$ is not assured.
Fig.4 shows some trajectories starting from points on a neighborhood of Ω. The trajectories starting from $x(0) = 0 \pm \delta$ and $\tilde{\theta}(0) > 0$, where $\delta > 0$ is a perturbation, indicate two critically different behaviors, which implies that the uniqueness and continuous dependence of the solution of the closed loop system are violated. In fact, $T_1 = \{(0, \tilde{\theta}) \mid \tilde{\theta} > 0\}$ is a *d-unstable* set, and $T_2 = \{(0, \tilde{\theta}) \mid \tilde{\theta} < 0\}$ is a stable invariant set. □

4 Conclusions

In this paper a solid stability analysis and control designs for an affine system with discontinuity are proposed. Since the nonsmooth system considered is not necessarily stabilized by any Lipschitz continuous controller, we introduce discontinuous controllers. If the upper and lower bounds of discontinuous uncertainty are available, a modifying L_gV structure-based switching controller can be applied, and the stability and the uniqueness of the Filippov solution are assured. In a case that the upper and lower bounds of the uncertainty are not available, we show an adaptive control approach to assure the strong stability and attractiveness of the state of the controlled system whereas a classical adaptive control methodology can not be applied due to the discontinuity. Throughout the paper, we consistently employ the concept of the Filippov solution.

References

1. J. Alvarez, I. Orlov, and L. Acho. An invariance principle for discontinuous dynamic systems with application to a coulomb friction oscillator. *Journal of Dynamic Systems, Measurement, and Control*, 122:687–690, 2000.
2. J. P. Aubin. *Differential Inclusions: Set-valued Maps and Viability Theory.* Springer-Verlag, New-York, 1984.
3. A. Bacciotti and F. Ceragioli. Stability and stabilization of discontinuous systems and nonsmooth lyapunov functions. *ESAIM Control, Optimisation and Calculus of Variations*, 4:361–376, 1999.
4. R. H. Clarke. *Optimization and Nonsmooth Analysis.* Society for Industrial and Applied Mathematics, Pennsylvania, 1990.
5. A. F. Filippov. Differential equations with discontinuous right-hand side. *American Mathematical Society Translations*, 42:199–231, 1964.
6. S-J. Kim and I-J. Ha. Existence of carathéodory solutions in nonlinear systems with discontinuous switching feedback controllers. *IEEE Transactions on Automatic Control*, 49:1167–1171, 2004.
7. Kanellakopoulos I. Krstic, M. and P Kokotovic. *Nonlinear and Adaptive Control Design.* John Wiley & Sons, New York, USA, 1995.
8. Y. Orlov, L. Aguilar, and J.C. Cadiou. Switched chattering control vs. backlash/friction phenomena in electrical servo-motors. *International Journal of Control*, 76:959–967, 2003.
9. B. E. Paden and S. S. Sastry. A calculus for computing filippov's differential inclusion with application to the variable structure control of robot manipulators. *IEEE Transactions on Circuits and Systems*, 34:73–82, 1987.
10. M. M. Polycarpou and P. A. Ioannou. On the existence and uniqueness of solutions in adaptive control systems. *IEEE Transactions on Automatic Control*, 38:474–479, 1993.
11. P. Sekhavat, Q. Wu, and N. Sepehri. Lyapunov-based friction compensation for accurate positioning of a hydraulic actuator. In *Proceeding of the 2004 American Control Conference*, pages 418–423, Boston, June 30- July 23, 2004.
12. D. Shevitz and B. E. Paden. Lyapunov stability theory of nonsmooth systems. *IEEE Transactions on Automatic Control*, 39:1910–1914, 1994.
13. G. Tao and F. L. Lewis. *Adaptive Control of Nonsmooth Dynamic Systems.* Springer-Verlag, London, 2001.

A Mathematical Preliminary

Consider the vector differential equation

$$\dot{x} = f(x,t), \quad x(t_0) = x_0 \tag{52}$$

where $x \in \mathbb{R}^n$ and $f : \mathbb{R}^n \times \mathbb{R} \to \mathbb{R}^n$ satisfies the following condition.

Condition B: f is defined almost everywhere and is measurable in an open region $D \subset R^{n+1}$. Further, for every compact set $Q \subset D$, there exists an integrable $B(t)$ such that $\|f(x,t)\| \leq B(t)$ a.e. in Q.

Under *Condition B*, the existence of the Filippov solution is assured [5].

Definition 1 (Filippov solution [5]). *A vector function $x(t)$ is called a solution of (52) on $[t_0, t_1]$ if $x(t)$ is absolutely continuous on $[t_0, t_1]$, and for almost all $t \in [t_0, t_1]$*

$$\dot{x} \in K[f](x,t) \tag{53}$$

where $K[\cdot]$ is the Filippov set defined by

$$K[f](x,t) = \bigcap_{\delta>0} \bigcap_{\mu N=0} \overline{co} f(B(x,\delta) - N, t). \tag{54}$$

$\bigcap_{\mu N=0}$ denotes the intersection over all sets N of Lebesgue measure zero, $B(x,\delta) = \{y \in R^n | \ \|y - x\| < \delta\}$ and $\overline{co}$ denotes the convex closure. □

Theorem 3 (Lyapunov[12]). *Let $\dot{x} = f(x,t)$ be essentially locally bounded and $0 \in K[f](0,t)$ in a region $\{x \in R^n \mid \|x\| < r\} \times \{t \mid t_0 \leq t < \infty\} \subset D$. Assume that the existence and uniqueness of the Filippov solution are assured. Let $V : R^n \times R \to R$ be a regular function (See [4] for definition) satisfying $V(0,t) = 0$ and $0 \leq V_1(\|x\|) \leq V(x,t) \leq V_2(\|x\|)$ in D for some $V_1, \ V_2 \in$ class $\mathcal{K}$. Now the set valued derivative $\dot{\tilde{V}}(x,t)$ is defined by*

$$\dot{\tilde{V}}(x(t),t) = \bigcap_{\xi \in \partial V(x(t),t)} \xi^T \begin{bmatrix} K[f](x(t),t) \\ 1 \end{bmatrix} \tag{55}$$

where $\partial V(x,t)$ is the Clarke's generalized gradient [4]. Then,
(1) $\dot{\tilde{V}}(x,t) \leq 0$ in D implies $x(t) \equiv 0$ is a uniformly stable solution.
(2) If in addition, there exists a class $\mathcal{K}$ function $w(x)$ in D with the property $\dot{\tilde{V}}(x,t) \leq -w(x) \leq 0$, then the solution $x(t) \equiv 0$ is uniformly asymptotically stable. Moreover, if $D = \mathbb{R}^n$ and V_1 is radially unbounded, then $x = 0$ is globally uniformly asymptotically stable. □

Non-smooth Finite-time Stabilization for A Class of Nonlinear Systems

Yiguang Hong and Jiankui Wang

Institute of Systems Science, Chinese Academy of Sciences, Beijing 100080, P.R.China yghong@iss.ac.cn

1 Introduction

Finite-time convergent control design via continuous time-invariant feedback laws has become the focal point of several recent studies. Based on the analysis of differential equations and homogeneity, continuous (and non-smooth) finite-time controllers via state or dynamic output feedback are proposed for the double integrator systems in the references such as [2, 6]. Moreover, finite-time control design for higher order nonlinear systems was considered in [5].

In practice, many stabilizable systems cannot be stabilized via smooth feedback. Recent results have been obtained for the nonlinear systems in p normal form, which includes some systems cannot be stabilized via smooth feedback, has been studied widely in recent years [3, 8]. It is not very easy to propose simple and effective global finite-time control strategies for nonlinear system in p normal form since some well-known methods like backstepping algorithms (via smooth feedback) in [7] may fail. Moreover, the non-smoothness in finite time design results from the demands of finite time convergence, in addition to possible under-actuated structures of the controlled systems.

This paper proposes a continuous finite-time control design method for a class of nonlinear systems. The organization of the paper is as follows. First, the problem formulation and preliminary results are given in Section 2. Then, in Section 3, continuous feedback laws are constructed for the global finite-time control of a class of nonlinear systems in p normal form.

2 Preliminaries

This paper considers the global finite-time control problem. Following the treatment in the literatures, finite-time stability can be introduced as follows.

Definition 1. *Consider a system*

$$\dot{x} = f(x), \qquad f(0) = 0, \quad x \in R^n, \tag{1}$$

where $f : U_0 \to R^n$ is continuous with respect to x on an open neighborhood U_0 of the origin $x = 0$. The equilibrium $x = 0$ of the system is (locally) finite-time stable if it is Lyapunov stable and finite-time convergent in a neighborhood $U \subseteq U_0$ of the origin. By 'finite-time convergence', we mean: if, for any initial condition $x_0 \in U$, there is a settling time $T > 0$, such that every solution $x(t; 0, x_0)$ of system (1) is defined with $x(t; 0, x_0) \in U/\{0\}$ for $t \in [0, T)$ and $\lim_{t\to T} x(t; 0, x_0) = 0$ and $x(t; 0, x_0) = 0$, $\forall t > T$. When $U = R^n$, the origin is a globally finite-time stable equilibrium.

The following result can be found in [2].

Lemma 1. *Consider the nonlinear system described in (1). Suppose there is a C^1 function $V(x)$ defined in a neighborhood $\hat{U} \subset U_0 \subset R^n$ of the origin, real numbers $c > 0$ and $0 < \alpha < 1$, such that $V(x)$ is positive definite on $\hat{U}$ and $\dot{V}(x) + cV^\alpha(x)$ is negative semidefinite (along the trajectory) on $\hat{U}$, then the origin of system (1) is locally finite-time stable. If $\hat{U} = R^n$ and $V(x)$ is also radially unbounded, the origin of system (1) is globally finite-time stable.*

The following two inequalities are quite well known ([1]).

1. Jensen's inequality:

$$S_{c_2}(x) \leq S_{c_1}(x), \quad 0 < c_1 < c_2, \tag{2}$$

where $S_c(x) = (\sum_{i=1}^n x_i^c)^{1/c}$, and $x_i \geq 0$, $1 \leq i \leq n$.

2. Young's inequality: For nonnegative numbers a, b, and $c > 0$, we have

$$ab \leq \frac{1+c}{a^{1+c}} + \frac{cb^{1+\frac{1}{c}}}{1+c} \leq a^{1+c} + b^{1+\frac{1}{c}} \tag{3}$$

The following inequality can be easily proved (referring to [8]): For any $x, y \in R$ and two odd integers $c \geq d > 0$ and $p = c/d \geq 1$, we have

$$|x - y|^p \leq 2^{p-1}|x^p - y^p| \tag{4}$$

A function is C^s with $s > 0$ means that the function admits a continuous sth-order derivatives, while a function is C^0 means that the function is continuous.

Lemma 2. *If $p > 0$ is a real number and $f(x)$ is a C^s ($s > 0$ an integer) function of x with $f(x) > 0$ for any $x \in R^n$, then $f(x)^p > 0$ is a C^s smooth function.*

In this paper, we will consider to construct a control law such that the following analytic nonlinear systems of the form:

$$\begin{cases} \dot{x}_1 = x_2^{m_1} + \sum_{i=0}^{m_1-1} x_2^i f_{1,i}(x_1) \\ \cdots \\ \dot{x}_{n-1} = x_n^{m_{n-1}} + \sum_{i=0}^{m_{n-1}-1} x_n^i f_{n-1,i}(x_1, ..., x_{n-1}) \\ \dot{x}_n = u \end{cases} \tag{5}$$

is finite-time stable, where $m_i, i = 1, ..., n-1$ are odd positive integers, and $f_{j-1}(x) = \sum_{i=0}^{m_{j-1}-1} x_j^i f_{j-1,i}$ are smooth with $f_j(0) = 0, j = 1, 2, ..., n-1$. For convenience, in the following analysis, we denote $f_n = 0$ and $u = x_{n+1}^{m_n}$ with $m_n = 1$.

Before further discussion, we take $k = \frac{p_0}{q_0} - 1 < 0$ ($p_0 > 0, q_0 > 0$ are odd integers) and

$$r_1 = 1, ..., r_i = \frac{r_{i-1} + k}{m_{i-1}}, \qquad r_i > -k, \quad i = 1, ..., n,$$

$$\beta_0 = r_2, \quad (\beta_i m_i + 1) r_{i+1} = (\beta_{i-1} m_{i-1} + 1) r_i > 0; \tag{6}$$

with

$$m_0 = 1, \quad \beta_i = \frac{p_i}{q_i},$$

and $p_i > 0, q_i > 0$ some odd integers.

Take

$$\Gamma_j(x) = (|x_1|^{c/r_1} + \cdots + |x_j|^{c/r_j})^{1/c}, \tag{7}$$

with $c > \max\{r_i, i = 1, ..., j\}$, $j = 1, ..., n$. In what follows we take $c = 1 + k + r_2$. Clearly, $\Gamma_i(x)^{1+k+r_2}$ is C^1. From (2) it is easy to see that

$$\Gamma_j(x)^d \le \sum_{i=1}^{j} |x_i|^{\frac{d}{r_i}}, \qquad 0 \le d \le c. \tag{8}$$

Lemma 3. *There is a C^∞ nonnegative function $b_j(x_1, ..., x_j)$ such that*

$$|f_j| \le \frac{1}{2} |x_{j+1}|^{m_j} + \Gamma_j^{r_j}(x) b_j(x_1, ..., x_j)$$

Proof: It is easy to get that

$$f_{j,i}(x_1, ..., x_j) = \int_0^1 df_{j,i}(\alpha x)$$

$$= \sum_{l=1}^{j} x_l \int_0^1 \frac{\partial f_{j,i}}{\partial \alpha x_l} d\alpha \triangleq \sum_{l=1}^{j} x_l \hat{f}_{j,i,l}(x_1, ..., x_j)$$

With (3) we have

$$|x_{j+1}^i f_{j,i}| \le \frac{|x_{j+1}|^{m_j}}{2m_j} + (2m_j)^{\frac{i}{m_j - i}} |f_{j+1,i}|^{\frac{m_j}{m_j - i}},$$

and then

$$|f_{j,i}|^{\frac{m_j}{m_j-i}} \le (\sum_{l=1}^{j} |x_l||\hat{f}_{j,i,l}(x_1,...,x_j)|)^{\frac{m_j}{m_j-i}}$$

$$\le j^{m_j} j \Gamma_j^{r_j} \sum_{l=1}^{j} (1+\hat{f}_{j,i,l}^{2m_j})(1+x_l^{2m_j}).$$

Therefore,

$$|f_j| = |\sum_{i=0}^{m_j-1} x_{j+1}^i f_{j,i}| \le \frac{1}{2}|x_{j+1}|^{m_j} + \Gamma_j^{r_j} b_j(x_1,...,x_j)$$

where

$$b_j = (2m_j j)^{m_j} j \sum_{i=0}^{m_j-1} \sum_{l=1}^{j} (1+\hat{f}_{j,i,l}^{2m_j})(1+x_l^{2m_j}) \ge 0$$

is a smooth function.

△

3 Finite-time stabilization

In this section, we will construct a finite time controller for system (5). To do this, we first give several notations. Set

$$W_1 = \frac{1}{1+r_2} x_1^{1+r_2} \tag{9}$$

and

$$W_j(x) \triangleq \int_{v_{j-1}(x_1,...,x_{j-1})^{\frac{1}{m_{j-1}}}}^{x_j} [s^{m_{j-1}\beta_{j-1}} - v_{j-1}(x_1,...,x_{j-1})^{\beta_{j-1}}]\, d\, s, \tag{10}$$

for $2 \le j \le n$, or equivalently

$$W_j = \frac{x_j^{m_{j-1}\beta_{j-1}+1} + m_{j-1}\beta_{j-1} v_{j-1}^{\frac{m_{j-1}\beta_{j-1}+1}{m_{j-1}}}}{m_{j-1}\beta_{j-1}+1} - x_j v_{j-1}^{\beta_{j-1}},$$

which is nonnegative and even positive when $x_j^{m_{j-1}} \ne v_{j-1}(x_1,...,x_{j-1})$. Define

$$w_1 = x_1^{r_2}, \tag{11}$$

$$w_j(x_1,...,x_j) \triangleq \frac{\partial W_j}{\partial x_j} = x_j^{m_{j-1}\beta_{j-1}} - v_{j-1}(w_1,...,w_{j-1})^{\beta_{j-1}},\ 2 \le j \le n, \tag{12}$$

where

$$v_j(x_1, ..., x_{j-1}) = -2l_j w_j^{\frac{r_j+k}{r_j m_{j-1}\beta_{j-1}}} \Phi_j, \quad 1 \leq j \leq n, \tag{13}$$

where $\Phi_j, 1 \leq j \leq n$ are smooth and nonnegative, which will be determined later.

Furthermore, we define

$$Q_j(w) = (|w_1|^{\frac{1+k+r_2}{r_2}} + |w_2|^{1+k+r_2} + ... + |w_j|^{\frac{(1+k+r_2)(\beta_{j-1}m_{j-1}+1)}{(1+r_2)\beta_{j-1}m_{j-1}}})^{\frac{1}{1+k+r_2}}. \tag{14}$$

Note that
$$\frac{(1+k+r_2)(\beta_{j-1}m_{j-1}+1)}{1+r_2} = \frac{1+k+r_2}{r_j} > 1,$$

for $j = 1, ..., n$, because $r_2 + k > 0$. It is easy to see that $Q_l \leq Q_i$ if $l \leq i$.

In what follows, we consider the finite time stabilization design for system (5), with a constructive procedure consistent with that given in [5].

Step 1: Consider the system

$$\dot{x}_1 = x_2^{m_1} + f_1(x_1, x_2) \tag{15}$$

with $f_1 = \sum_{i=0}^{m_1-1} x_2^i f_{1,i}(x_1)$. From Lemma 3, we have $|f_1| \leq |x_2|^{m_1}/2 + |x_1|b_1(x_1)$

Take $v_1 = -2x_1^{\frac{1}{\beta_1}} \Phi_1(x_1)$ with

$$\Phi_1 = [\frac{2+k}{2} - \frac{k}{2}x_1^2]b_1(x_1) + L_1,$$

which is C^∞ with respect to x_1, according to Lemma 2.

Note that $\beta_1 = \frac{1}{r_2 m_1} = \frac{1}{1+k}$ by (6). Take a function $V_1 = W_1$ defined in (9), and we have

$$\dot{V}_1 = \frac{\partial V_1}{\partial x_1}(x_2^{m_1} + f_1) \leq x_1^{r_2} x_2^{m_1} + |\frac{x_1^{r_2} x_2^{m_1}}{2}| - \frac{w_1 v_1}{2} + \frac{w_1 v_1}{2} + |x_1|^{1+r_2} b_1$$

$$\leq -L_1 x_1^{1+k+r_2} + \frac{3}{2}|w_1||x_2^{m_1} - v_1| \tag{16}$$

It is obvious that $v_1(x_1)^{\beta_1}$ is C^∞ smooth with respect to x_1.

In addition, next lemma is proposed for the following analysis.

Lemma 4. *There is a C^∞ nonnegative function $\rho_{1,1}$ such that*

$$|\frac{\partial v_1^{\beta_1}}{\partial x_1}| \leq \rho_{1,1}(x_1).$$

Proof:

$$|\frac{\partial v_1^{\beta_1}}{\partial x_1}| = 2^{\beta_1}|\Phi_1^{\beta_1} + \beta_1 x_1 \Phi_1^{\beta_1-1}\frac{\partial \Phi_1}{\partial x_1}| \leq 2^{\beta_1}\{\Phi_1^{\beta_1} + \frac{\beta_1 \Phi_1^{\beta_1-1}}{2}[1 + x_1^2(\frac{\partial \Phi_1}{\partial x_1})^2]\}$$

which leads to the conclusion.

△

After Step $j-1$: For system

$$\begin{cases} \dot{x}_1 = x_2^{m_1} + \sum_{i=0}^{m_1-1} x_2^i f_{1,i}(x_1) \\ \cdots \\ \dot{x}_{j-2} = x_{j-1}^{m_{j-2}} + \sum_{i=0}^{m_{j-2}-1} x_{j-1}^i f_{j-2,i}(x_1,...,x_{j-1}) \\ \dot{x}_{j-1} = x_j^{m_{j-1}} + f_{j-1}(x) \end{cases} \tag{17}$$

we already have the function $V_{j-1} = \sum_{i=1}^{j-1} W_i$ such that

$$\dot{V}_{j-1} \le -L_{j-1} Q_{j-1}(w_1, w_2, ..., w_{j-1})^{1+k+r_2} + \frac{3}{2}|w_{j-1}||x_j^{m_j} - v_{j-1}| \tag{18}$$

The construction of v_{j-1}, w_{j-1}, and V_{j-1}, based on v_{j-2}, w_{j-2}, and V_{j-2}, will be introduced in detail in Step j.

We assume that we already have

(i) $v_i^{\beta_i}$ defined in (13) is C^1 and so is W_i, $1 \le i \le j-1$ ($j \le n$); $\Phi_i(x)$ is positive and C^∞ smooth with respect to x;

(ii) There is a C^∞ function $\rho_{i,l}$ for any given $1 \le i \le l \le j-1$ such that

$$|\frac{\partial v_l^{\beta_l}}{\partial x_i}| \le Q_l^{(r_l+k)\beta_l - r_i}(x)\rho_{i,l}(x).$$

These assumptions will be validated in next step.

Step j: It is time to consider system

$$\begin{cases} \dot{x}_1 = x_2^{m_1} + f_1 \\ \cdots \\ \dot{x}_{j-1} = x_j^{m_{j-1}} + f_{j-1} \\ \dot{x}_j = x_{j+1}^{m_j} + f_j(x) \end{cases} \tag{19}$$

with $f_k = \sum_{i=0}^{m_k-1} x_{k+1}^i f_{k,i}(x_1, x_2, ..., x_k)$, $(1 \le k \le j)$.

Then construct C^1 positive definite function

$$V_j = W_j(x) + V_{j-1}(w_1, ..., w_{j-1}), \tag{20}$$

and W_j is defined in (10).

At first, we give a lemma for the construction of our proposed control law.

Lemma 5. *There is a* C^∞ *positive function* $\pi_j(x_1, ..., x_j)$ *such that* $\Gamma_j^{r_j} \le Q_j^{r_j}\pi_j$.

Proof: In order to prove the inequality $\Gamma_j^{r_j} \le Q_j^{r_j}\pi_j$, we only need to prove the following inequality:

$$\sum_{i=1}^{j}|x_i|^{\frac{1+k+r_2}{r_i}} \le \sum_{i=1}^{j}|w_i|^{\frac{1+k+r_2}{r_i m_{i-1}\beta_{i-1}}} \pi_j^{\frac{1+k+r_2}{r_j}} \tag{21}$$

For $i = 1$, the result is obvious because $|x_1|^{1+k+r_2} = |w_1|^{\frac{1+k+r_2}{r_2}}$. Thus, we consider the case when $2 \le i \le j$. In this case, we have

$$|x_i|^{\frac{1+k+r_2}{r_i}} = |x_i^{m_{i-1}\beta_{i-1}}|^{\frac{1+k+r_2}{r_i m_{i-1}\beta_{i-1}}} \le [|w_i| + |v_{i-1}|^{\beta_{i-1}}]^{\frac{1+k+r_2}{r_i m_{i-1}\beta_{i-1}}}$$

$$\le 2^{\frac{k+r_i}{r_i m_{i-1}\beta_{i-1}}}|w_i|^{\frac{r_j}{r_i m_{i-1}\beta_{i-1}}} + 2^{\frac{k+r_i+(1+k+r_2)\beta_{i-1}}{r_i m_{i-1}\beta_{i-1}}} \Phi_{i-1}^{\frac{1+k+r_2}{r_i m_{i-1}}}|w_{i-1}|^{\frac{1+k+r_2}{r_{i-1}m_{i-2}\beta_{i-2}}}.$$

In this way,

$$\sum_{i=1}^{j}|x_i|^{\frac{1+k+r_2}{r_i}} \le [1+\sum_{i=1}^{j-1}2^{\frac{k+r_{i+1}+(1+k+r_2)\beta_i}{r_{i+1}m_i\beta_i}}\Phi_i^{\frac{1+k+r_2}{r_{i+1}m_i}}]\sum_{i=1}^{j}|w_i|^{\frac{1+k+r_2}{r_i m_{i-1}\beta_{i-1}}}$$

$$\triangleq \pi_j^{\frac{1+k+r_2}{r_j}}\sum_{i=1}^{j}|w_i|^{\frac{1+k+r_2}{r_i m_{i-1}\beta_{i-1}}}$$

Based on Lemma 2 and Assumption (i) we can get:

$$\pi_j = [1+\sum_{i=1}^{j-1}2^{\frac{k+r_{i+1}+(1+k+r_2)\beta_i}{r_{i+1}m_i\beta_i}}\Phi_i^{\frac{1+k+r_2}{r_{i+1}m_i}}]^{\frac{r_j}{1+k+r_2}}.$$

△

Let us go back to consider the derivative of V_j:

$$\dot V_j \le -L_{j-1}Q_{j-1}^{1+k+r_2}+\frac{3}{2}|w_{j-1}||x_j^{m_{j-1}}-v_{j-1}|+w_j[x_{j+1}^{m_j}+f_j]+\sum_{i=1}^{j-1}\frac{\partial W_j}{\partial x_i}[x_{i+1}^{m_i}+f_i]$$

Then we analyze each term on the right hand side of this inequality about $\dot V_j$.

By (3), we first have

$$\frac{3}{2}|w_{j-1}||x_j^{m_{j-1}}-v_j| \le 3\times 2^{-\frac{1}{\beta_j}}|w_{j-1}||w_j|^{\frac{1}{\beta_{j-1}}}$$

$$\le \frac{L_{j-1}}{3n}Q_{j-1}^{1+k+r_2} + l_j^0|w_j|^{\frac{1+k+r_2}{r_j\beta_{j-1}m_{j-1}}}.$$

where $l_j^0 = 3^{\frac{1+k+r_2}{r_j m_{j-1}}}2^{-\frac{1+k+r_2}{r_j m_{j-1}\beta_{j-1}}} \times \frac{r_{j-1}+k}{1+k+r_2}\left(\frac{L_{j-1}(1+k+r_2)}{3n(1+r_2-r_{j-1})}\right)^{-\frac{1+r_2-r_{j-1}}{r_{j-1}+k}}$.

The following lemma is given to facilitate the related analysis.

Lemma 6. $|w_j|^{\frac{1}{r_j m_{j-1}\beta_{j-1}}} \le h_j(x)$, *where* $h_j(x)$ *are positive and* C^∞.

Proof: The proof is conducted by induction. For $i=1$, $|w_1|^{\frac{1}{r_2}} \le \frac{1+x_1^2}{2} \triangleq h_1(x)$, where $h_1(x)$ obviously is a positive smooth function. For $i=l$, we assume

$$|w_l|^{\frac{1}{r_l m_{l-1}\beta_{l-1}}} \le h_l(x),$$

where $h_l(x)$ is a positive C^∞ function. Then, for $i=l+1\le j$, based on (2), we have

$$|w_{l+1}|^{\frac{1}{r_{l+1}m_l\beta_l}} \le (\frac{1+x_{l+1}^2}{2})^{\frac{1}{r_{l+1}}} + |v_l|^{\frac{1}{r_{l+1}m_l}}$$

and

$$|v_l|^{\frac{1}{r_{l+1}m_l}} = 2^{\frac{1}{r_{l+1}m_l}}\Phi_l^{\frac{1}{r_{l+1}m_l}}|w_l|^{\frac{1}{r_l m_{l-1}\beta_{l-1}}} \le 2^{\frac{1}{r_{l+1}m_l}}\Phi_l^{\frac{1}{r_{l+1}m_l}}h_l(x)$$

According to the construction of Φ_l and Lemma 2, if we define

$$h_{l+1}(x) \triangleq (\frac{1+x_{l+1}^2}{2})^{\frac{1}{r_{l+1}}} + 2^{\frac{1}{r_{l+1}m_l}}\Phi_l^{\frac{1}{r_{l+1}m_l}}h_l(x)$$

then we can get the conclusion by induction.

△

Moreover, in a similar way, we have

Lemma 7. *For $\forall j\ge 2$, there exists a C^∞ nonnegative function $\tilde{\psi}_j$ such that*

$$|\sum_{i=1}^{j-1}\frac{\partial W_j}{\partial x_i}(x_{i+1}^{m_i}+f_i)| \le \frac{L_{j-1}}{3n}Q_j^{1+k+r_2} + |w_j|^{\frac{1+k+r_2}{r_j m_{j-1}\beta_{j-1}}}\tilde{\psi}_j(x). \qquad (22)$$

Proof: Note that using (4),

$$|v_{j-1}^{1/m_{j-1}} - x_j| \le 2^{1-\frac{1}{m_{j-1}\beta_{j-1}}}|w_j|^{\frac{1}{\beta_{j-1}m_{j-1}}}, \quad i\le j-1.$$

Therefore, for any $i\le j-1$,

$$|\frac{\partial W_j}{\partial x_i}| = |v_{j-1}^{1/m_{j-1}} - x_j||\frac{\partial v_{j-1}^{\beta_{j-1}}}{\partial x_i}| \le 2^{1-\frac{1}{m_{j-1}\beta_{j-1}}}|w_j|^{\frac{1}{\beta_{j-1}m_{j-1}}}|\frac{\partial v_{j-1}^{\beta_{j-1}}}{\partial x_i}|. \qquad (23)$$

By Assumption (ii) in Step $j-1$, we can obtain

$$|\frac{\partial v_{j-1}^{\beta_{j-1}}}{\partial x_i}| \le Q_{j-1}^{(r_{j-1}+k)\beta_{j-1}-r_i}\rho_{i,j-1}. \qquad (24)$$

Recalling Lemmas 3 and 5, we have

$$|x_{i+1}^{m_i}+f_i| \le \frac{3}{2}|x_{i+1}^{m_i}-v_i| + \frac{3}{2}|v_i| + |f_i|$$

$$\le 3\times 2^{-\frac{1}{\beta_i}}|w_{i+1}^{\frac{1}{\beta_i}}| + 3|w_i^{\frac{r_i+k}{r_i m_{i-1}\beta_{i-1}}}\Phi_i| + \Gamma_i^{r_i}b_i$$

$$\leq (3 \times 2^{-\frac{1}{\beta_i}} + 3\Phi_i)Q_{i+1}^{r_i+k} + Q_i^{r_i+k}(\Gamma_i^{-k}\pi_i^{\frac{r_i+k}{r_i}} b_i). \tag{25}$$

Then based on $Q_i \leq Q_j$, $i \leq j$, inequalities (23), (24), (25), (3) and $1 \leq \frac{1+k+r_2}{r_j}$, we have

$$|\sum_{i=1}^{j-1} \frac{\partial W_j}{\partial x_i}[x_{i+1}^{m_i} + f_i]| \leq \frac{L_{j-1}}{3n} Q_j^{1+k+r_2}$$

$$+|w_j|^{\frac{1+k+r_2}{r_j\beta_{j-1}m_{j-1}}} (\frac{L_{j-1}(1+k+r_2)}{3n(1+k+r_2-r_j)})^{\frac{r_j-1-k-r_2}{r_j}}$$

$$\times \frac{r_j(2^{1-\frac{1}{m_{j-1}\beta_{j-1}}})^{\frac{1+k+r_2}{r_j}}}{1+k+r_2}\{\sum_{i=1}^{j-1}\rho_{i,j-1}[3\times 2^{-\frac{1}{\beta_i}} + 3\Phi_i + \Gamma_i^{-k}\pi_i^{\frac{r_i+k}{r_i}} b_i]\}^{\frac{1+k+r_2}{r_j}}. \tag{26}$$

Note that, by (8) and (3),

$$\Gamma_i^{-k} \leq \sum_{l=1}^{i} |x_l|^{-\frac{k}{r_l}} \leq \sum_{l=1}^{i}(\frac{1+x_l^2}{2})^{-\frac{k}{r_l}}$$

Then, if we define $\tilde{\psi}_j$ as:

$$\tilde{\psi}_j \triangleq (\frac{L_{j-1}(1+k+r_2)}{3n(1+k+r_2-r_j)})^{\frac{r_j-1-k-r_2}{r_j}} \times \frac{r_j(2^{1-\frac{1}{m_{j-1}\beta_{j-1}}})^{\frac{1+k+r_2}{r_j}}}{1+k+r_2}$$

$$\times\{\sum_{i=1}^{j-1}\rho_{i,j-1}[3\times 2^{-\frac{1}{\beta_i}} + 3\Phi_i + \sum_{l=1}^{i}(\frac{1+x_l^2}{2})^{-\frac{k}{r_l}}\pi_i^{\frac{r_i+k}{r_i}} b_i]\}^{\frac{1+k+r_2}{r_j}}.$$

the conclusion follows.

△

In addition, with Lemma 3 and $w_j v_j \leq 0$, where v_j will be determined later,

$$w_j(x_{j+1}^{m_j} + f_j) \leq w_j x_{j+1}^{m_j} + |w_j||\frac{x_{j+1}^{m_j}}{2}| + |w_j|\Gamma_j(x)^{r_j} b_j$$

$$\leq \frac{3}{2}|w_j||x_{j+1}^{m_j} - v_j| + \frac{w_j v_j}{2} + \frac{L_{j-1}}{3n}Q_j^{1+k+r_2} + |w_j|^{\frac{1+k+r_2}{r_j\beta_{j-1}m_{j-1}}}\bar{b}_j(x_1, ..., x_j)$$

for some nonnegative function $\bar{b}_j$:

$$\bar{b}_j = (\frac{L_{j-1}}{3n})^{\frac{-r_j}{1+k+r_2-r_j}}|w_j|^{\frac{-k(1+k+r_2)}{(1+k+r_2-r_j)(1+r_2-r_j)}}(\pi_j b_j)^{\frac{1+k+r_2}{1+k+r_2-r_j}}.$$

According to Lemma 6 select

$$\tilde{b}_j = (\frac{L_{j-1}}{3n})^{\frac{-r_j}{1+k+r_2-r_j}}|h_j|^{\frac{-k(1+k+r_2)}{(1+k+r_2-r_j)}}(\pi_j b_j)^{\frac{1+k+r_2}{1+k+r_2-r_j}}.$$

which is a C^∞ nonnegative function by Lemma 2. Then we have

$$w_j(x_{j+1}^{m_j} + f_j) \le \frac{3}{2}|w_j||x_{j+1}^{m_j} - v_j| + w_j v_j/2$$

$$+\frac{L_{j-1}}{3n}Q_j^{1+k+r_2} + |w_j|^{\frac{1+k+r_2}{r_j\beta_{j-1}m_{j-1}}}\tilde{b}_j(x_1, ..., x_j).$$

Construct

$$v_j = -2w_j^{\frac{r_j+k}{r_j m_{j-1}\beta_{j-1}}}\Phi_j, \tag{27}$$

where, for $j \ge 2$,

$$\Phi_j(x) = \left(l_j + \frac{2L_{j-1}}{3n} + l_j^0 + \tilde{\psi}_j(x) + \tilde{b}_j(x)\right).$$

Then

$$\frac{w_j v_j}{2} + |w_j|^{\frac{1+k+r_2}{r_j\beta_{j-1}m_{j-1}}}\left(\frac{2L_{j-1}}{3n} + l_j^0 + \tilde{\psi}_j + \tilde{b}_j\right)$$

$$\le -l_j|w_j|^{\frac{1+k+r_2}{r_j\beta_{j-1}m_{j-1}}}$$

Therefore,

$$\dot{V}_j(x) \le -L_j Q_j^{1+k+r_2} + \frac{3}{2}|w_j||x_{j+1}^{m_j} - v_j| \tag{28}$$

where $L_j = \min\{\frac{n-1}{n}L_{j-1}, l_j\}$.

Before the end of Step j, we need to verify the assumptions listed in Step $j-1$. It is easy to see that $v_j^{\beta_j}$ is C^1. Therefore, we only need to confirm Assumption (ii) for Step j.

Lemma 8. *There is a C^∞ smooth nonnegative function $\rho_{i,j}$ for any given $i \le j$ for $j \ge 1$ such that*

$$|\frac{\partial v_j^{\beta_j}}{\partial x_i}| \le Q_j^{(r_j+k)\beta_j - r_i}\rho_{i,j}(x).$$

Proof: For $j = 1$, by Lemma 4, the conclusion is obvious. Therefore, we only consider the case when $j \ge 2$. Note that

$$|\frac{\partial v_j^{\beta_j}}{\partial x_i}| \le 2^{\beta_j} w_j^{\frac{r_j - r_{j+1}}{r_j m_{j-1}\beta_{j-1}}}\Phi_j^{\beta_j - 1}\{\frac{r_{j+1}m_j\beta_j}{r_j m_{j-1}\beta_{j-1}}|\frac{\partial w_j}{\partial x_i}|\Phi_j + \beta_j|w_j||\frac{\partial \Phi_j}{\partial x_i}|\}$$

Note that

$$|\frac{\partial w_j}{\partial x_i}| = \begin{cases} m_{j-1}\beta_{j-1}|x_j|^{m_{j-1}\beta_{j-1}-1}, & if \quad i = j \\ |\frac{\partial v_{j-1}^{\beta_{j-1}}}{\partial x_i}|, & if \quad i < j \end{cases}$$

In the case of $i = j$, by Lemma 5 we have

$$|\frac{\partial v_j^{\beta_j}}{\partial x_j}|$$

$$\leq Q_j^{(r_j+k)\beta_j - r_j}\{2^{\beta_j}\Phi_j^{\beta_j-1}\beta_j[\frac{r_{j+1}m_j}{r_j}\pi_j^{m_{j-1}\beta_{j-1}-1}\Phi_j + |w_j|^{\frac{1}{m_{j-1}\beta_{j-1}}}|\frac{\partial \Phi_j}{\partial x_j}|]\}$$

$$\leq Q_j^{(r_j+k)\beta_j - r_j}\{2^{\beta_j}\Phi_j^{\beta_j-1}\beta_j[\frac{r_{j+1}m_j}{r_j}\pi_j^{m_{j-1}\beta_{j-1}-1}\Phi_j + 1 + h_j(1+(\frac{\partial \Phi_j}{\partial x_j})^2)^{\frac{1}{r_j}}]\}$$

$$\triangleq Q_j^{(r_j+k)\beta_j - r_j}\rho_{i,j}(x)$$

Then the conclusion follows.

In the case of $i < j$, with Assumption (iii) given in Step $j-1$,

$$|\frac{\partial w_j}{\partial x_i}| = |\frac{\partial v_{j-1}^{\beta_{j-1}}}{\partial x_i}| \leq Q_{j-1}^{(r_{j-1}+k)\beta_{j-1}-r_i}\rho_{i,j-1}(x).$$

Then similar to the discussion for $i = j$, we can construct $\rho_{i,j}$ and get the conclusion.

$\triangle$

Up to Step n: The control law $u = v_n(x)$.

According to the construction of $V_n = \sum_{j=1}^{n} W_j$, we have

$$V_n = \sum_{j=1}^{n}\int_{v_{j-1}^{\frac{1}{m_{j-1}}}}^{x_j}[s^{m_{j-1}\beta_{j-1}} - v_{j-1}^{\beta_{j-1}}]\,d\,s \leq \sum_{j=1}^{n} 2|w_j|^{\frac{r_2+1}{r_j m_{j-1}\beta_{j-1}}}.$$

Then, by (2), we have

$$V_n^{\frac{1+k+r_2}{1+r_2}} \leq 2Q_n^{1+k+r_2}. \tag{29}$$

Thus, there is $L_n^* > 0$ such that

$$\dot{V}_n \leq -L_n^* V_n^{\frac{1+k+r_2}{1+r_2}}. \tag{30}$$

By Lemma 1, the closed-loop system is finite-time stable.

Therefore, we have our main result.

Theorem 1. *The finite-time control law for system (5) can be constructed as*

$$u = v_n = -2w_n^{\frac{r_n+k}{r_n m_{n-1}\beta_{n-1}}}\Phi_n \tag{31}$$

with w_n and Φ_n defined as above.

Remark 1. If we denote $sig(y)^\alpha = |y|^\alpha sgn(y)$, then we need not assume that p_i, q_i are odd. In this case, in a similar way, the finite-time control law can be constructed in the form of

$$v_j = -2sig(w_j)^{\frac{r_j+k}{r_j m_{j-1} \beta_{j-1}}} \Phi_j,$$

where $\Phi_j(x)$ are positive functions for $j \geq 1$. Moreover, we can extend the result to the discontinuous case, where $r_n + k = 0$. If so, the control law will be in a discontinuous form of $u = v_n(x) = -2sgn(w_n)\Phi_n(x)$ for some function $\Phi_n > 0$ constructed in a way similar to that given in the above design procedure. However, the solutions to the discontinuous closed-loop differential equations have to be understood or analyzed in the sense of Filippov [4].

4 Conclusions

In this paper, global finite-time stabilization problem for a large class of nonlinear control systems in the p-normal form is considered. An iterative design approach is given based on Lyapunov function. The finite-time stabilizing control laws are constructed in the form of non-smooth time-invariant feedback.

Acknowledgement

This work was supported partially by NSF of China under Grants 60221301 and 10472129, and partially by National Key Lab of Security Control and Simulation of Power Systems and Large-scale Generating Installations, Tsinghua University.

References

1. Beckenbach, E., Bellman, R. (1971), *Inequalities*, Springer-Verlag, Berlin.
2. Bhat, S., Bernstein, D. (1998), Continuous finite-time stabilization of the translational and rotational double integrators, *IEEE Trans. Automatic Control*, vol. 43, 678-682.
3. Cheng, D., Lin, W. (2003), On p normal form of nonlinear systems, *IEEE Trans. Automatic Contr.*, vol. 48, 1242-1248.
4. Filippov, A. (1988), *Differential Equations with Discontinuous Right Hand Side*, Kluwer, Dordrecht.
5. Hong, Y. (2002), Finite-time stabilization and stabilizability of a class of controllable systems, *Systems & Control Letters*, vol. 46, 231-236.
6. Hong, Y., Huang, J., Xu, Y. (2001), On an output feedback finite-time stabilization problem, *IEEE Trans Automatic Control*, vol. 46, 305-309.
7. Krstic, M., Kanellakopoulos, I., Kokotovic, P. (1995), *Nonlinear and Adaptive Control Systems Design*, John Wiley, New York.
8. Qian, C., Lin, W. (2001), A continuous feedback approach to global strong stabilization of nonlinear systems, *IEEE Trans. Automatic Control*, vol. 46, 1061-1079.

$\mathcal{L}_2$-gain Analysis and Control Synthesis for Discrete-time Switched Systems with State Delay

Guangming Xie, Qi Fu and Long Wang

Center for Systems and Control, LTCS and Department of Mechanics and Engineering Science, Peking University, Beijing 100871, China
xiegming@mech.pku.edu.cn

1 Introduction

In recent years, the study of switched systems has received growing attention in control theory and its applications. Switched systems are an important class of hybrid dynamic systems consisting of a family of linear time-invariant subsystems and a switching law specifying the switching between them. There have been many works for switched linear systems, primarily on stability analysis and stabilization synthesis. A survey can be seen in [1].

Time delays are very common phenomena in many real control systems which is a great source of instability and poor performance, the study of control systems with time delay has also received considerable attention during the past decades (see [16] and the references therein).

Analysis and control problems for switched systems with time delay haven been studied by the authors and their cooperators. [2]-[5] established the necessary and sufficient criteria for controllability and reachability of switched linear systems with delayed control input in continuous time domain. The results for the discrete time case can be seen in [6]-[9]. Stability analysis and stabilization synthesis problems for switched linear systems with state-delays were studied in continuous time domain [10] and in discrete time domain [11], respectively. [12] discussed the quadratic stability and stabilization for discrete-time switched linear systems with state delay.

There are some other results on switched systems with time delay. [13] studied the stability and $\mathcal{L}_2$ gain analysis for switched linear systems with time delay, but their model was limited to symmetric type systems.

In this paper, switched linear systems with delayed state in the discrete-time domain are considered. The $\mathcal{L}_2$-gain analysis and control synthesis problems are investigated for this class of switched systems. The approach followed in such paper looks at the existence of a switched quadratic Lyapunov function to check the quadratical stability and $\mathcal{L}_2$ property of the switched system under consideration. Two different conditions for the existence of such a Lyapunov function are given. The conditions are expressed in terms of LMI which are suitable for application [14]. Similar to the analysis result, the design of the switched state feedback controller is derived as well.

This paper is organized as follows. In Section 2, the problems are formulated and some preliminaries are presented. Section 3 recalls the results of quadratic stability analysis and stabilization synthesis for switched linear systems without disturbance input case. Section 4 is on $\mathcal{L}_2$-gain analysis and Section 5 is on control synthesis. In Section 6 an illustrating example is given. Section 7 concludes the whole paper.

Notations: M^T is the transpose of the matrix M. $M > 0$ (resp. < 0) means that M is positive definite (resp. negative definite). I and $\mathbf{0}$ denote the identity and zero matrices of appropriate size respectively. When being symmetric, the symbol $*$ denotes the corresponding symmetric block.

2 Problem Formulation and Preliminaries

Consider a discrete-time switched system with state delay given by

$$\begin{aligned} x(t+1) &= A_{r(t),1}x(t) + A_{r(t),2}x(t-\tau) + B_{r(t),1}w(t) \\ &\quad + B_{r(t),2}u(t) \\ z(t) &= C_{r(t),1}x(t) + C_{r(t),2}x(t-\tau) + D_{r(t),1}w(t) \\ &\quad + D_{r(t),2}u(t) \\ x(t) &= 0, t \le 0 \end{aligned} \tag{1}$$

where $x(t) \in \mathbb{R}^{q_x}$ is the state, $w(t) \in \mathbb{R}^{q_w}$ is the exogenous disturbance input, $u(t) \in \mathbb{R}^{q_u}$ is the control input, $z(t) \in \mathbb{R}^{q_z}$ is the controlled output, $r(t)$ is the switching signal which takes its value in the finite set $\mathcal{I} = \{1, \ldots, N\}$. Moreover, $r(t) = i$ means that the subsystem $(A_{i,1}, A_{i,2}, B_{i,1}, B_{i,2}, C_{i,1}, C_{i,2}, D_{i,1}, D_{i,2})$ is activated. The positive integer τ is the fixed delay.

In this work, we are interested in the $\mathcal{L}_2$-gain analysis and control synthesis problems for this class of switched systems. $\mathcal{L}_2$-gain analysis is to establish a condition under which the autonomous switched system

$$\begin{aligned} x(t+1) &= A_{r(t),1}x(t) + A_{r(t),2}x(t-\tau) + B_{r(t),1}w(t) \\ z(t) &= C_{r(t),1}x(t) + C_{r(t),2}x(t-\tau) + D_{r(t),1}w(t) \\ x(t) &= 0, t \le 0 \end{aligned} \tag{2}$$

is quadratically stable with an $\mathcal{L}_2$-gain smaller than a positive scalar γ. By control synthesis, we mean to design a switched state feedback control law as

$$u(t) = K_{r(t)}x(t) \tag{3}$$

ensuring the corresponding closed-loop system

$$\begin{aligned}x(t+1) &= \widehat{A}_{r(t),1}x(t) + A_{r(t),2}x(t-\tau) + B_{r(t),1}w(t)\\ z(t) &= \widehat{C}_{r(t),1}x(t) + C_{r(t),2}x(t-\tau) + D_{r(t),1}w(t)\\ x(t) &= 0, t \le 0\end{aligned} \tag{4}$$

to be quadratically stable with an $\mathcal{L}_2$-gain smaller than a positive scalar γ, where

$$\widehat{A}_{i,1} = A_{i,1} + B_{i,2}K_i$$

and

$$\widehat{C}_{i,1} = C_{i,1} + D_{i,2}K_i, i \in \mathcal{I}.$$

Next, we present some definitions and lemmas as preliminaries of this paper.

Definition 1. *The $\mathcal{L}_2$-gain* g *of the switched system (2) under arbitrary switching signal is defined as*

$$\begin{aligned}\mathrm{g} := \ inf\Big\{\gamma \ge 0 : \sum_{t=0}^{\infty} z^T(t)z(t) \le \gamma^2 \sum_{t=0}^{\infty} w^T(t)w(t),\\ \forall r(t), \forall w \in \mathcal{L}_2 = \{w(t) : \sum_{t=0}^{\infty} w^T(t)w(t) \ is\ finite\}.\Big\}\end{aligned} \tag{5}$$

This definition is also valid for the system (4).

Definition 2. *[12] Switched system $x(t+1) = A_{r(t),1}x(t) + A_{r(t),2}x(t-\tau)$ is said to be quadratically stable if there exists a quadratic positive definite Lyapunov function $V(t, x(t))$, a constant $\epsilon > 0$, such that under arbitrary switching signal,*

$$\begin{aligned}\triangle V(t, x(t)) &= V(t+1, x(t+1)) - V(t, x(t))\\ &\le \epsilon\, x(t)^T x(t), \quad \forall x(t) \ne 0.\end{aligned}$$

Definition 3. *Switched system (2) is said to be quadratically stable with an $\mathcal{L}_2$-gain smaller than a positive scalar γ if the following conditions hold:*
(i) $x(t+1) = A_{r(t),1}x(t) + A_{r(t),2}x(t-\tau)$ is quadratically stable;
(ii) for all nonzero disturbances input signals $w \in \mathcal{L}_2$,

$$\sum_{t=0}^{\infty} z^T(t)z(t) < \gamma^2 \sum_{t=0}^{\infty} w^T(t)w(t).$$

Lemma 1. *Given matrices P, Q, M of compatible dimensions with P, Q symmetric positive definite, then*

$$\begin{bmatrix} P & M \\ M^T & Q \end{bmatrix} > 0$$

if and only if

$$\begin{bmatrix} P & -M \\ -M^T & Q \end{bmatrix} > 0$$

Proof. It directly follows by

$$\begin{bmatrix} -I & 0 \\ 0 & I \end{bmatrix} \begin{bmatrix} P & M \\ M^T & Q \end{bmatrix} \begin{bmatrix} -I & 0 \\ 0 & I \end{bmatrix} = \begin{bmatrix} P & -M \\ -M^T & Q \end{bmatrix}.$$

Lemma 2. *Given matrices T, M of compatible dimensions with T symmetric positive definite, then the following two statements are equivalent.*
i) there exists symmetric positive definite matrix S satisfying

$$\begin{bmatrix} S & SM^T \\ MS & T \end{bmatrix} > 0; \tag{6}$$

ii) there exist symmetric positive definite matrix S and matrix G satisfying

$$\begin{bmatrix} G + G^T - S & G^T M^T \\ MG & T \end{bmatrix} > 0. \tag{7}$$

Proof. See Appendix A.

3 Recall of Quadratic Stability and Stabilization

In this section, we first recall the results of quadratic stability analysis and stabilization synthesis for switched linear system without disturbance input w.

Proposition 1. *[12] The following statements are equivalent:*
i) There exists a piecewise quadratic Lyapunov function of the form

$$V(t, x(t)) = x^T(t) P_{r(t)} x(t) + \sum_{i=1}^{\tau} x^T(t-i) Q_{r(t-i)} x(t-i) \tag{8}$$

whose difference is negative definite, proving asymptotic stability of (2) with $w = 0$.
ii) There exist symmetric matrices $P_1, \ldots, P_N, Q_1, \ldots, Q_N$, satisfying

$$\begin{bmatrix} P_j & P_j A_{i,1} & P_j A_{i,1} & 0 \\ A_{i,1}^T P_j & P_i & 0 & Q_i \\ A_{i,2} D^T P_j & 0 & Q_k & 0 \\ 0 & Q_i & 0 & Q_i \end{bmatrix} > 0, \forall\, i, j, k \in \mathcal{I}. \tag{9}$$

iii) There exist symmetric matrices $S_1, \ldots, S_N, T_1, \ldots, T_N$ and matrices $G_1, \cdots, G_N$, satisfying

$$\begin{bmatrix} G_i + G_i^T - S_i & G_i^T A_{i,1}^T & G_i^T & 0 \\ A_{i,1} G_i & S_j & 0 & A_{i,2} T_k \\ G_i & 0 & T_i & 0 \\ 0 & T_k A_{i,2}^T & 0 & T_k \end{bmatrix} > 0, \forall\, i, j, k \in \mathcal{I}. \tag{10}$$

Proposition 2. *[12] For system (1), there exist switched state feedback controller (3) and switched quadratic Lyapunov function (8) proving the stability of the closed-loop system (4) with $w = 0$ if and only if there exist symmetric matrices S_i, T_i, matrices X_i, $i \in \mathcal{I}$ such that $\forall\, i, j, k \in \mathcal{I}$*

$$\begin{bmatrix} S_i & (A_{i,1}S_i + B_{i,1}X_i)^T & S_i & 0 \\ A_{i,1}S_i + B_{i,1}X_i & S_j & 0 & A_{i,2}T_k \\ S_i & 0 & T_i & 0 \\ 0 & T_kA_{i,2}^T & 0 & T_k \end{bmatrix} > 0 \tag{11}$$

Moreover, if (11) is feasible, the state feedback control matrices can be taken as

$$K_i = X_iS_i^{-1}, i \in \mathcal{I}. \tag{12}$$

Proposition 3. *[12] For system (1), there exist switched state feedback controller (3) and switched quadratic Lyapunov function (8) proving the stability of the closed-loop system (4) if and only if there exist symmetric matrices S_i, T_i, matrices G_i and X_i, $i \in \mathcal{I}$ such that $\forall\, i, j, k \in \mathcal{I}$*

$$\begin{bmatrix} G_i + G_i^T - S_i & G_i^T A_{i,1}^T + X_i^T B_{i,1}^T & G_i^T & 0 \\ A_{i,1}G_i + B_{i,1}X_i & S_j & 0 & A_{i,2}T_k \\ G_i & 0 & T_i & 0 \\ 0 & T_kA_{i,2}^T & 0 & T_k \end{bmatrix} > 0 \tag{13}$$

Moreover, if (13) is feasible, the state feedback control matrices can be taken as

$$K_i = X_iG_i^{-1}, \ \forall\, i \in \mathcal{I} \tag{14}$$

4 $\mathcal{L}_2$-gain Analysis Result

In this section, we investigate the $\mathcal{L}_2$-gain property of the autonomous switched system (2). First, we give a fundamental lemma.

Lemma 3. *Switched system (2) is quadratically stable with an $\mathcal{L}_2$-gain smaller than a positive scalar γ if there exists a function $V(t, x(t)) : \mathbb{Z}^+ \times \mathbb{R}^{q_x} \to \mathbb{R}^+$ satisfying*
(i) $V(t, x(t))$ is quadratic positive definite;
(ii) the difference of $V(t, x(t))$ along the solution of switched system (2) satisfies

$$\Delta V(t, x(t)) + z^T(t)z(t) - \gamma^2 w^T(t)w(t) < 0 \tag{15}$$

Proof. When assuming the zero input, (15) implies $\Delta V(t, x(t)) < 0$. By Definition 2, switched system $x(t+1) = A_{r(t),1}x(t) + A_{r(t),2}x(t-\tau)$ is quadratically stable. On the other hand, we sum (15) from 0 to $m(\forall m \in \mathbb{Z}^+)$, then get

$$V(m+1,x(m+1)) - V(0,x(0)) + \sum_{t=0}^{m} z^T(t)z(t)$$

$$< \gamma^2 \sum_{t=0}^{m} w^T(t)w(t)$$

Since $V(m+1,x(m+1)) \geq 0$ and $x(0) = 0$, this implies

$$\sum_{t=0}^{m} z^T(t)z(t) < \gamma^2 \sum_{t=0}^{m} w^T(t)w(t).$$

Hence, by Definition 3, switched system (2) is quadratically stable with an $\mathcal{L}_2$-gain smaller than a positive scalar γ.

We consider the following switched Lyapunov function

$$V(t,x(t)) = x^T(t)P_{r(t)}x(t) + \sum_{l=1}^{\tau} x^T(t-l)Q_{r(t-l)}x(t-l) \tag{16}$$

with $P_1, Q_1, \cdots, P_N, Q_N$ being positive definite matrices.

Theorem 1. *For a given $\gamma > 0$, the following statements are equivalent:*
(i) There exist a switched Lyapunov function of the form (16) whose difference along the solution of system (2) satisfies (15), proving that system (2) is quadratically stable with an $\mathcal{L}_2$-gain smaller than γ;
(ii) There exist positive definite matrices $P_1, Q_1, \cdots, P_N, Q_N$ satisfying

$$\begin{bmatrix} P_j & P_jA_{i,1} & P_jA_{i,2} & P_jB_{i,1} & 0 \\ * & P_i - Q_i & 0 & 0 & C_{i,1}^T \\ * & * & Q_k & 0 & C_{i,2}^T \\ * & * & * & \gamma^2 I & D_{i,1}^T \\ * & * & * & * & I \end{bmatrix} > 0, \forall i,j,k \in \mathcal{I} \tag{17}$$

The Lyapunov function is then given by (16);
(iii) there exist positive definite matrices $S_1, T_1, \cdots, S_N, T_N$ and matrices $G_1, \cdots, G_N$ satisfying

$$\begin{bmatrix} G_i + G_i^T - S_i & G_i^T A_{i,1}^T & G_i^T & 0 & 0 & G_i^T C_{i,1}^T \\ * & S_j & 0 & A_{i,2}T_k & B_{i,1} & 0 \\ * & * & T_i & 0 & 0 & 0 \\ * & * & * & T_k & 0 & T_k C_{i,2}^T \\ * & * & * & * & \gamma^2 I & D_{i,1}^T \\ * & * & * & * & * & I \end{bmatrix}$$

$$> 0, \quad \forall i,j,k \in \mathcal{I} \tag{18}$$

The Lyapunov function is then given by

$$V(t,x(t)) = x^T(t)S_{r(t)}^{-1}x(t) + \sum_{i=1}^{\tau} x^T(t-i)T_{r(t-i)}^{-1}x(t-i).$$

Proof. • *To prove i) ⇒ ii)*: suppose that there exists a Lyapunov function of the form (16), whose difference satisfies (15), then

$$\begin{aligned} &\triangle V(t,x(t)) + z^T(t)z(t) - \gamma^2 w^T(t)w(t) \\ &= x^T(t+1)P_{r(t+1)}x(t+1) \\ &\quad + \sum_{l=1}^{\tau} x^T(t+1-l)Q_{r(t+1-l)}x(t+1-l) \\ &\quad - x^T(t)P_{r(t)}x(t) - \sum_{l=1}^{\tau} x^T(t-l)Q_{r(t-l)}x(t-l) \\ &\quad + z^T(t)z(t) - \gamma^2 w^T(t)w(t) \\ &= [x^T(t), x^T(t-\tau), w^T(t)]\triangle P_{r(t)} \begin{bmatrix} x(t) \\ x(t-\tau) \\ w(t) \end{bmatrix} < 0 \end{aligned} \tag{19}$$

holds under arbitrary switching signal, where

$$\begin{aligned} &\triangle P_{r(t)} \\ &= \begin{bmatrix} A^T_{r(t),1}P_{r(t+1)}A_{r(t),1} - P_{r(t)} + Q_{r(t)} + C^T_{r(t),1}C_{r(t),1} \\ * \\ * \end{bmatrix. \\ &\qquad A^T_{r(t),1}P_{r(t+1)}A_{r(t),2} + C^T_{r(t),1}C_{r(t),2} \\ &\qquad A^T_{r(t),2}P_{r(t+1)}A_{r(t),2} - Q_{r(t-\tau)} + C^T_{r(t),2}C_{r(t),2} \\ &\qquad * \\ &\qquad \left. \begin{matrix} A^T_{r(t),1}P_{r(t+1)}B_{r(t),1} + C^T_{r(t),1}D_{r(t),1} \\ A^T_{r(t),2}P_{r(t+1)}B_{r(t),1} + C^T_{r(t),2}D_{r(t),1} \\ B^T_{r(t),1}P_{r(t+1)}B_{r(t),1} + D^T_{r(t),1}D_{r(t),1} - \gamma^2 I \end{matrix} \right]. \end{aligned}$$

It follows that (19) holds for all $\begin{bmatrix} x(t) \\ x(t-\tau) \\ w(t) \end{bmatrix} \neq 0$ and the special configuration $r(t+1) = j$, $r(t) = i$, $r(t-\tau) = k$. Hence we have

$$\begin{bmatrix} A^T_{i,1}P_jA_{i,1} - P_i + Q_i + C^T_{i,1}C_{i,1} & A^T_{i,1}P_jA_{i,2} + C^T_{i,1}C_{i,2} & A^T_{i,1}P_jB_{i,1} + C^T_{i,1}D_{i,1} \\ * & A^T_{i,2}P_jA_{i,2} - Q_k + C^T_{i,2}C_{i,2} & A^T_{i,2}P_jB_{i,1} + C^T_{i,2}D_{i,1} \\ * & * & B^T_{i,1}P_jB_{i,1} + D^T_{i,1}D_{i,1} - \gamma^2 I \end{bmatrix} < 0, \forall\, i,j,k \in \mathcal{I},$$

which is equivalent, by the Schur complement, to

$$\begin{bmatrix} A_{i,1}^T P_j A_{i,1} - P_i + Q_i & A_{i,1}^T P_j A_{i,2} & A_{i,1}^T P_j B_{i,1} & C_{i,1}^T \\ * & A_{i,2}^T P_j A_{i,2} - Q_k & A_{i,2}^T P_j B_{i,1} & C_{i,2}^T \\ * & * & B_{i,1}^T P_j B_{i,1} - \gamma^2 I & D_{i,1}^T \\ * & * & * & -I \end{bmatrix} < 0, \forall\, i,j,k \in \mathcal{I},$$

which is equivalent, by the Schur complement, to

$$\begin{bmatrix} -P_j & P_j A_{i,1} & P_j A_{i,2} & P_j B_{i,1} & 0 \\ * & Q_i - P_i & 0 & 0 & C_{i,1}^T \\ * & * & -Q_k & 0 & C_{i,2}^T \\ * & * & * & -\gamma^2 I & D_{i,1}^T \\ * & * & * & * & -I \end{bmatrix} < 0, \forall i,j,k \in \mathcal{I},$$

that is

$$\begin{bmatrix} P_j & -P_j A_{i,1} & -P_j A_{i,2} & -P_j B_{i,1} & 0 \\ * & P_i - Q_i & 0 & 0 & -C_{i,1}^T \\ * & * & Q_k & 0 & -C_{i,2}^T \\ * & * & * & \gamma^2 I & -D_{i,1}^T \\ * & * & * & * & I \end{bmatrix} > 0, \forall i,j,k \in \mathcal{I},$$

which is equivalent, by Lemma 1, to (17).

• *To prove ii) ⇒ i)*, suppose that (17) is satisfied for all $i,j,k \in \mathcal{I}$. Then we have

$$\begin{bmatrix} P_i - A_{i,1}^T P_j A_{i,1} - Q_i & -A_{i,1}^T P_j A_{i,2} & -A_{i,1}^T P_j B_{i,1} & C_{i,1}^T \\ * & Q_k - A_{i,2}^T P_j A_{i,2} & -A_{i,2}^T P_j B_{i,1} & C_{i,2}^T \\ * & * & \gamma^2 I - B_{i,1}^T P_j B_{i,1} & D_{i,1}^T \\ * & * & * & I \end{bmatrix} > 0, \forall\, i,j,k \in \mathcal{I}.$$

which is equivalent to

$$\begin{bmatrix} P_i - A_{i,1}^T P_j A_{i,1} - Q_i - C_{i,1}^T C_{i,1} & -A_{i,1}^T P_j A_{i,2} - C_{i,1}^T C_{i,2} & -A_{i,1}^T P_j B_{i,1} - C_{i,1}^T D_{i,1} \\ * & Q_k - A_{i,2}^T P_j A_{i,2} - C_{i,2}^T C_{i,2} & -A_{i,2}^T P_j B_{i,1} - C_{i,2}^T D_{i,1} \\ * & * & \gamma^2 I - B_{i,1}^T P_j B_{i,1} - D_{i,1}^T D_{i,1} \end{bmatrix} > 0, \forall\, i,j,k \in \mathcal{I}.$$

Then, for any switching signal $r(t)$, we have

$$\triangle P_{r(t)} = \begin{bmatrix} A^T_{r(t),1}P_{r(t+1)}A_{r(t),1} - P_{r(t)} + Q_{r(t)} + C^T_{r(t),1}C_{r(t),1} & A^T_{r(t),1}P_{r(t+1)}A_{r(t),2} + C^T_{r(t),1}C_{r(t),2} & A^T_{r(t),1}P_{r(t+1)}B_{r(t),1} + C^T_{r(t),1}D_{r(t),1} \\ * & A^T_{r(t),2}P_{r(t+1)}A_{r(t),2} - Q_{r(t-\tau)} + C^T_{r(t),2}C_{r(t),2} & A^T_{r(t),2}P_{r(t+1)}B_{r(t),1} + C^T_{r(t),2}D_{r(t),1} \\ * & * & B^T_{r(t),1}P_{r(t+1)}B_{r(t),1} + D^T_{r(t),1}D_{r(t),1} - \gamma^2 I \end{bmatrix} < 0.$$

It follows that $\forall \begin{bmatrix} x(t) \\ x(t-\tau) \\ w(t) \end{bmatrix} \neq 0,$

$$\triangle V(t, x(t)) + z^T(t)z(t) - \gamma^2 w^T(t)w(t) < 0.$$

- *To prove ii) $\Leftrightarrow$ iii)*, let $S_i = P_i^{-1}, T_i = Q_i^{-1}, \forall\, i \in \mathcal{I}$, then (17) is equivalent to

$$\begin{bmatrix} P_j & P_jA_{i,1} & P_jA_{i,2} & P_jB_{i,1} & 0 & 0 \\ * & P_i & 0 & 0 & C^T_{i,1} & I \\ * & * & Q_k & 0 & C^T_{i,2} & 0 \\ * & * & * & \gamma^2 I & D^T_{i,1} & 0 \\ * & * & * & * & I & 0 \\ * & * & * & * & * & T_i \end{bmatrix} > 0, \forall i, j, k \in \mathcal{I},$$

which is equivalent to

$$\begin{bmatrix} S_j & A_{i,1}S_i & A_{i,2}T_k & B_{i,1} & 0 & 0 \\ * & S_i & 0 & 0 & S_iC^T_{i,1} & S_i \\ * & * & T_k & 0 & T_kC^T_{i,2} & 0 \\ * & * & * & \gamma^2 I & D^T_{i,1} & 0 \\ * & * & * & * & I & 0 \\ * & * & * & * & * & T_i \end{bmatrix} > 0, \forall i, j, k \in \mathcal{I},$$

which is equivalent to

$$\begin{bmatrix} S_i & S_iA^T_{i,1} & S_i & 0 & 0 & S_iC^T_{i,1} \\ * & S_j & 0 & A_{i,2}T_k & B_{i,1} & 0 \\ * & * & T_i & 0 & 0 & 0 \\ * & * & * & T_k & 0 & T_kC^T_{i,2} \\ * & * & * & * & \gamma^2 I & D^T_{i,1} \\ * & * & * & * & * & I \end{bmatrix} > 0, \forall i, j, k \in \mathcal{I},$$

which is equivalent, by Lemma 2, to (18).

Remark 1. It is easy to see that LMIs (17) holds implies that LMIs (9) holds, and LMIs (18) holds implies that LMIs (10) holds. This fact shows that the condition in Theorem 1 is sufficient for quadratic stability.

5 Control Synthesis Result

In this section, we give LMI-type conditions to build a switched state feedback controller as (3) ensuring the closed-loop system (4) is quadratically stable with an $\mathcal{L}_2$-gain smaller than a positive scalar γ.

Theorem 2. *For a given $\gamma > 0$, if there exist positive definite matrices $S_1, T_1, \cdots, S_N, T_N$ and matrices $F_1, \cdots, F_N$ such that $\forall i, j, k \in \mathcal{I}$*

$$\begin{bmatrix} S_i & S_i^T A_{i,1}^T + F_i^T B_{i,2}^T & S_i^T & 0 & 0 & S_i^T C_{i,1}^T + F_i^T D_{i,2}^T \\ * & S_j & 0 & A_{i,2} T_k & B_{i,1} & 0 \\ * & * & T_i & 0 & 0 & 0 \\ * & * & * & T_k & 0 & T_k C_{i,2}^T \\ * & * & * & * & \gamma^2 I & D_{i,1}^T \\ * & * & * & * & * & I \end{bmatrix} > 0 \quad (20)$$

then system (4) is quadratically stable with an $\mathcal{L}_2$-gain smaller than γ, and the switched state feedback controller can be taken as

$$u(t) = F_{r(t)} S_{r(t)}^{-1} x(t).$$

Proof. Replacing $S_i, A_{i,1}, C_{i,1}$ with $S_i^{-1}, A_{i,1} + B_{i,2}K_i, C_{i,1} + D_{i,2}K_i$ respectively in Theorem 1 and performing the change of variable $F_i = K_i S_i$ complete the proof.

Theorem 3. *For a given $\gamma > 0$, if there exist positive definite matrices $S_1, T_1, \cdots, S_N, T_N$ and matrices $G_1, F_1, \cdots, G_N, F_N$ such that $\forall i, j, k \in \mathcal{I}$*

$$\begin{bmatrix} G_i + G_i^T - S_i & G_i^T A_{i,1}^T + F_i^T B_{i,2}^T & G_i^T & 0 & 0 & G_i^T C_{i,1}^T + F_i^T D_{i,2}^T \\ * & S_j & 0 & A_{i,2} T_k & B_{i,1} & 0 \\ * & * & T_i & 0 & 0 & 0 \\ * & * & * & T_k & 0 & T_k C_{i,2}^T \\ * & * & * & * & \gamma^2 I & D_{i,1}^T \\ * & * & * & * & * & I \end{bmatrix} > 0 \quad (21)$$

then system (4) is quadratically stable with an $\mathcal{L}_2$-gain smaller than γ, and the switched state feedback controller can be taken as

$$u(t) = F_{r(t)} G_{r(t)}^{-1} x(t).$$

Proof. Replacing $A_{i,1}, C_{i,1}$ with $A_{i,1} + B_{i,2}K_i, C_{i,1} + D_{i,2}K_i$ respectively in Theorem 1 and performing the change of variable $F_i = K_i G_i$ complete the proof.

Remark 2. It is easy to see that LMIs (20) holds implies that LMIs (11) holds, and LMIs (21) holds implies that LMIs (13) holds. This fact shows that the conditions in Theorems 2 and 3 are sufficient for quadratic stabilization.

By Theorems 2 and 3, a switched state feedback controller can be designed to ensure that the closed-loop system (4) is quadratically stable with an $\mathcal{L}_2$-gain as small as possible under our condition via solving one of the following eigenvalue problems.

Problem 1. If there exists a solution to the following eigenvalue problem in the variables $S_i, T_i, G_i, F_i, \gamma$

$$\begin{aligned} &\text{minimize } \gamma^2 \\ &\text{subject to } S_i > 0, T_i > 0, \gamma > 0 (\forall i \in \mathcal{I}) \\ &\qquad\qquad \text{and (20).} \end{aligned}$$

then there exists a switched state feedback controller as (3) with

$$K_i = F_i S_i^{-1}, \, i \in \mathcal{I}$$

such that the closed-loop system (4) is quadratically stable with an $\mathcal{L}_2$-gain smaller than γ.

Or equivalently,

Problem 2. If there exists a solution to the following eigenvalue problem in the variables $S_i, T_i, G_i, F_i, \gamma$

$$\begin{array}{l}\text{minimize } \gamma^2 \\ \text{subject to } S_i > 0, T_i > 0, \gamma > 0 (\forall i \in \mathcal{I}) \\ \qquad\qquad \text{and (21).}\end{array}$$

then there exists a switched state feedback controller as (3) with

$$K_i = F_i G_i^{-1},\ i \in \mathcal{I}$$

such that the closed-loop system (4) is quadratically stable with an $\mathcal{L}_2$-gain smaller than γ.

6 Illustrating Example

In this section, we give a numerical example to illustrate the utilization of our results.

Example 1. Consider system (1) with $\mathcal{I} = \{1, 2\}$ and

$$A_{11} = \begin{bmatrix} 0.1 & 0 \\ 0 & 0.6 \end{bmatrix}, A_{12} = \begin{bmatrix} 0.1 & 0 \\ 0 & 0.01 \end{bmatrix},$$

$$B_{1,1} = \begin{bmatrix} 0.2 \\ 0 \end{bmatrix}, B_{1,2} = \begin{bmatrix} 1 \\ 0 \end{bmatrix},$$

$$C_{11} = \begin{bmatrix} 0.3 & 0 \\ 0 & 0.4 \end{bmatrix}, C_{12} = \begin{bmatrix} 0.04 & 0 \\ 0 & -0.5 \end{bmatrix},$$

$$D_{1,1} = \begin{bmatrix} 0.1 \\ 0 \end{bmatrix}, D_{1,2} = \begin{bmatrix} 2 \\ 0 \end{bmatrix};$$

$$A_{21} = \begin{bmatrix} -0.7 & 0 \\ 0 & -0.6 \end{bmatrix}, A_{22} = \begin{bmatrix} -0.02 & 0 \\ 0 & -0.5 \end{bmatrix},$$

$$B_{2,1} = \begin{bmatrix} 0 \\ 0.1 \end{bmatrix}, B_{2,2} = \begin{bmatrix} 0 \\ 2 \end{bmatrix},$$

$$C_{21} = \begin{bmatrix} -0.5 & 0 \\ 0 & 0.4 \end{bmatrix}, C_{22} = \begin{bmatrix} 0.4 & 0 \\ 0 & 0.03 \end{bmatrix},$$

$$D_{2,1} = \begin{bmatrix} 0 \\ 0.2 \end{bmatrix}, D_{2,2} = \begin{bmatrix} 0 \\ 1 \end{bmatrix};$$

By solving Problem 2, we can take the switched state feedback control law (3) with

$$K_1 = \begin{bmatrix} -0.1198 & 0 \end{bmatrix}, \quad K_2 = \begin{bmatrix} 0 & 0.1474 \end{bmatrix}$$

to ensure that the closed-loop system (4) is quadratically stable with an $\mathcal{L}_2$-gain smaller than $\gamma = 0.2717$.

7 Conclusion

This paper has addressed the $\mathcal{L}_2$-gain analysis and control synthesis problems for switched linear systems with state delay in the discrete-time domain. For the $\mathcal{L}_2$-gain analysis problem, two LMI(Linear Matrices Inequality)-based conditions have been obtained such that the autonomous switched system with state delay is quadratically stable with an $\mathcal{L}_2$-gain smaller than a positive scalar γ. For the control synthesis problem, two LMI-based conditions have been derived to build a switched state feedback controller ensuring the closed-loop system is quadratically stable with an $\mathcal{L}_2$-gain smaller than a positive scalar γ. All the obtained results are delay independent. The future work includes finding delay dependent conditions and considering the case with both state delay and input delay, or time-varying delay, and so on.

Appendix A

The following result is needed for the proof of Lemma 2.

Lemma 4. *[15] Let the matrices $U \in \mathbb{C}^{n\times m}, W \in \mathbb{C}^{k\times n}$, and $\Phi = \Phi^T \in \mathbb{C}^{n\times n}$, be given. Then the following statements are equivalent:*
i) There exists a matrix $V \in \mathbb{C}^{m\times k}$ satisfying

$$UVW + (UVW)^T + \Phi < 0$$

ii) The following two conditions hold

$$\mathcal{N}_U \Phi \mathcal{N}_U^T < 0 \quad or \quad UU^T > 0$$

$$\mathcal{N}_W^T \Phi \mathcal{N}_W < 0 \quad or \quad W^T W > 0$$

where $\mathcal{N}_U$ and $\mathcal{N}_W^T$ are respectively orthogonal complement of U and W^T; that is

$$\mathcal{N}_U U = 0$$
$$\mathcal{N}_W^T W^T = 0.$$

Proof (Proof of Lemma 2). By Lemma 1, (7) is equivalent to

$$\begin{bmatrix} G + G^T - S & -G^T M^T \\ -MG & T \end{bmatrix} > 0$$

that is

$$\begin{bmatrix} S - G - G^T & G^T M^T \\ MG & -T \end{bmatrix} < 0$$

which can be written as

$$\begin{bmatrix} S & 0 \\ 0 & -T \end{bmatrix} + \begin{bmatrix} -I \\ M \end{bmatrix} G[\, I \;\; 0\,] + \left\{ \begin{bmatrix} -I \\ M \end{bmatrix} G[\, I \;\; 0\,] \right\}^T < 0$$

Noticing $Q > 0$, we get the following equivalent inequality by Lemma 4

$$[\,M\ \ I\,]\begin{bmatrix} S & 0 \\ 0 & -T \end{bmatrix}\begin{bmatrix} M^T \\ I \end{bmatrix} = -T + MSM^T < 0$$

which is equivalent, by using the Schur complement, to (6) since $S > 0$.

References

1. D. Liberzon and A. S. Morse, "Basic problems in stability and design of switched systems," *IEEE Control Systems*, 19(5), 1999, p 59-70.
2. G. Xie, L. Wang, Q. Ye, "Controllability of a class of hybrid dynamic systems (I) - Basic properties and preliminary results", *APPL MATH MECH-ENGL*, 24(9), Sep., 2003, p 1041-1050.
3. G. Xie, L. Wang, Q. Ye, "Controllability of a class of hybrid dynamic systems (II) - Single time-delay case", *APPL MATH MECH-ENGL*, 24(9), Sep., 2003, p 1051-1062.
4. G. Xie, L. Wang, Q. Ye, "Controllability of a class of hybrid dynamic systems (III) - Multiple time-delay case", *APPL MATH MECH-ENGL*, 24(9), Sep., 2003, p 1063-1074.
5. Y. Wang, G. Xie, L. Wang, "Controllability of switched time-delay systems under constrained switching", *J. MATH. ANAL. APPL.*, 286 (2): 397-421, OCT 15 2003.
6. G. Xie, L. Wang, "Controllability of Switched Linear Discrete-time Systems with Time Delay", *Proceedings of the American Control Conference*, v 4, 2003, p 2797-2802.
7. Y. Wang, G. Xie, L. Wang, "Controllability of Switched Linear Discrete-time Systems with Multiple Time Delays", *Proceedings of the European Control Conference*, September 2003, Cambridge, UK.
8. G. Xie, L. Wang, "On Reachability of Switched Linear Discrete-time Systems with State Delays", accepted by*IEEE International Symposium on Intelligent Control*, ISIC, 2004.
9. G. Xie, L. Wang, "Reachability of switched linear discrete-systems with state delays", accepted by *Dynamics of Continuous, Discrete and Impulsive Systems, Series B*, 2004.
10. G. Xie, L. Wang, "Stability and stabilization of switched linear systems with state delay: continuous-time case", *the 16th Mathematical Theory of Networks and Systems Conference*, MTNS 2004.
11. G. Xie, L. Wang, "Stability and stabilization of switched linear systems with state delay: discrete-time case", *the 16th Mathematical Theory of Networks and Systems Conference*, MTNS 2004.
12. G. Xie, L. Wang, "Quadratic Stability and Stabilization of Discrete-time Switched Systems with State Delay", accepted by IEEE Conference on Decision and Control, CDC 2004.
13. G. Zhai, Y. Sun, X. Chen, A. N. Michel, "Stability and L_2 gain analysis for switched symmetric systems with time delay", *Proceedings fo the 2003 American Control conference*, Denver, Jun., 2003, pp 4-6.

14. S. Boyd, L. E. Ghaou, E. Feron and V. Balakrishnan, *Linear matrix inequalities in systems and control theory*. SIAM. Philadelphia; 1993.
15. R.E.Skelton, T.Iwasaki, and K.Grigoriadis, "A unified approach to linear control design", Taylor and Francis series in Systems and Control; 1997.
16. J. P. Richard, "Time-delay systems: an overview of some recent advances and open problems", *Automatic*, 39, 2003, pp 1667-1694.

Robust Control for Discrete-time Switched Systems with Uncertainty

Songlin Chen, Yu Yao and Fenghua He

Control and Simulation Centor, Harbin Institute of Technology, Harbin, 150080, P.R.China `[songlin, yaoyu, hefenghua]@hit.edu.cn`

1 Introduction

By a switched systems, we mean a hybrid dynamic system that consists of a family of subsystems described by differential or difference equations and a switching rule that orchestrates switching between these subsystems. Switched systems and switched multi-controller systems have numerous applications in control of mechanical systems, process control, automotive industry, power systems, aircraft and traffic control, and many other fields.

In recent years, the study of switched systems has received increasing attention (see [2, 3, 4, 7, 10, 12, 13]). A survey of basic problem in stability and design of switched systems was given in [10] where many contributions were summarized. Most of these contributions dealt with stability of switched systems. The methodologies used in studying the qualitative properties of switched systems are very diverse. In [3], multiple Lyapunov functions were introduced and a result for the stability of linear time-invariant switched system was established. Linear matrix inequality (LMI) techniques (see [8, 11, 13]) prove to be a very good tool to study the stability and robustness of switched system. In [11], LMI problems were formulated for the stability analysis of switched systems consisting of linear subsystems. For robust stabilization of polytopic uncertain switched systems, a quadratic stabilizing switching law was designed for polytopic uncertain switched systems based on LMI techniques in [13].

In [5], switched Lyapunov functions were used to study the state feedback and static output feedback control for discrete-time switched systems . Based on results in [5], design problem of dynamic output feedback for discrete-time switched system was studied from a γ-performance point of view, and the designed switched output feedback control ensured that the L_2 gain of the closed loop system is less than or equal to some specified positive level γ in [6]. However, [5, 6] did not explicitly consider uncertainty in the model. Uncertainty in the plant model is one of the main challenges in control theory, and it is of practical importance to deal with systems with uncertain parameters.

Motivated by the above existing results, this paper intend to study the robust control problem of uncertain switched systems. For the discrete-time switched linear systems with norm-bounded time-varying uncertainty both in state and input matrices, the switched state feedback controller and dynamic output feedback controller are designed individually to guarantee the robust stability and disturbance attenuation properties (γ-performance) of the corresponding closed loop system under arbitrary switching. In the case of state feedback, LMI-based conditions are presented. In the case of dynamic output feedback, if some real numbers are given in advance, the conditions can be dealt with as LMIs which can be tested with efficient algorithms.

This paper is organized as follows. In section 2, the stability analysis and γ-performance definition of discrete-time switched system are given. Stabilizing controller and γ-performance controller via state feedback are designed in section 3. Main results for stabilizing controller and γ-performance controller via dynamic output feedback are proposed in section 4. Some illustrative examples along with simulation are presented in Section 5. Some conclusions are provided in the end.

2 Preliminaries

2.1 Robust Stability Analysis

Consider the uncertain discrete-time switched system given by

$$x(k+1) = \underbrace{(A_\sigma + \Delta A_\sigma)}_{\bar{A}_\sigma} x(k), \tag{1}$$

where $x(k) \in \mathbb{R}^n$ is the state and $\sigma \in \Xi = \{1, 2, \cdots, \mathrm{N}\}$ is a switching rule. Moreover, $\sigma = i$ implies that the ith subsystem is activated. The uncertain matrice ΔA_i is assumed to have the forms as follows

$$\Delta A_i = M_i F_i(k) N_i, \quad \forall i \in \Xi. \tag{2}$$

A_i are constant matrices of compatible dimensions. M_i and N_i are constant matrices of compatible dimensions which characterize the structure of uncertainty, and $F_i(k)$ are the norm-bound time-varying matrices which satisfy

$$F_i(k)' F_i(k) \leq I, \quad \forall i \in \Xi, \tag{3}$$

where I is an identity matrix of compatible dimensions. This kind of uncertainty was extensively studied in the literature [9].

The following lemma gives a stability sufficient condition for system (1) without uncertainty.

Lemma 1 [6] *The autonomous switched system (1) with* $\Delta A_i = 0$ *is asymptotically stable if there exist symmetric matrices* $S_i \in \mathbb{R}^{n\times n}$ *and matrices* $G_i \in \mathbb{R}^{n\times n}$ $(\forall i \in \Xi)$ *such that*

$$\begin{bmatrix} G_i + G_i' - S_i & G_i'A_i' \\ A_iG_i & S_j \end{bmatrix} > 0, \quad \forall\{i,j\} \in \Xi \times \Xi. \tag{4}$$

In order to establish the robust stability criteria for uncertain switched system, we introduce the following lemma.

Lemma 2 *Let Y, M and N be given matrices of compatible dimensions, where Y is a symmetric matrix, then*

$$Y + MFN + N'F'M' > 0$$

holds for any matrix F satisfying $F^TF < I$ if and only if there exists a constant $\varepsilon > 0$ such that

$$Y + MM' - \varepsilon^{-1}N'N > 0.$$

Theorem 1 *The uncertain autonomous switched system (1) is asymptotically stable for any uncertainty described by (2) and (3), if there exist symmetric matrices $S_i \in \mathbb{R}^{n\times n}$ and $G_i \in \mathbb{R}^{n\times n}$ ($\forall i \in \Xi$) and real numbers $\varepsilon_{ij} > 0$ ($\forall\{i,j\} \in \Xi \times \Xi$) such that*

$$\begin{bmatrix} G_i + G_i' - S_i & G_i'A_i' & G_i'N_i' \\ A_iG_i & S_j - \varepsilon_{ij}M_iM_i' & 0 \\ N_iG_i & 0 & \varepsilon_{ij}I \end{bmatrix} > 0, \quad \forall\{i,j\} \in \Xi \times \Xi. \tag{5}$$

Proof By Lamma 1 and Lamma 2, the proof is easy and we omit it here.

2.2 γ-performance

Consider the uncertain discrete-time switched system given by

$$\begin{cases} x(k+1) = \underbrace{(A_\sigma + \Delta A_\sigma)}_{\bar{A}_\sigma} x(k) + \underbrace{(B_\sigma^w + \Delta B_\sigma)}_{\bar{B}_\sigma^w} w(k) \\ z(k) = C_\sigma^z x(k) + D_\sigma^{zw} w(k) \end{cases} \tag{6}$$

where $x(k) \in \mathbb{R}^n$ is the state with $x(0) = 0$, $w(k) \in \mathbb{Z}^q$ is the exogenous disturbance which belong to $L_2[0,\infty)$, and $z(k) \in \mathbb{R}^p$ is the controlled output. σ is a switching rule defined as previously. A_i, B_i^w, C_i^z, D_i^{zw} are constant matrices of compatible dimensions which describe the nominal system. The uncertain matrices ΔA_i and ΔB_i are described by

$$\Delta A_i = M_i^A F_i^A(k) N_i^A, \quad \Delta B_i = M_i^B F_i^B(k) N_i^B, \quad \forall i \in \Xi, \tag{7}$$

where M_i^A, N_i^A, M_i^B, N_i^B are constant matrices of compatible dimensions which distribute uncertainty on rows and columns of the corresponding matrices. The time-varying matrices $F_i^A(k)$, $F_i^B(k)$ satisfy

$$F_i^A(k)'F_i^A(k) \le I, \quad F_i^B(k)'F_i^B(k) \le I, \quad \forall i \in \Xi. \tag{8}$$

Given a constant $\gamma > 0$, the following definition concerns the γ-performance criterion for discrete-time switched systems.

Definition 1 *The autonomous system (6) is said to have a γ-performance if it is asymptotically stable and*

$$\gamma^{-1}\sum_{k=0}^{\infty} z(k)^2 < \gamma \sum_{k=0}^{\infty} w(k)^2, \quad \forall \sum_{k=0}^{\infty} w(k)^2 > 0.$$

The following lemma gives a sufficient condition to check if the system (6) without uncertainty has a γ-performance.

Lemma 3 [6] *The autonomous system (6) with $\Delta A_i = \Delta B_i = 0$ has a γ-performance if there exist symmetric matrices $S_i \in \mathbb{R}^{n\times n}$ and matrices $G_i \in \mathbb{R}^{n\times n}$ ($\forall i \in \Xi$) such that*

$$\begin{bmatrix} G_i + G_i' - S_i & * & * & * \\ 0 & \gamma I & * & * \\ A_i G_i & B_i^w & S_j & * \\ C_i^z G_i & D_i^{zw} & 0 & \gamma I \end{bmatrix} > 0, \quad \forall \{i,j\} \in \Xi \times \Xi. \tag{9}$$

where $*$ denotes that the block can be obtained easily by symmetry.

The following theorem gives a sufficient condition to check if the system (6) with uncertain description (7) and (8) has a γ-performance.

Theorem 2 *The uncertain autonomous switched system (6) has a γ-performance for any uncertainty described by (7) and (8) if there exist symmetric matrices $S_i \in \mathbb{R}^{n\times n}$ and $G_i \in \mathbb{R}^{n\times n}$ ($\forall i \in \Xi$), and real numbers $\varepsilon_{ij} > 0$ ($\forall \{i,j\} \in \Xi \times \Xi$) such that*

$$\begin{bmatrix} G_i + G_i' - S_i & * & * & * & * & * \\ 0 & \gamma I & * & * & * & * \\ A_i G_i & B_i^w & S_j - \varepsilon_{ij}[M_i^A (M_i^A)' + M_i^B (M_i^B)'] & * & * & * \\ C_i^z G_i & D_i^{zw} & 0 & \gamma I & * & * \\ N_i^A G_i & 0 & 0 & 0 & \varepsilon_{ij} I & * \\ 0 & N_i^B & 0 & 0 & 0 & \varepsilon_{ij} I \end{bmatrix} > 0, \tag{10}$$

$\forall \{i,j\} \in \Xi \times \Xi.$

Proof By lamma 3, if there exist symmetric matrices $S_i \in \mathbb{R}^{n\times n}$ and matrices $G_i \in \mathbb{R}^{n\times n}$ ($\forall i \in \Xi$) such that

$$\begin{bmatrix} G_i + G_i' - S_i & * & * & * \\ 0 & \gamma I & * & * \\ \bar{A}_i G_i & \bar{B}_i^w & S_j & * \\ C_i^z G_i & D_i^{zw} & 0 & \gamma I \end{bmatrix} > 0, \quad \forall \{i,j\} \in \Xi \times \Xi, \tag{11}$$

then the uncertain switched system (6) has a γ-performance. The formula (11) can be written as

$$Y_{ij} + \tilde{M}_i \tilde{F}_i(k) \tilde{N}_i + \tilde{N}_i' \tilde{F}_i(k)' \tilde{M}_i' > 0, \quad \forall \{i, j\} \in \Xi \times \Xi, \tag{12}$$

where

$$Y_{ij} = \begin{bmatrix} G_i + G_i' - S_i & * & * & * \\ 0 & \gamma I & * & * \\ A_i G_i & B_i^w & S_j & * \\ C_i^z G_i & D_i^{zw} & 0 & \gamma I \end{bmatrix}, \quad \tilde{M}_i = \begin{bmatrix} 0 & 0 \\ 0 & 0 \\ M_i^A & M_i^B \\ 0 & 0 \end{bmatrix},$$

$$\tilde{N}_i = \begin{bmatrix} N_i^A G_i & 0 & 0 & 0 \\ 0 & N_i^B & 0 & 0 \end{bmatrix}, \quad \tilde{F}_i(k) = diag\left[F_i^A(k) \; F_i^B(k) \right]. \tag{13}$$

It follows from (7) that $\tilde{F}_i(k)$ satisfy $\tilde{F}_i(k)' \tilde{F}_i(k) \leq I$ ($\forall i \in \Xi$). By Lemma 2, inequalities (12) hold if and only if there exist real numbers $\varepsilon_{ij} > 0$ ($\forall \{i, j\} \in \Xi \times \Xi$) such that

$$Y_{ij} - \varepsilon_{ij} \tilde{M}_i \tilde{M}_i' - \varepsilon_{ij}^{-1} \tilde{N}_i' \tilde{N}_i > 0, \quad \forall \{i, j\} \in \Xi \times \Xi. \tag{14}$$

Substitute (12) into formula (14) and expand formula (14), we get

$$\begin{bmatrix} \Phi_{ij} & * & * & * \\ 0 & \gamma I - \varepsilon_{ij}^{-1} (N_i^B)' N_i^B & * & * \\ A_i G_i & B_i^w & S_j - \varepsilon_{ij} [M_i^A (M_i^A)' + M_i^B (M_i^B)'] & * \\ C_i^z G_i & D_i^{zw} & 0 & \gamma I \end{bmatrix} > 0, \tag{15}$$

$\forall \{i, j\} \in \Xi \times \Xi$, where $\Phi_{ij} = G_i + G_i' - S_i - \varepsilon_{ij}^{-1} G_i' (N_i^A)' N_i^A G_i$. By Schur complement theorem [1], the formula (15) is equivalent to (10), which ends the proof. □

Remark 1 With the similar technique in Theorem 2, the same uncertainty in other matrices (e.g., C_σ^z and D_σ^{zw}, etc.) can be incorporated without destroying the tractability of results.

3 State Feedback Design

In this section, consider the following discrete-time switched system given by

$$\begin{cases} x(k+1) = \underbrace{(A_\sigma + \Delta A_\sigma)}_{\bar{A}_\sigma} x(k) + \underbrace{(B_\sigma + \Delta B_\sigma)}_{\bar{B}_\sigma} u(k) + B_\sigma^w w(k) \\ z(k) = C_\sigma^z x(k) + D_\sigma^z u(k) + D_\sigma^{zw} w(k) \end{cases} \tag{16}$$

where $x(k) \in \mathbb{R}^n$ is the state, $w(k) \in \mathbb{R}^q$ is the exogenous disturbance which belong to $L_2[0, \infty)$, $u(k) \in \mathbb{R}^q$ and $z(k) \in \mathbb{R}^p$ are control input and controlled output respectively. σ is defined as previously. The switching rule σ is a prior unknown and we assume that its instantaneous value is available in real time such that the right control can be switched simultaneously. A_i, B_i, B_i^w, C_i^z, D_i^z, and D_i^{zw} are constant matrices of compatible dimensions. The uncertain matrices ΔA_i and ΔB_i are described by (7) and (8).

3.1 Stabilizing State Feedback Design

The design problem of stabilizing switched state feedback conrol is considered in this subsection. We aim to find

$$u(k) = K_\sigma x(k), \tag{17}$$

such that the corresponding closed loop system

$$x(k+1) = (\bar{A}_\sigma + \bar{B}_\sigma K_\sigma)x(k) \tag{18}$$

is asymptotically stable. A sufficient condition for existence of such controller is given by the following theorem.

Theorem 3 *The closed loop system (18) is asymptotically stable for any uncertainty described by (7) and (8), if there exist symmetric positive definite matrices $S_i \in \mathbb{R}^{n\times n}$, matrices $G_i \in \mathbb{R}^{n\times n}$, $R_i \in \mathbb{R}^{m\times n}$ ($\forall i \in \Xi$), and real numbers $\varepsilon_{ij} > 0$ ($\forall\{i,j\} \in \Xi \times \Xi$) such that*

$$\begin{bmatrix} G_i + G_i' - S_i & * & * & * \\ A_iG_i + B_iR_i & S_j - \varepsilon_{ij}[M_i^A(M_i^A)' + M_i^B(M_i^B)'] & * & * \\ N_i^AG_i & 0 & \varepsilon_{ij}I & * \\ N_i^BR_i & 0 & 0 & \varepsilon_{ij}I \end{bmatrix} > 0, \tag{19}$$

$\forall\{i,j\} \in \Xi \times \Xi$, then the switched state feedback controller stabilizing the uncertain switched system (16) is given by (17) with $K_i = R_iG_i^{-1}$.

3.2 γ-Performance State Feedback Design

In this subsection, the design problem of γ-performance switched state feedback control for uncertain switched system (16) is approached. Our objective is to build the switched state feedback controller (17) such that the corresponding closed loop system

$$\left\{ \begin{aligned} x(k+1) &= (\bar{A}_\sigma + \bar{B}_\sigma K_\sigma)x(k) + B_\sigma^w w(k) \\ z(k) &= (C_\sigma^z + D_\sigma^z K_\sigma)x(k) + B_\sigma^{zw} w(k) \end{aligned} \right. \tag{20}$$

has a γ-performance. The following theorem gives a sufficient condition to build such controller.

Theorem 4 *The closed loop system (20) has a γ-performance for any uncertainty described by (7) and (8), if there exist symmetric positive definite matrices $S_i \in \mathbb{R}^{n\times n}$, matrices $G_i \in \mathbb{R}^{n\times n}$, $R_i \in \mathbb{R}^{q\times n}$ ($\forall i \in \Xi$), and real numbers $\varepsilon_{ij} > 0$ ($\forall\{i,j\} \in \Xi \times \Xi$) such that*

$$\begin{bmatrix} G_i + G_i' - S_i & * & * & * & * & * \\ 0 & \gamma I & * & * & * & * \\ A_iG_i + B_iR_i & B_i^w & S_j - \varepsilon_{ij}[M_i^A(M_i^A)' + M_i^B(M_i^B)'] & * & * & * \\ C_i^zG_i + D_i^zR_i & D_i^{zw} & 0 & \gamma I & * & * \\ N_i^AG_i & 0 & 0 & 0 & \varepsilon_{ij}I & * \\ N_i^BR_i & 0 & 0 & 0 & 0 & \varepsilon_{ij}I \end{bmatrix} > 0, \tag{21}$$

$\forall (i,j) \in (\Xi \times \Xi)$, then the γ-performance switched state feedback controller is given by (17) with $K_i = R_i G_i^{-1}$.

Proof With $R_i = K_i G_i$, the proof is similar to Theorem 2, and we omit it here. □

4 Dynamic Output Feedback (DOF)

Consider the class of uncertain discrete-time switched system as follows

$$\begin{cases} x(k+1) = \underbrace{(A_\sigma + \Delta A_\sigma)}_{\bar{A}_\sigma} x(k) + \underbrace{(B_\sigma + \Delta B_\sigma)}_{\bar{B}_\sigma} u(k) + B_\sigma^w w(k) \\ z(k) = C_\sigma^z x(k) + D_\sigma^z u(k) + D_\sigma^{zw} w(k) \\ y(k) = C_\sigma x(k) + D_\sigma^{yw} w(k) \end{cases} \tag{22}$$

where $x(k) \in \mathbb{R}^n$ is the state, $u(k) \in \mathbb{R}^m$ is the input, $w(k) \in \mathbb{Z}^q$ is the exogenous disturbance which belong to $L_2[0,\infty)$, $y(k)$ and $z(k)$ are the measured output and the controlled output respectively. σ is a switching rule defined as in section 3. A_i, B_i, B_i^w, C_i^z, D_i^z, D_i^{zw}, C_i, D_i^{yw} are constant matrices of compatible dimensions. The uncertain matrices ΔA_i and ΔB_i are described by (7) and (8).

4.1 Stabilization DOF Control Design

In this section, a switched dynamic output feedback control

$$\begin{cases} \eta(k+1) = A_\sigma^c \eta(k) + B_\sigma^c y(k) \\ u(k) = C_\sigma^c \eta(k) + D_\sigma^c y(k) \end{cases} \tag{23}$$

are to construct to ensure the stability of the corresponding closed loop system

$$\begin{bmatrix} x(k+1) \\ \eta(k+1) \end{bmatrix} = \begin{bmatrix} \bar{A}_\sigma + \bar{B}_\sigma D_\sigma^c C_\sigma & \bar{B}_\sigma C_\sigma^c \\ B_\sigma^c C_\sigma & A_\sigma^c \end{bmatrix} \begin{bmatrix} x(k) \\ \eta(k) \end{bmatrix} \tag{24}$$

The switched dynamic feedback control (23) and the switched system (22) share the same rule σ, which means that at an arbitrary instant k if the system switches from (A_i, B_i, C_i) to (A_j, B_j, C_j), the control switches simultaneously from $(A_i^c, B_i^c, C_i^c, D_i^c)$ to $(A_j^c, B_j^c, C_j^c, D_j^c)$.

To obtain the main result, we describe the closed loop system (24) as the form of system (1) as follows

$$\tilde{x}(k+1) = (\tilde{A}_\sigma + \tilde{M}_\sigma \tilde{F}_\sigma \tilde{N}_\sigma)\tilde{x}(k), \tag{25}$$

where $\tilde{x}(k) = \begin{bmatrix} x(k) \ \eta(k) \end{bmatrix}'$, and the corresponding matrices are

$$\tilde{A}_\sigma = \begin{bmatrix} A_\sigma + B_\sigma D_\sigma^c C_\sigma & B_\sigma C_\sigma^c \\ B_\sigma^c C_\sigma & A_\sigma^c \end{bmatrix}, \quad \tilde{M}_\sigma = \begin{bmatrix} M_\sigma^A & M_\sigma^B \\ 0 & 0 \end{bmatrix},$$

$$\tilde{F}_\sigma = \begin{bmatrix} F_\sigma^A & 0 \\ 0 & F_\sigma^B \end{bmatrix}, \quad \tilde{N}_\sigma = \begin{bmatrix} N_\sigma^A & 0 \\ N_\sigma^B D_\sigma^c C_\sigma & N_\sigma^B C_\sigma^c \end{bmatrix}. \tag{26}$$

The following theorem gives a sufficient condition to build such a stabilizing DOF control.

Theorem 5 *If there exist matrices X, W, Y, symmetric matrices S_i, R_i, matrices Q_i, P_i, L_i, H_i, E_i ($\forall(i \in \Xi)$), and real numbers $\varepsilon_{ij} > 0$ ($\forall\{i,j\} \in \Xi \times \Xi$) such that*

$$\begin{bmatrix} X+X'-S_i & * & * & * & * & * & * & * \\ I+W-P_i & Y+Y'-R_i & * & * & * & * & * & * \\ A_iX+B_iL_i & A_i+B_iH_iC_i & S_j & * & * & * & * & * \\ Q_i & YA_i+E_iC_i & P_j & R_j & * & * & * & * \\ N_i^AX & N_i^A & 0 & 0 & \varepsilon_{ij}I & * & * & * \\ N_i^BL_i & N_i^BH_iC_i & 0 & 0 & 0 & \varepsilon_{ij}I & * & * \\ 0 & 0 & (M_i^A)' & (M_i^A)'Y' & 0 & 0 & \varepsilon_{ij}^{-1}I & * \\ 0 & 0 & (M_i^B)' & (M_i^B)'Y' & 0 & 0 & 0 & \varepsilon_{ij}^{-1}I \end{bmatrix} > 0, \tag{27}$$

$\forall\{i,j\} \in \Xi \times \Xi$, then the switched dynamic output feedback control given by (23) with

$$\begin{aligned} A_i^c &= V^{-1}\left[Q_i - Y(A_i+B_iH_iC_i)X - VB_i^cC_iX - Y_i, B_iC_i^cU\right]U^{-1}, \\ B_i^c &= V^{-1}(E_i - YB_iH_i), \\ C_i^c &= (L_i - H_iC_iX)U_i^{-1}, \\ D_i^c &= H_i, \\ VU &= W - YX, \end{aligned} \tag{28}$$

$\forall i \in \Xi$, stabilizes the system (22) for any uncertainty described by (7) and (8).

Proof By Theorem 1, if there exist symmetric matrices $\tilde{S}_i$ ($\forall i \in \Xi$), matrix $\tilde{G}$, and real numbers $\varepsilon_{ij} > 0$ ($\forall\{i,j\} \in \Xi \times \Xi$) such that

$$\begin{bmatrix} G+G'-S_i & G'A_i' & G'N_i' \\ A_iG & S_j-\varepsilon_{ij}M_iM_i' & 0 \\ N_iG & 0 & \varepsilon_{ij}I \end{bmatrix} > 0, \quad \forall\{i,j\} \in \Xi \times \Xi, \tag{29}$$

then the corresponding closed loop system (25) is asymptotically stable. Furthermore, by Schur complement theorem, (29) is equivalent to

$$\begin{bmatrix} \tilde{G}+\tilde{G}'-\tilde{S}_i & * & * & * \\ \tilde{A}_i\tilde{G} & \tilde{S}_j & * & * \\ \tilde{N}_i\tilde{G} & 0 & \varepsilon_{ij}^I & * \\ 0 & \tilde{M}_i' & 0 & \varepsilon_{ij}^{-1}I \end{bmatrix} > 0, \quad \forall\{i,j\} \in \Xi \times \Xi. \tag{30}$$

The condition (30) is not linear with respect to unknown matrices variables $\tilde{G}$, $\tilde{S}_i$, A_i^c, B_i^c, C_i^c and D_i^c. To linearize the condition and get a well tractable condition in terms of LMIs, we employed the method of variable substitution presented in [6]. Matrices $\tilde{G}$ and $\tilde{G}^{-1}$ are defined as follows

$$\tilde{G} = \begin{bmatrix} X & \bullet \\ U & \bullet \end{bmatrix}, \quad \tilde{G}^{-1} = \begin{bmatrix} Y' & \bullet \\ V' & \bullet \end{bmatrix}, \tag{31}$$

where $\bullet$ denote blocks in these matrices which don't appear in the final result. Define transformation matrix

$$T = \begin{bmatrix} I & 0 \\ Y & V \end{bmatrix}. \tag{32}$$

Multiplying (30) on the left by $\tilde{T} = diag\begin{bmatrix} T & T & I & I \end{bmatrix}$ and on the right by $\tilde{T}'$, we get

$$\begin{bmatrix} T(\tilde{G} + \tilde{G}' - \tilde{S}_i)T' & * & * & * \\ T\tilde{A}_i\tilde{G}T' & T\tilde{S}_jT' & * & * \\ \tilde{N}_i\tilde{G}T' & 0 & \varepsilon_{ij}I & * \\ 0 & \tilde{M}_i'T' & 0 & \varepsilon_{ij}^{-1}I \end{bmatrix} > 0, \quad \forall\{i,j\} \in \Xi \times \Xi. \tag{33}$$

The formula above is equivalent to (30). Letting

$$T\tilde{S}_jT' = \begin{bmatrix} S_j & P_j' \\ P_j & R_j \end{bmatrix}, \tag{34}$$

and expanding each block in (33), we get

$$T(\tilde{G} + \tilde{G}')T' = \begin{bmatrix} X + X' & I + W' \\ I + W & Y + Y' \end{bmatrix},$$

$$T\tilde{A}_i\tilde{G}T' = \begin{bmatrix} A_iX + B_iL_i & A_i + B_iH_iC_i \\ Q_i & YA_i + E_iC_i \end{bmatrix},$$

$$\tilde{N}_i\tilde{G}T' = \begin{bmatrix} N_i^AX & N_i^A \\ N_i^BL_i & N_i^BH_iC_i \end{bmatrix},$$

$$\tilde{M}_i'T' = \begin{bmatrix} (M_i^A)' & (M_i^A)'Y' \\ (M_i^B)' & (M_i^B)'Y' \end{bmatrix}, \tag{35}$$

where

$$\begin{aligned} Q_i &\triangleq Y(A_i + B_iD_i^cC_i)X + VB_i^cC_iX + YB_i, C_i^cU + VA_i^cU, \\ L_i &\triangleq D_i^cC_iX + C_i^cU, \\ E_i &\triangleq YB_iD_i^c + VB_i^c, \\ H_i &\triangleq D_i^c, \\ W &\triangleq YX + VU. \end{aligned} \tag{36}$$

Inequalities (17) can be obtained by substituting (34) and (35) into (33). The proof for reversibility from (36) to (27) is similar to the proof of Theorem 4 in reference [6]. □

4.2 γ-Performance Switched DOF Control

The design problem for γ-erformance dynamic output feedback control is considered in this subsection. Our objective is to build a switched DOF control (23) such that the corresponding closed loop system

$$\begin{cases} \begin{bmatrix} x(k+1) \\ \eta(k+1) \end{bmatrix} = \begin{bmatrix} \bar{A}_\sigma + B_\sigma D^c_\sigma C_\sigma & \bar{B}_\sigma C^c_\sigma \\ B^c_\sigma C_\sigma & A^c_\sigma \end{bmatrix} \begin{bmatrix} x(k) \\ \eta(k) \end{bmatrix} + \begin{bmatrix} B^w_\sigma + \bar{B}_\sigma D^c_\sigma D^{yw}_\sigma \\ B^c_\sigma D^{yw}_\sigma \end{bmatrix} w(k) \\ z(k) = \begin{bmatrix} C^z_\sigma + D^z_\sigma D^c_\sigma C_\sigma & D^z_\sigma C^c_\sigma \end{bmatrix} \begin{bmatrix} x(k) \\ \eta(k) \end{bmatrix} + \begin{bmatrix} D^{zw}_\sigma + D^z_\sigma D^c_\sigma D^{yw}_\sigma \end{bmatrix} w(k) \end{cases} \tag{37}$$

has a γ-performance. We describe the closed loop system (37) as the form of system (6) as follows

$$\begin{cases} \tilde{x}(k+1) = (\tilde{A}_\sigma + \tilde{M}^A_\sigma \tilde{F}^A_\sigma \tilde{N}^A_\sigma) + (\tilde{B}_\sigma + \tilde{M}^B_\sigma F^B_\sigma \tilde{N}^B_\sigma) w(k) \\ z(k) = \tilde{C}_\sigma \tilde{x}(k) + \tilde{D}_\sigma w(k) \end{cases} \tag{38}$$

where $\tilde{x}(k) = \begin{bmatrix} x(k) \ \eta(k) \end{bmatrix}'$, and associated matrices are

$$\tilde{A}_\sigma = \begin{bmatrix} A_\sigma + B_\sigma D^c_\sigma C_\sigma & B_\sigma C^c_\sigma \\ B^c_\sigma C_\sigma & A^c_\sigma \end{bmatrix}, \quad \tilde{B}_\sigma = \begin{bmatrix} B^w_\sigma + B_\sigma D^c_\sigma D^{yw}_\sigma \\ B^c_\sigma D^{yw}_\sigma \end{bmatrix},$$

$$\tilde{M}^A_\sigma = \begin{bmatrix} M^A_\sigma & M^B_\sigma \\ 0 & 0 \end{bmatrix}, \quad \tilde{F}^A_\sigma = \begin{bmatrix} F^A_\sigma & 0 \\ 0 & F^B_\sigma \end{bmatrix}, \quad \tilde{N}^A_\sigma = \begin{bmatrix} N^A_\sigma & 0 \\ N^B_\sigma D^c_\sigma C_\sigma & N^B_\sigma C^c_\sigma \end{bmatrix},$$

$$\tilde{M}^B_\sigma = \begin{bmatrix} M^B_\sigma \\ 0 \end{bmatrix}, \quad \tilde{F}^B_\sigma = \begin{bmatrix} F^B_\sigma \end{bmatrix}, \quad \tilde{N}^B_\sigma = \begin{bmatrix} N^B_\sigma D^c_\sigma D^{yw}_\sigma \end{bmatrix},$$

$$\tilde{C}_\sigma = \begin{bmatrix} C^z_\sigma + D^z_\sigma D^c_\sigma C_\sigma & D^z_\sigma C^c_\sigma \end{bmatrix}, \quad \tilde{D}_\sigma = \begin{bmatrix} D^{zw}_\sigma + D^z_\sigma D^c_\sigma D^{yw}_\sigma \end{bmatrix}. \tag{39}$$

A sufficient condition to build such DOF control for uncertain switched system (22) is given by the following theorem.

Theorem 6 *If there exists matrices X, W, Y, symmetric matrices S_i, R_i, matrices Q_i, P_i, L_i, H_i, E_i ($\forall i \in \Xi$) and real numbers $\varepsilon_{ij} > 0$ ($\forall \{i,j\} \in \Xi \times \Xi$) such that*

$$\begin{bmatrix} \Phi_{11} & \Phi'_{21} \\ \Phi_{21} & \Phi_{22} \end{bmatrix} > 0, \quad \forall \{i,j\} \in (\Xi \times \Xi), \tag{40}$$

where

$$\Phi_{11} = \begin{bmatrix} X + X' - S_i & * & * & * & * & * \\ I + W - P_i & Y + Y' - R_i & * & * & * & * \\ 0 & 0 & \gamma I & * & * & * \\ A_i X + B_i L_i & A_i + B_i H_i C_i & B^w_i + B_i H_i D^{yw}_i & S_j & * & * \\ Q_i & Y A_i + E_i C_i & E_i D^{yw}_i + Y B^w_i & P_j & R_j & * \\ C^z_i X + D^z_i L_i & C^z_i + D^z_i H_i C_i & D^{zw}_i + D^z_i H_i D^{yw}_i & 0 & 0 & \gamma I \end{bmatrix},$$

$$\Phi_{21} = \begin{bmatrix} N_i^A X & N_i^A & 0 & 0 & 0 & 0 \\ N_i^B L_i & N_i^B H_i C_i & 0 & 0 & 0 & 0 \\ 0 & 0 & N_i^B H_i D_i^{yw} & 0 & 0 & 0 \\ 0 & 0 & 0 & (M_i^A)' & (M_i^A)'Y' & 0 \\ 0 & 0 & 0 & (M_i^B)' & (M_i^B)'Y' & 0 \\ 0 & 0 & 0 & (M_i^B)' & (M_i^B)'Y' & 0 \end{bmatrix},$$

$$\Phi_{22} = diag\left[\varepsilon_{ij}I \ \varepsilon_{ij}I \ \varepsilon_{ij}I \ \varepsilon_{ij}^{-1}I \ \varepsilon_{ij}^{-1}I\right],$$

then the switched dynamic output feedback control given by (23) with (28) ensures that the closed loop system (37) has γ-performance for any uncertainty described by (7) and (8).

Proof By Theorem 2, if there exist symmetric matrices $\tilde{S}_i$ ($\forall i \in \Xi$), matrix $\tilde{G}$, and real number $\varepsilon_{ij} > 0$ ($\forall \{i,j\} \in \Xi \times \Xi$) such that

$$\begin{bmatrix} \tilde{G}+\tilde{G}-\tilde{S}_i & * & * & * & * & * \\ 0 & \gamma I & * & * & * & * \\ \tilde{A}_i\tilde{G} & \tilde{B}_i & \tilde{S}_j - \varepsilon_{ij}[\tilde{M}_i^A(\tilde{M}_i^A)' - \tilde{M}_i^B(\tilde{M}_i^B)'] & * & * & * \\ \tilde{C}_i\tilde{G} & \tilde{D}_i & 0 & \gamma I & * & * \\ \tilde{N}_i^A\tilde{G} & 0 & 0 & 0 & \varepsilon_{ij}I & * \\ 0 & \tilde{N}_i^B & 0 & 0 & 0 & \varepsilon_{ij}I \end{bmatrix} > 0, \quad (41)$$

$\forall \{i,j\} \in \Xi \times \Xi$, then the closed loop system (37) has γ-performance for any uncertainty described by (7) and (8). By Schur complement theorem, (41) is equivalent to

$$\begin{bmatrix} \tilde{G}+\tilde{G}-\tilde{S}_i & * & * & * & * & * & * & * \\ 0 & \gamma I & * & * & * & * & * & * \\ \tilde{A}_i\tilde{G} & \tilde{B}_i & \tilde{S}_j & * & * & * & * & * \\ \tilde{C}_i\tilde{G} & \tilde{D}_i & 0 & \gamma I & * & * & * & * \\ \tilde{N}_i^A\tilde{G} & 0 & 0 & 0 & \varepsilon_{ij}I & * & * & * \\ 0 & \tilde{N}_i^B & 0 & 0 & 0 & \varepsilon_{ij}I & * & * \\ 0 & 0 & (\tilde{M}_i^A)' & 0 & 0 & 0 & \varepsilon_{ij}^{-1}I & * \\ 0 & 0 & (\tilde{M}_i^B)' & 0 & 0 & 0 & 0 & \varepsilon_{ij}^{-1}I \end{bmatrix} > 0 \quad (42)$$

$\forall \{i,j\} \in \Xi \times \Xi$. In order to transform the condition (43) into a tractable condition, the similar method of change of variables defined in (36) can be used again. Multiplying (43) on the left by $\tilde{T} = diag\left[T\ I\ T\ I\ I\ I\ I\ I\right]$ and on the right by $\tilde{T}'$, the condition (40) can be gained. □

Remark 2 If the real numbers $\varepsilon_{ij} > 0$ ($\forall \{i,j\} \in \Xi \times \Xi$) are given in advance, the inequalities (27) and (40) are linear with respect to other unknown matrices variables, and which can be checked with efficient and reliable numerical algorithms available.

Remark 3 Switch from subsystem i to subsystem j is described by pair (i,j). Under arbitrary switching, the proposed conditions have to be satisfied for all ordered pairs (i,j). However, if not all switch are allowed, the proposed condition can be mortified and satisfied only for all possible ordered pairs, moreover the calculation burden and conservativeness of the results can be reduced.

5 Illustrative Examples

In this section, two examples are presented to illustrate the proposed synthesis method above.

Example 1 Consider the uncertain switched system (16) composed of two subsystems as follows

$$\bar{A}_1 = \begin{bmatrix} 1.8 + \Delta_{11}^A(k) & -1 \\ 1 & -2 + \Delta_{12}^A(k) \end{bmatrix}, \quad \bar{B}_1 = \begin{bmatrix} -1.2 + \Delta_{11}^B(k) \\ 0.8 + \Delta_{12}^B(k) \end{bmatrix},$$

$$\bar{A}_2 = \begin{bmatrix} -1.4 & 0.5 \\ 0 & 1.6 + \Delta_{21}^A(k) \end{bmatrix}, \quad \bar{B}_2 = \begin{bmatrix} 1.2 \\ -1.3 \end{bmatrix},$$

$$B_i^w = \begin{bmatrix} 1\,0 \\ 0\,1 \end{bmatrix}, \quad C_i^z = \begin{bmatrix} 1\,0 \\ 0\,1 \\ 0\,0 \end{bmatrix}, \quad D_i^z = \begin{bmatrix} 0 \\ 0 \\ 1 \end{bmatrix}, \quad D_i^{zw} = \begin{bmatrix} 0\,0 \\ 0\,0 \\ 0\,0 \end{bmatrix}, \quad i = 1, 2 \tag{43}$$

where $|\Delta_{11}^A(k)| \leq 0.3$, $|\Delta_{12}^A(k)| \leq 0.1$, $|\Delta_{21}^A(k)| \leq 0.4$ and $|\Delta_{1j}^B(k)| \leq 0.2$ $(j = 1, 2)$. Converting the uncertain matrices $\bar{A}_i$ and $\bar{B}_i$ into the forms as (7) and (8), we get $M_1^A = diag\,[0.3\ 0.1]$, $N_1^A = diag\,[1\ 1]$, $M_1^B = [0.2\ 0.2]'$, $N_1^B = [1]$, $M_2^A = [0\ 0.4]'$, $N_2^A = [0\ 1]$, $M_2^B = [0\ 0]'$ and $N_2^B = [0]$. It should be noticed that the two subsystems are unstable. First, the design of the switched state feedback controller stabilizing the system (43) is considered. By the LMI toolbox for Matlab 6.5, the LMIs (18) is feasible for system (43), and a stabilizing state feedback controller is

$$K_1 = [0.3242\ 0.5881], \quad K_2 = [0.4529\ 0.5809].$$

Secondly, the design of the γ-performance switched state feedback controller for the system (43) is considered. We minimize γ subject to (21) via LMI toolbox. The optimal value of γ is 30.6382 and the corresponding γ-performance state feedback controller is

$$K_1 = [0.3456\ 0.5659], \quad K_2 = [0.4634\ 0.5750].$$

Example 2 Consider the uncertain switched systems (22) composed of two subsystems as follows

$$\bar{A}_1 = \begin{bmatrix} 1.3 + \Delta_{11}^A(k) & 0.2 \\ 0 + \Delta_{12}^A(k) & -1.2 \end{bmatrix}, \quad \bar{B}_1 = \begin{bmatrix} -2 + \Delta_1^B(k) \\ 1.5 \end{bmatrix}, \quad C_1 = [2\ 1],$$

$$\bar{A}_2 = \begin{bmatrix} -1.4 & 0.2 + \Delta_{21}^A(k) \\ 0 & 1.1 + \Delta_{22}^A(k) \end{bmatrix}, \quad \bar{B}_2 = \begin{bmatrix} 1.6 \\ -2.1 + \Delta_2^B(k) \end{bmatrix}, \quad C_2 = \begin{bmatrix} 1.5 & 1.2 \end{bmatrix},$$

$$B_i^w = \begin{bmatrix} 1 & 0 \\ 0 & 1 \end{bmatrix}, \quad C_i^z = \begin{bmatrix} 1 & 0 \\ 0 & 1 \\ 0 & 0 \end{bmatrix}, \quad D_i^z = \begin{bmatrix} 0 \\ 0 \\ 1 \end{bmatrix},$$

$$D_i^{zw} = \begin{bmatrix} 0 & 0 \\ 0 & 0 \\ 0 & 0 \end{bmatrix}, \quad D_i^{yw} = \begin{bmatrix} 0 & 0 \end{bmatrix}, \quad i = 1, 2 \tag{44}$$

where $|\Delta_{ij}^A(k)| \le 0.5$ $(i, j = 1, 2)$, $|\Delta_1^B(k)| \le 0.2$, and $|\Delta_2^B(k)| \le 0.3$. Converting the uncertain matrices $\bar{A}_i$ and $\bar{B}_i$ into the forms as (6) and (7), we get the corresponding coefficient matrices $M_1^A = \begin{bmatrix} 0.5 & 0.5 \end{bmatrix}'$, $N_1^A = \begin{bmatrix} 1 & 0 \end{bmatrix}$, $N_1^B = \begin{bmatrix} 1 \end{bmatrix}$, $M_1^B = \begin{bmatrix} 0 & 0.2 \end{bmatrix}'$, $M_2^A = \begin{bmatrix} 0.5 & 0 \end{bmatrix}'$, $N_2^A = \begin{bmatrix} 1 & 1 \end{bmatrix}$, $M_2^B = \begin{bmatrix} 0 & 0.4 \end{bmatrix}'$ and $N_2^B = \begin{bmatrix} 1 \end{bmatrix}$. First, the design of stabilizing switched DOF controller is considered. Given $\varepsilon_{ij} = 1, (i, j = 1, 2)$, a DOF controller stabilizing system (45) is obtained as follows:

$$A_1^c = \begin{bmatrix} 0.0685 & -0.2665 \\ 0.2878 & -0.0361 \end{bmatrix}, \quad B_1^c = \begin{bmatrix} 0.0674 \\ -0.1183 \end{bmatrix},$$

$$C_1^c = \begin{bmatrix} -0.0124 & -0.0333 \end{bmatrix}, \quad D_1^c = \begin{bmatrix} 0.2464 \end{bmatrix},$$

$$A_2^c = \begin{bmatrix} -0.2505 & 0.3398 \\ -0.3317 & 0.1384 \end{bmatrix}, \quad B_2^c = \begin{bmatrix} -0.1101 \\ 0.0449 \end{bmatrix},$$

$$C_2^c = \begin{bmatrix} 0.0130 & -0.0040 \end{bmatrix}, \quad D_2^c = \begin{bmatrix} 0.2191 \end{bmatrix}.$$

Secondly, the design of the γ-performance switched DOF controller for system (45) is considered. Setting real numbers $\varepsilon_{ij} = 1, (i, j = 1, 2)$, we minimize γ subject to (40) via LMI toolbox. The optimal value of γ is 45.7945 and the derived γ-performance switched DOF controller is as follows

$$A_1^c = \begin{bmatrix} 1.9002 & 0.6946 \\ -3.1042 & -1.1345 \end{bmatrix}, \quad B_1^c = \begin{bmatrix} -0.1668 \\ -0.3305 \end{bmatrix},$$

$$C_1^c = \begin{bmatrix} -0.0585 & -0.0214 \end{bmatrix}, \quad D_1^c = \begin{bmatrix} 0.2748 \end{bmatrix}$$

$$A_2^c = \begin{bmatrix} -0.3223 & -0.1188 \\ -1.3887 & -0.5053 \end{bmatrix}, \quad B_2^c = \begin{bmatrix} 0.1160 \\ 0.5511 \end{bmatrix},$$

$$C_2^c = \begin{bmatrix} -0.0192 & -0.0070 \end{bmatrix}, \quad D_2^c = \begin{bmatrix} 0.2339 \end{bmatrix}. \tag{45}$$

Given $\Delta_{ij}^A(k) = 0.5$ $(i, j = 1, 2), \Delta_1^B(k) = 0.2, \Delta_2^B(k) = 0.3$, and initial state $x(0) = \begin{bmatrix} -6 & 4 \end{bmatrix}$, the simulation result under the random switching signal is shown in Fig.1 and Fig.2.

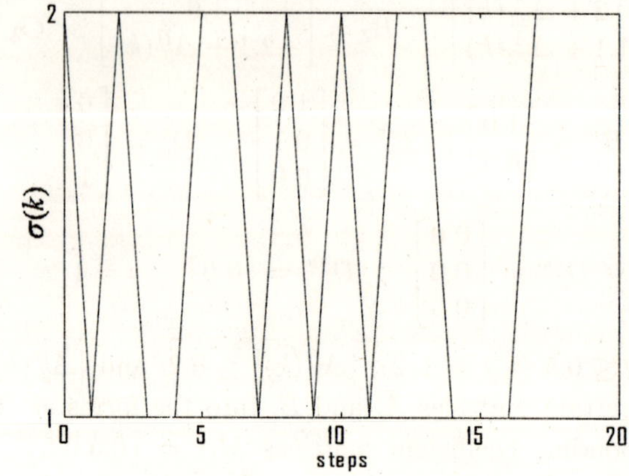

Fig.1. random switching signal

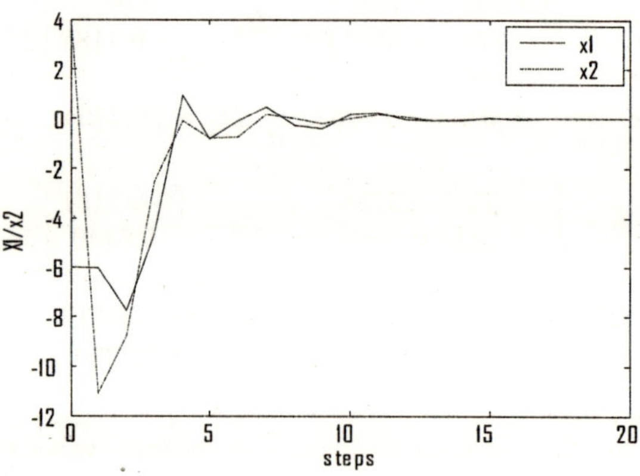

Fig.2. State responses of system (44) with controller (45)

6 Conclusion

In this paper the robust control design for uncertain discrete-time switched system is studied. For the switched systems with norm-bounded time-varying uncertainty, sufficient conditions are presented to realize the stabilization and γ-performance via state feedback and dynamic output feedback under arbitrary switching. In the case of state feedback, the conditions are expressed in terms of LMIs, and in the case of dynamic output feedback, the conditions can be dealt with as LMIs if some real numbers are given in advance. Thus these results can be applied easily by efficient algorithms for solving LMIs.

References

1. S. Boyd, L. El Ghaoui, E. Feron, et al,. *Linear Matrix Inequalities in System and Control Theory.* Philadelphia: SIAM, 1994.
2. M. S. Branicky. Multiple Lyapunov functions and other analysis tools for switched and hybrid systems. *IEEE Transactions on Automatic Control*, 43:475-482, 1998.
3. M. S. Branicky. Stability of switched and hybrid systems. *Proceedings of the 33rd IEEE Conference on Decision and Control*, pages 3498-3503, 1994.
4. D. Cheng. Stabilization of planar switched systems *Systems and Control Letters*, 51:79-88, 2004.
5. J. Daafouz, P. Riedinger, and C. Iung Stability analysis and control synthesis for switched systems: a switched Lyapunov function approach. *IEEE Trans on Automatic Control*, 47:1883-1887, 2002.
6. J. Daafouz, and J. Bernussou. Robust dynamic output feedback control for switched system. *Proceeding of the 41st IEEE Conference on Decision and Control*, pages 4389-4394, 2002.
7. R. A. Decarlo, M. S. Branicky, and S. Pettersson. Perspectives and results on the stability and stabilizability of hybrid systems. *Proceedings of the IEEE*, 88:1069-1083, 2000.
8. M. Johansson and A. Rantzer. Computation of piecewise quadratic Lyapunov functions for hybrid systems. *IEEE Transactions on Automatic Control* 43:555-559, 1998.
9. P. P. Khargonekar, I. R. Petersen, and K. Zhou. Robust stabilization of uncertain linear systems: quadratic stabilizability and H∞ control theory. *IEEE Transactions on Automatic Control*, 35:356-361, 1990.
10. D. Liberzon, and A. S. Morse. Basic problems in stability and design of switched systems. *IEEE Control Systems Magazine*, 19:59-70, 1999.
11. S. Pettersson and B. Lennartson. Stability and robustness for hybrid systems. *Proceedings of the 35th IEEE Conference on Decision and Control*, pages 1202-1207, 1996.
12. H. Ye, A. N. Michel, and L. Hou. Stability theory for hybrid dynamical systems. *IEEE Transactions on Automatic Control*, 43:461-474, 1998.
13. G. S. Zhai, H. Lin, and P. J. Antsaklis. Quadratic stabilizability of switched linear systems with polytopic uncertainties. *International Journal of Control*, 76:747-753, 2003.

Part 3

Adaptive Control and Estimation

A New Framework For Adaptive Control

Adaptive Robust Control Approach

Kang-Zhi Liu

Department of Electrical and Mechanical Engineering, Chiba University, 1-33 Yayoi-cho, Inage-ku, Chiba 263-8255, Japan kzliu@faculty.chiba-u.jp

1 Introduction

As a powerful technology of dealing with parametric uncertainty in feedback system design, adaptive control has been studied extensively since 1950's. The main stream of approaches is transfer function based and the fundamental scheme in the proof of stability is Lyapunov stability/hyperstability theories[1, 2, 9]. This classic approach has achieved many fruitful results and has a good record of industrial applications[1]. However, this classic approach also embodies some inherent limitations such as design complexity stemming from high relative degree and the minimal phase condition coming from model-matching.

Meanwhile, the state space approach was explored since late 1970's as is represented by the research of Kreisselmier. In his landmark paper[4], Kreissekmier proposed a smart structure of full-order observer for SISO uncertain parametric systems and showed that the state of uncertain systems could be reconstructed by a set of filters and an adaptive law for uncertain parameters. This observer and the related state regulation problem were investigated in a series of papers[4, 5, 6]. Unfortunately, only local results on state regulation problem were obtained except for that the PE requirement is put on the system input.

This observer structure was extended to a class of SISO nonlinear systems in the adaptive observer backstepping[7] in which global boundedness of states and asymptotic tracking of output were achieved.

However, to the knowledge of the author no global exponential adaptive stabilization method has been established for linear parametric systems. This is possibly because all approaches up to now are Lyapunov based. So in this paper, a totally different approach is explored. This approach is motivated by the ISS small-gain theory of [3] and Kreisselmier observer. In the present approach, the adaptive system has a structure of constant state feedback, Kreisselmier observer and adaptive law. The constant state feedback robustly stabilizes the uncertain parametric system, Kreisselmier observer reconstructs the state of plant and the adaptive law improves the parameter estimate. An attractive feature of this approach is that the designs of these ingredients are independent. They are unified by a newly developed ISS small-gain condition to yield global exponential stability of the states of plant and observer, with explicit convergence rate. Roughly speaking, what is achieved by robust constant state feedback may be recovered by this adaptive control method. No requirement is necessary other than robust controllability and observability.

2 Preliminaries

In this section, Gronwell-Bellman Lemma is extended first. Then an upper bound for solutions to 2 coupled integral inequalities is derived based on it and it is shown that the exponential convergence is guaranteed by a certain small-gain condition. This result forms the foundation for the stability analysis of cascaded systems and the adaptive design of this paper.

2.1 Extended Gronwell-Bellman Lemma

Lemma 1. *Let $\eta : [a,b] \mapsto \mathbb{R}$ be continuous and $c,\ \lambda > 0$. If a function $y : [a,b] \mapsto \mathbb{R}$ satisfies*

$$y(t) \le \eta(t) + c\int_a^t y(s)ds - ce^{-\lambda t}\int_a^t e^{\lambda s}y(s)ds \tag{1}$$

for $a \le t \le b$, then on the same interval there holds

$$y(t) \le \eta(t) + g(t) * \eta(t) \tag{2}$$

in which $g(t)$ is the nonnegative function below

$$g(t) = \frac{c\lambda(e^{-(\lambda-\sqrt{\lambda^2+4c\lambda})t/2} - e^{-(\lambda+\sqrt{\lambda^2+4c\lambda})t/2})}{\sqrt{\lambda^2+4c\lambda}}.$$

Proof. This lemma can be proved by solving a 2nd order ODE about

$$z(t) = c\int_a^t y(s)ds - ce^{-\lambda t}\int_a^t e^{\lambda s}y(s)ds.$$

□

2.2 Coupled Integral Inequalities

Theorem 1. *Suppose 2 functions $x, y : [t_0, \infty) \mapsto \mathbb{R}$ satisfy*

$$x(t) \leq k_1 e^{-\lambda_1(t-t_0)} + a\int_{t_0}^{t} e^{-\lambda_1(t-\tau)}y(\tau)d\tau \tag{3}$$

$$y(t) \leq k_2 e^{-\lambda_2(t-t_0)} + b\int_{t_0}^{t} e^{-\lambda_2(t-\tau)}x(\tau)d\tau \tag{4}$$

in which $k_1, k_2, a, b > 0$ *and* $\lambda_2 > \lambda_1 > 0$. *If*

$$g := \frac{ab}{\lambda_1\lambda_2} < 1 \tag{5}$$

is true, then both $x(t)$ *and* $y(t)$ *are exponentially convergent functions with a convergence rate no less than*

$$(\lambda_1 + \lambda_2 - \sqrt{(\lambda_2 - \lambda_1)^2 + 4ab})/2 > 0.$$

Proof. Let $\lambda = \lambda_2 - \lambda_1 > 0$ and define $z(t) = e^{\lambda_1(t-t_0)}x(t)$. Substitution of (4) into (3) gives

$$z(t) \leq \eta(t) + c\int_{t_0}^{t} z(s)ds - ce^{-\lambda t}\int_{t_0}^{t} e^{\lambda s}z(s)ds \tag{6}$$

in which $c = ab/\lambda$, $\eta(t) = k_1 + ak_2(1 - e^{-\lambda(t-t_0)})/\lambda$. Note $k_1 + ak_2/\lambda \neq 0$ generically. It is easy to see that $z(t)$ is bounded from above by an exponential function with a rate

$$\beta = \frac{-\lambda + \sqrt{\lambda^2 + 4c\lambda}}{2} = \frac{-\lambda + \sqrt{\lambda^2 + 4ab}}{2}.$$

Hence, $x(t)$ is bounded from above by an exponential function with a rate

$$\beta - \lambda_1 = \frac{-(\lambda_1 + \lambda_2) + \sqrt{(\lambda_2 - \lambda_1)^2 + 4ab}}{2} \tag{7}$$

which is negative iff condition (5) holds. The convergence rate of $y(t)$ follows from this conclusion and (4). □

Since the 2nd term of (3) is less than $a/\lambda_1 \times \max_{\tau\in[t_0,t]} y(\tau)$, a/λ_1 can be intepreted as an ISS gain for $x(t)$. Similarly, b/λ_2 is an ISS gain for $y(t)$. Therefore, (5) is an ISS small-gain condition. The upper bound for $x(t)$, $y(t)$ may be computed according to Lemma 1 if necessary.

3 Stability of Cascaded Systems

Let $\theta = [\theta_1 \ \cdots \ \theta_r]^T$ be an uncertain time-invariant parameter vector taking value in a bounded and positive polyhedral set S hereafter:

$$S = \{\theta | \theta_i \in [\underline{\theta}_i,\ \overline{\theta}_i],\ \underline{\theta}_i > 0,\ i = 1, \ldots, r\}. \tag{8}$$

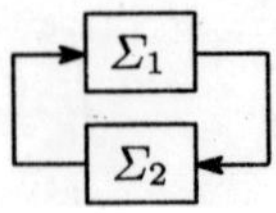

Fig. 1. Cascaded system

Consider the cascaded systems Σ_1 and Σ_2

$$\Sigma_1 : \dot{x}_1 = A_1(\theta)x_1 + B_1(\theta)y_2, y_1 = C_1(\theta)x_1 \tag{9}$$
$$\Sigma_2 : \dot{x}_2 = A_2(\theta)x_2 + B_2(\theta)y_1, y_2 = C_2(\theta)x_2. \tag{10}$$

Theorem 2. *Assume that* $\exists\lambda_1, \lambda_2, k_1, k_2, a, b > 0$ *such that for all* $\theta \in S$

$$\|e^{A_1 t}\| \le k_1 e^{-\lambda_1 t}, \ \|e^{A_1 t}B_1 C_2\| \le ae^{-\lambda_1 t} \tag{11}$$
$$\|e^{A_2 t}\| \le k_2 e^{-\lambda_2 t}, \ \|e^{A_2 t}B_2 C_1\| \le be^{-\lambda_2 t} \tag{12}$$

hold and $\lambda_2 > \lambda_1$. *Then the closed system is exponetially stable for all* $\theta \in S$ *if*

$$\frac{ab}{\lambda_1\lambda_2} < 1 \tag{13}$$

holds. The convergence rate is no less than

$$(\lambda_1 + \lambda_2 - \sqrt{(\lambda_2 - \lambda_1)^2 + 4ab})/2.$$

Proof. Obviously,

$$\|x_1(t)\| \le k_1 e^{-\lambda_1 t}\|x_1(t_0)\| + a\int_{t_0}^{t} e^{-\lambda_1(t-\tau)}\|x_2(\tau)\|d\tau$$

and

$$\|x_2(t)\| \le k_2 e^{-\lambda_2 t}\|x_2(t_0)\| + b\int_{t_0}^{t} e^{-\lambda_2(t-\tau)}\|x_1(\tau)\|d\tau$$

hold. Then the conlusion is immediately implied by Theorem 1. □

4 Robust State Feedback Design

Let the MIMO uncertain parametric plant be given by

$$\dot{x} = A(\theta)x + B(\theta)u \tag{14}$$
$$y = Cx \tag{15}$$

in which $x \in \mathbb{R}^n$, $u \in \mathbb{R}^p$, $y \in \mathbb{R}^q$ and $A(\theta)$, $B(\theta)$ are matrices affine in $\theta \in S$. It is assumed that

Assumption 1 $(A(\theta),\ B(\theta))$ is controllable for all $\theta \in S$ (Robust controllability)

The design problem in this section is to find a *constant feedback matrix* F such that

$$u = Fx \tag{16}$$

stabilizes system (14) exponentially with a convergence rate $\lambda_1 > 0$. That is, in the closed loop system

$$\dot{x} = (A(\theta) + B(\theta)F)x := A_c(\theta)x \tag{17}$$

the largest real part among all eigenvalues of matrix $A_c(\theta)$ satisfies

$$\max_i \mathrm{Re}\lambda_i(A(\theta)) \leq -\lambda_1,\ \forall\theta \in S. \tag{18}$$

This design problem is well studied. For example, [10] showed that for a certain class of systems this stabilization can be achieved no matter how large the uncertain parameters vary. And in the context of quadratic stabilization, F can be designed by solving some LMIs. For systems with less than 2 uncertain parameters, one may even use the root locus technique in the design. So this issue will not be pursued further in this paper. It is just assumed that a solution F has been found. Hence there exist $k_1, a > 0$ such that $\forall\theta \in S$

$$\|e^{A_c(\theta)t}\| \leq k_1 e^{-\lambda_1 t},\ \|e^{A_c(\theta)t}B(\theta)F\| \leq ae^{-\lambda_1 t} \tag{19}$$

hold.

When the state is estimated by some sort of observer and the estimate is $\hat{x}$, then the state feedback changes into $u = F\hat{x}$ and the closed loop becomes

$$\dot{x} = A_c(\theta)x + B(\theta)Fe \tag{20}$$

in which e denotes the estimation error of state

$$e = \hat{x} - x. \tag{21}$$

Therefore

$$\|x(t)\| \leq k_1 e^{-\lambda_1 t}\|x_1(t_0)\| + a\int_{t_0}^{t} e^{-\lambda_1(t-\tau)}\|e(\tau)\|d\tau \tag{22}$$

holds.

5 Kreisselmier Observer

Kreisselmier observer is used to reconstruct the state from input and measured output. The assumption made here is

Assumption 2 $(C, A(\theta))$ is observable for all $\theta \in S$ (Robust observability)

The full-order case and the minimal-order case are discussed separately below.

5.1 Full-Order Observer

In the following full-order observer

$$\dot{\hat{x}} = (A(\theta) + L(\theta)C)\hat{x} + B(\theta)u - L(\theta)y, \tag{23}$$

it is assumed that there is an observer gain $L(\theta)$ *affine in* θ such that

$$T := A(\theta) + L(\theta)C \tag{24}$$

is a constant matrix independent of θ and its eigenvalue can be assigned arbitrarily. This is possible at least in the so-called observer canonical form[8] in which

$$\begin{aligned} A &= A_0 + K_0(\theta)DC_0, \ C = DC_0 \\ A_0 &= \operatorname{diag}(A_{11} \ \cdots \ A_{qq}), \ C_0 = \operatorname{diag}(c_{11} \ \cdots \ c_{qq}) \\ A_{ii} &= \begin{bmatrix} 0 & 0 \\ I_{\sigma_i - 1} & 0 \end{bmatrix}, \ c_{ii} = \begin{bmatrix} 0 \cdots 0 \ 1 \end{bmatrix} \in \mathbb{R}^{1 \times \sigma_i} \end{aligned}$$

and D is a lower triangular matrix with all diagonal entries equal to 1, $K_0(\theta)$ is a nonzero matrix. Here σ_i $(i = 1, \ldots, q)$ is the so-called observability index.

Assumption 3 D is independent of θ and $K_0(\theta)$ is affine in θ.

This assumption is satisfied for almost all physical systems. Subject to this assumption

$$L(\theta) = -K_0(\theta) + L^* D^{-1} \tag{25}$$

gives the desired matrix because the eigenvalues of matrix $T = A_0 + L^*C_0$ can be arbitrarily assigned by suitable L^*.

Next express $L(\theta), B(\theta)$ as

$$L(\theta) = L_0 + \sum_{i=1}^{r} \theta_i L_i, \ B(\theta) = B_0 + \sum_{i=1}^{r} \theta_i B_i \tag{26}$$

and prepare the following filters $(i = 0, 1 \ldots, r)$

$$\dot{\zeta}_i = T\zeta_i + B_i u + L_i y, \ \zeta_i(0) = 0. \tag{27}$$

Then

$$x = \zeta_0 + \theta_1\zeta_1 + \cdots + \theta_r\zeta_r - \epsilon \tag{28}$$

holds in which ϵ is the estimation error due to nonzero initial state $x(0)$ and satisfies

$$\dot{\epsilon} = T\epsilon, \ \epsilon(0) = -x(0). \tag{29}$$

Let the estimate of parameter θ_i be denoted by $\hat{\theta}_i$, then the estimated state is given by

$$\hat{x} = \zeta_0 + \hat{\theta}_1\zeta_1 + \cdots + \hat{\theta}_r\zeta_r \tag{30}$$

and the estimation error turns out to be

$$e = \tilde{\theta}_1\zeta_1 + \cdots + \tilde{\theta}_r\zeta_r + \epsilon,\ \tilde{\theta}_i = \hat{\theta}_i - \theta_i. \tag{31}$$

Here $\tilde{\theta}_i$ denotes the estimation error of parameter. Further, in order to analyze the norm bound of e, define the following notations for simplicity

$$\begin{aligned} &\zeta = [\zeta_1^T \ \cdots \ \zeta_r^T]^T \\ &\tilde{\Theta} = [\tilde{\theta}_1 I_n \ \cdots \ \tilde{\theta}_r I_n],\ J = \text{diag}(T \ \cdots \ T) \\ &G = [B_1^T \ \cdots \ B_r^T]^T,\ H = [L_1^T \ \cdots \ L_r^T]^T. \end{aligned} \tag{32}$$

Then e can be written as

$$e = \tilde{\Theta}\zeta + \epsilon. \tag{33}$$

Since $y = Cx$ and $u = Fx + Fe$,

$$\dot{\zeta} = J\zeta + (HC + GF)x + GFe \tag{34}$$

holds. Now due to the stability of matrix T, there exist $\lambda_2(>\lambda_1), k_2, b, c > 0$ such that

$$\begin{aligned} &\|e^{Tt}\| \le k_2 e^{-\lambda_2 t},\ \|e^{Jt}(HC+GF)\| \le be^{-\lambda_2 t} \\ &\|e^{Jt}GF\| \le ce^{-\lambda_2 t} \end{aligned} \tag{35}$$

hold. Further, define a constant as

$$\mu = c\|\tilde{\Theta}\|_{[t_0,\infty)} \ge 0 \tag{36}$$

in which $\|\tilde{\Theta}\|_{[t_0,\infty)}$ is a norm defined below

$$\|\tilde{\Theta}\|_{[t_0,\infty)} = \max_{t\in[t_0,\infty)} \sqrt{\sum_{i=1}^{r} \tilde{\theta}_i(t)^2}. \tag{37}$$

By using

$$\|e^{Jt}\| = \|e^{Tt}\|,\ \|\epsilon(t)\| = \|e^{Tt}\epsilon(t_0)\| \le k_2 e^{\lambda_2(t-t_0)}\|\epsilon(t_0)\|$$

and Gronwell-Bellman Lemma, the following norm bound is derived from the response of ζ:

$$\|\zeta(t)\| \le k_2\|\zeta(t_0)\|e^{-(\lambda_2-\mu)(t-t_0)} + b\int_{t_0}^{t} e^{-(\lambda_2-\mu)(t-\tau)}\|x(\tau)\|d\tau. \tag{38}$$

Theorem 3. *The closed loop system composed of robust constant state feedback and full-order Kreisselmier observer is exponentially stable for all $\theta \in S$ and $t > t_0$ if*

$$\frac{ab\|\tilde{\Theta}\|_{[t_0,\infty)}}{\lambda_1(\lambda_2-\mu)} < 1 \tag{39}$$

and the covergence rate is at least no less than

$$(\lambda_1+\lambda_2-\mu-\sqrt{(\lambda_2-\lambda_1-\mu)^2+4ab\|\tilde{\Theta}\|_{[t_0,\infty)}})/2.$$

Proof. The exponential stability of x, ζ follows from Theorem 1 immediately if one notes that x has a norm bound

$$\|x(t)\| \le \left(k_1\|x(t_0)\| + \frac{ak_2}{\lambda_2-\lambda_1}\|\epsilon(t_0)\|\right)e^{-\lambda_1(t-t_0)} + a\|\tilde{\Theta}\|_{[t_0,\infty)}\int_{t_0}^{t} e^{-\lambda_1(t-\tau)}\|\zeta(\tau)\|d\tau \tag{40}$$

in this case. Since $e = \tilde{\Theta}\zeta + \epsilon$ in which $\tilde{\Theta}$ is bounded (guaranteed by projection algorithm) and ϵ is exponentially stable with a convergence rate λ_2, e is expoenentially stable with the convergence rate given in the theorem. This completes the proof. □

5.2 Minimal-Order Observer

Without loss of generality, it can be assumed that

$$C = [I_q \;\; 0]. \tag{41}$$

Further, matrices $A(\theta)$ and $B(\theta)$ are decomposed as

$$A(\theta) = \begin{bmatrix} A_{11}(\theta) & A_{12}(\theta) \\ A_{21}(\theta) & A_{22}(\theta) \end{bmatrix}, \; B(\theta) = \begin{bmatrix} B_1(\theta) \\ B_2(\theta) \end{bmatrix} \tag{42}$$

accordingly. Due to the robust observability of $(C, A(\theta))$, the pair $(A_{12}(\theta), A_{22}(\theta))$ is also robustly observable. So it can be assumed that $(A_{12}(\theta), A_{22}(\theta))$ is in the observer canonical form via similarity transformation on x_2. It is assumed that

A4 A_{11}, A_{12}, B_1 are constant matrices and $A_{21}(\theta)$, $B_2(\theta)$ are affine in θ.
A5 There exists a matrix $L(\theta)$ affine in θ such that

$$T = A_{22}(\theta) + L(\theta)A_{12} \tag{43}$$

is independent of θ and its eigenvalues can be assigned arbitrarily.

The minimal-order observer is constructed according to Gopinath's algorithm. As in the standard procedure, one calculates the following matrix

$$V = [0 \;\; I_{n-q}] + L[I_q \;\; 0] = [L \;\; I_{n-q}].$$

Then the minimal-order observer is given by

$$\dot{z} = Tz + VBu + (A_{21} - TL)y \tag{44}$$

and the state x is recovered as

$$x = \begin{bmatrix} C \\ V \end{bmatrix}^{-1} \begin{bmatrix} y \\ z \end{bmatrix} = \begin{bmatrix} y \\ z - Ly \end{bmatrix} \tag{45}$$

if the initial state $x_2(0)$ as well as the parameter θ are known and $z(0)$ is set as $x_2(0) + Ly(0)$.

Now express $L(\theta), A_{21}(\theta), B_2(\theta)$ as

$$L(\theta) = L_0 + \sum_{i=1}^{r} \theta_i L_i, \ B_2(\theta) = B_{20} + \sum_{i=1}^{r} B_{2i}$$

$$A_{21}(\theta) = A_{210} + \sum_{i=1}^{r} \theta_i A_{21i} \tag{46}$$

and prepare the following filters

$$\dot{\zeta}_i = T\zeta_i + (B_{2i} + L_0 B_1)u + (A_{21i} - TL_i)y \tag{47}$$
$$\zeta_i(0) = 0, \ i = 0, 1 \ldots, r.$$

Then

$$z = \zeta_0 + \sum_{i=1}^{r} \theta_i \zeta_i \tag{48}$$

holds. Let the estimate of parameter θ_i be denoted by $\hat{\theta}_i$. Then the estimate $\hat{x}_2$ is obtained as

$$\hat{x}_2 = \hat{z} - L(\hat{\theta})y = \zeta_0 + \sum_{i=1}^{r} \hat{\theta}_i \zeta_i - L(\hat{\theta})y. \tag{49}$$

and the estimation error

$$e = \hat{x}_2 - x_2$$

turns out to be

$$e = \hat{z} - z - [L(\hat{\theta}) - L(\theta)]y + \epsilon$$
$$= \sum_{i=1}^{r} \tilde{\theta}_i(\zeta_i - L_i y) + \epsilon. \tag{50}$$

Here $\tilde{\theta}_i$ denotes the estimation error of parameter $\tilde{\theta}_i = \hat{\theta}_i - \theta_i$ and ϵ is the estimation error due to nonzero initial state $x_2(0)$ and satisfies

$$\dot{\epsilon} = T\epsilon, \ \epsilon(0) = -x_2(0). \tag{51}$$

Further, in order to analyze the norm bound of e, let us look at the dynamics of

$$\bar{\zeta}_i = \zeta_i - L_i y,\ i = 1, \ldots, r. \tag{52}$$

Noting $\dot{y} = C\dot{x} = A_{11}y + A_{12}(\theta)x_2 + B_1 u$, some easy calculation gives

$$\dot{\bar{\zeta}}_i = T\bar{\zeta}_i + (A_{21i} - L_i A_{11})y - L_i A_{12}(\theta)x_2 + B_{2i}u.$$

Now let $F = [F_1 \ \ F_2]$ be a partition compatible with that of state x. Then $u = F\hat{x} = F_1 y + F_2 x_2 + F_2 e$ and this state equation reduces to

$$\dot{\bar{\zeta}}_i = T\bar{\zeta}_i + H_i x + G_i e \tag{53}$$

in which

$$\begin{aligned} G_i &= B_{2i}F_2 \\ H_i &= [A_{21i} - L_i A_{11} + B_{2i}F_2, \ \ B_{2i}F_2 - L_i A_{12}]. \end{aligned}$$

As in the full-order case, define the following notations for simplicity

$$\begin{aligned} \bar{\zeta} &= [\bar{\zeta}_1^T \ \ \cdots \ \ \bar{\zeta}_r^T]^T \\ \tilde{\Theta} &= [\tilde{\theta}_1 I_{n-q} \ \ \cdots \ \ \tilde{\theta}_r I_{n-q}],\ J = \text{diag}(T \ \ \cdots \ \ T) \\ G &= [G_1^T \ \ \cdots \ \ G_r^T]^T,\ H = [H_1^T \ \ \cdots \ \ H_r^T]^T. \end{aligned} \tag{54}$$

Then

$$e = \tilde{\Theta}\bar{\zeta} + \epsilon \tag{55}$$

$$\dot{\bar{\zeta}} = J\bar{\zeta} + Ge + Hx \tag{56}$$

hold.

Note in the present case $\hat{x} - x = [0 \ \ e^T]^T$, the closed loop of state feedback reduces to

$$\dot{x} = A_c(\theta)x + B(\theta)F_2 e. \tag{57}$$

Further, there exist $\lambda_1,\ k_1,\ a > 0$ and $\lambda_2,\ k_2,\ b > 0$ such that for all $\theta \in \Theta$

$$\begin{aligned} &\|e^{A_c(\theta)t}\| \le k_1 e^{-\lambda_1 t},\ \|e^{A_c(\theta)t}B(\theta)F_2\| \le ae^{-\lambda_1 t} \\ &\|e^{Tt}\| \le k_2 e^{-\lambda_2 t},\ \|e^{Jt}G\| \le ce^{-\lambda_2 t} \|e^{Jt}H\| \le be^{-\lambda_2 t} \end{aligned} \tag{58}$$

hold. Moreover, let constants μ and $\|\tilde{\Theta}\|_{[t_0,\infty)}$ be as defined by (36), (37) respectively. Then the theorem below holds whose proof is identical to that of Theorem 3.

Theorem 4. *The closed loop system composed of robust constant state feedback and minimal-order Kreisselmier observer is exponentially stable for all* $\theta \in S$ *and* $t > t_0$ *if*

$$\frac{ab\|\tilde{\Theta}\|_{[t_0,\infty)}}{\lambda_1(\lambda_2 - \mu)} < 1 \tag{59}$$

and the covergence rate is at least no less than

$$(\lambda_1 + \lambda_2 - \mu - \sqrt{(\lambda_2 - \lambda_1 - \mu)^2 + 4ab\|\tilde{\Theta}\|_{[t_0,\infty)}})/2.$$

6 Improving Convergence Rate by Parameter Adaptation

If the parameter error bound $\|\tilde{\Theta}\|_{[t_0,\infty)}$ is reduced by parameter adaptation, the convergence rate can be improved. So in this section, the design of adaptive laws is discussed.

Full-Order Observer

Define

$$\xi_i = C\zeta_i, \; i = 1, \ldots, r. \tag{60}$$

Also denote the estimated output by $\hat{y} = C\hat{x}$. Then the output estimation error $e_y = \hat{y} - y = C(\hat{x} - x)$ can be expressed as

$$e_y = \sum_{i=1}^{r} \tilde{\theta}_i \xi_i + C\epsilon. \tag{61}$$

This error signal is in the conventional linear form in parameter estimation error, hence any adaptation methods may be used to derive the adaptive law.

Minimal-Order Observer

In this case, a virual output error has to be introduced in order to activate adaptation because the output is not estimated. Such virtual output is generated in the following way. Note the state equation about y is

$$\dot{y} = A_{11}y + A_{12}x_2 + B_1 u.$$

Let $\hat{y}$ denote the (virtual) output generated from

$$\dot{\hat{y}} = A_{11}y + A_{12}\hat{x}_2 + B_1 u. \tag{62}$$

The corresponding output error is denoted by $e_y = \hat{y} - y$. Then

$$\dot{e}_y = A_{12}(\hat{x}_2 - x_2) = \sum_{i=1}^{r} \tilde{\theta}_i A_{12}\bar{\zeta}_i + A_{12}\epsilon \tag{63}$$

holds. Although $\dot{e}_y$ has the property required for error signal used in adaptation, it is not available. So in order to construct the adaptive law, a filtered output error signal is introduced

$$\xi(t) = \frac{s}{s+\alpha} e_y(t)$$

which has a dynamics given by

$$\dot{\xi} = -\alpha\xi + \dot{y}_y$$
$$= -\alpha\xi + \sum_{i=1}^{r} \tilde{\theta}_i A_{12}\bar{\zeta}_i + A_{12}\epsilon. \quad (64)$$

Then the adaptive law proposed for the minimal-order Kreisselmier observer is

$$\dot{\hat{\theta}}_i = -\gamma_i (A_{12}\bar{\zeta}_i)^T \xi, \ \gamma_i > 0, \ i = 1, \ldots, r. \quad (65)$$

7 Concluding Remarks

In this paper, a totally new adaptive approach has been proposed in the state space for linear parametric systems. The ingredients are robust constant state feedback, Kreisselmier observer and adaptive laws. The key in unifying these ingredients together to achieve global exponential stability is a newly developed ISS small-gain condition.

This approach has made the following breakthroughs as compared with the classical transfer function and Lyapunov based approach: (1) No minimal phase requirement (2) No design complexity due to high relative degree (3) Separated designs of robust state feedback, Kreisselmier observer and adative law (4) Global exponential stability guarantee for all states (5) It is an MIMO theory.

In this paper, only stabilization problem is treated. When the internal model is put into the loop, asymptotic reference tracking problem can be solved as a stabilization problem. However, it remains unclear how to extend this approach to handle performance problems as measured by input/output norm. Further, this approach has a fundamental limitation: it is not able to surpass the performance of robust constant state feedback! To get over this limitation, estimated parameter dependent state feedback needs to be investigated.

Finally, it is also an interesting theme to see whether this approach can be extended to deal with some classes of nonlinear systems.

References

1. K. J. Åström and B. Wittenmark (1995), *Adaptive control, 2nd version*, Addison Wesley.
2. P. Ioannou and J. Sun (1996), *Robust adaptive control*, Prentice-Hall.
3. Z. Jiang and I. M. Y. Mareels (1997), A small-gain control method for nonlinear cascaded systems with dynamic uncertainties, *IEEE Trans. on Automatic Control*, Vol. 42, No. 3, 292-308.
4. G. Kreisselmier (1977), Adaptive observers with exponential rate of convergence, *IEEE Trans. on Automatic Control*, Vol. 22, No. 1, 2-8.

5. G. Kreisselmier (1980), Adaptive control via adaptive observation and asymptotic feedback matrix synthesis, *IEEE Trans. on Automatic Control*, Vol. 25, No.4, 717-722.
6. G. Kreisselmier (1982), On adaptive state regulation, *IEEE Trans. on Automatic Control*, Vol. 27, No. 1, 3-17.
7. M. Krstić and et al. (1995), *Nonlinear and adaptive control design*, Wiley-Interscience.
8. N. Suda (1993), *Linear systems theory (in Japanese)*, Asakura-Shoten.
9. T. Suzuki (2001), *Adaptive control (in Japanese)*, Corona Publishing Co., Ltd..
10. K. Wei (1994), Stabilization of linear time-invariant inteval systems via constant state feedback control, *IEEE Trans. on Automatic Control*, Vol. 39, No. 1, 22-32.

Adaptive Control of Nonholonomic Systems

Inverse Optimality and Application to Mobile Robot

Yoshihiko Miyasato

The Institute of Statistical Mathematics, Tokyo 106-8569, Japan
miyasato@ism.ac.jp

1 Introduction

Nonholonomic systems are known as dynamical systems with nonintegrable constraints, and cannot be stabilized via continuous feedbacks. Recently, control problems of nonholonomic systems attract much attentions in the fields of motion control of several mechanical systems such as mobile robots, space robots, and underactuated robots. Among those, certain nonholonomic systems are described in the forms of chained systems, and adaptive and nonlinear controller designs for chained forms have been studied extensively by several researchers [1][2]. In the present paper, we propose design methods of nonlinear adaptive control for uncertain nonholonomic systems based on the notion of inverse optimality [3][4]. The proposed methodologies are applied to chained systems of high orders, and the controller designs by both state and output feedbacks are shown. It is seen that the resulting control strategies are optimal to certain meaningful cost functionals. Simulation studies are also performed, where control of mobile robots is considered.

2 State Feedback Case

Consider the chained system of the following form.

$$\begin{cases} \dot{x}_0 = u_0 + \phi_0(x_0)^T\theta, \\ \dot{x}_i = u_0 x_{i+1} + \phi_i(u_0, \bar{x}_i)^T\theta, \quad (1 \le i \le n-1), \\ \dot{x}_n = u_1 + \phi_n(u_0, x)^T\theta, \end{cases} \tag{1}$$

$$\bar{x}_i \equiv [x_0, \cdots, x_i]^T, \quad (x = \bar{x}_n \equiv [x_0, \cdots, x_n]^T), \tag{2}$$

where $x_0(0) \neq 0$, and θ is an unknown parameter. $\phi_0(x_0)$, $\phi_1(u_0, \bar{x}_1) \sim \phi_n(u_0, x)$ are smooth nonlinear functions satisfying

$$\begin{cases} \phi_0(x_0) = x_0 \cdot \varphi_0(x_0), \\ \phi_i(u_0, \bar{x}_i) = \sum_{j=1}^{i} x_j \cdot \varphi_{ij}(u_0, \bar{x}_i), \end{cases} \tag{3}$$

where $\varphi_0(x_0)$, $\varphi_{11}(u_0, \bar{x}_1) \sim \varphi_{nn}(u_0, x)$ are also nonlinear smooth functions. The control objective is to stabilize the system by utilizing control signals u_0 and u_1. The design procedure is based on the integrator backstepping composed of **Step 0** $\sim$ **Step n**.

(Step 0) First, stabilize x_0 by utilizing u_0.

$$u_0 = x_0 g_0(x_0, \hat{\theta}_0) + v_0, \tag{4}$$

$$g_0(x_0, \hat{\theta}_0) = -\varphi_0(x_0)^T \hat{\theta}_0 - \sqrt{k_0^2 + \{\varphi_0(x_0)^T \hat{\theta}_0\}^2}, \quad (k_0 > 0), \tag{5}$$

where v_0 is a stabilizing signal to be determined later. $\hat{\theta}_0$ is an estimate of θ at **Step 0**, and is determined by the following adaptive law.

$$\dot{\hat{\theta}}_0 = \tau_0 \equiv \Gamma_0 \phi_0(x_0) x_0, \qquad (\Gamma_0 = \Gamma_0^T > 0). \tag{6}$$

Here define a positive function W_0

$$W_0 = \frac{1}{2} x_0^2 + \frac{1}{2} \tilde{\theta}_0^T \Gamma_0^{-1} \tilde{\theta}_0, \qquad (\tilde{\theta}_0 = \theta - \hat{\theta}_0), \tag{7}$$

and take the time derivative of it.

$$\dot{W}_0 = -\sqrt{k_0^2 + \{\varphi_0(x_0)^T \hat{\theta}_0\}^2}\; x_0^2 + x_0 v_0. \tag{8}$$

For that relation, the following virtual system is introduced.

$$\dot{x}_0 = -\sqrt{k_0^2 + \{\varphi_0(x_0)^T \hat{\theta}_0\}^2}\; x_0 + v_0 \equiv \bar{f}_0 + \bar{g}_0 v_0. \tag{9}$$

We are to stabilize that virtual system via v_0 by utilizing optimal control strategy. For that purpose, consider Hamilton-Jacobi (HJ) equation and its solution V_0.

$$\mathcal{L}_{\bar{f}_0} V_0 - \frac{1}{4r_0} (\mathcal{L}_{\bar{g}_0} V_0)^2 + q_0 = 0, \tag{10}$$

$$V_0 = \frac{1}{2} x_0^2. \tag{11}$$

q_0 and r_0 are positive functions, and those are determined from HJ equation (10) based on inverse optimality for given solution V_0 (11). The substitution of V_0 (11) into HJ equation (10) yields

$$-\sqrt{k_0^2 + \{\varphi_0(x_0)^T \hat{\theta}_0\}^2}\; x_0^2 - \frac{1}{4r_0} x_0^2 + q_0 = 0. \tag{12}$$

Then, q_0 and r are given by

$$q_0 = \sqrt{k_0^2 + \{\varphi_0(x_0)^T \hat{\theta}_0\}^2}\; x_0^2 + \frac{1}{4r_0} x_0^2, \qquad r_0 > 0, \tag{13}$$

and the control signal v_0^* is obtained as a solution for the corresponding optimal control problem.

$$v_0 = v_0^* \equiv -\frac{1}{2r_0}\mathcal{L}_{\bar{g}_0} V_0 = -\frac{1}{2r_0} x_0. \tag{14}$$

For the original process (x_0), $\dot{W}_0$ is evaluated as follows:

$$\dot{W}_0 = -q_0 + \frac{1}{4r_0}x_0^2 + x_0 v_0 = -(q_0 + r_0 v_0^2) + r_0\left(v_0 + \frac{1}{2r_0}x_0\right)^2. \tag{15}$$

Then, $v_0 = v_0^*$ is an optimal control solution which minimizes the following cost functional J_0.

$$J_0(t) = \int_0^t (q_0 + r_0 v_0^2)d\tau + W_0(t). \tag{16}$$

For simplicity of notation, u_0 is described as follows:

$$u_0 = \left(g_0 - \frac{1}{2r_0}\right)x_0 \equiv \tilde{g}_0(x_0, \hat{\theta}_0)x_0, \qquad (\tilde{g}_0(x_0, \hat{\theta}_0) \neq 0). \tag{17}$$

(State scaling transformation) [5] The next discontinuous state scaling transformation is introduced.

$$z_i \equiv \frac{x_i}{x_0^{n-i}}, \quad (1 \leq i \leq n). \tag{18}$$

Then we have

$$\begin{cases} \dot{z}_i = \tilde{g}_0(x_0, \hat{\theta}_0)z_{i+1} + f_i(x_0, z_i, \hat{\theta}_0) + \psi_i(x_0, \bar{z}_i, \hat{\theta}_0)^T\theta, & (1 \leq i \leq n-1), \\ \dot{z}_n = u_1 + \psi_n(u_0, x)^T\theta, \end{cases} \tag{19}$$

$$\begin{cases} \psi_i(x_0, \bar{z}_i, \hat{\theta}_0) \equiv \frac{\phi_i(u_0,\bar{x}_i)}{x_0^{n-i}} - \frac{(n-i)z_i\phi_0(x_0)}{x_0}, \\ \psi_n(u_0, x) \equiv \phi_n(u_0, x), \end{cases} \tag{20}$$

$$f_i(x_0, z_i, \hat{\theta}_0) \equiv -(n-i)\tilde{g}_0(x_0, \hat{\theta}_0)z_i, \tag{21}$$

$$\bar{z}_i \equiv [z_1, \cdots, z_i]^T. \tag{22}$$

The design procedure is continued for the new coordinate $[z_1, \cdots, z_n]$.
(Step 1) Define $e_1 = z_1$, $e_2 = z_2 - \alpha_1$, then we have

$$\dot{e}_1 = \tilde{g}_0(e_2 + \alpha_1) + f_1 + \psi_1^T\theta. \tag{23}$$

α_1 is a virtual input and is determined so as to stabilize e_1.

$$\alpha_1 = \frac{1}{\tilde{g}_0}(-k_1 e_1 - f_1 - \psi_1^T\hat{\theta}), \qquad (k_1 > 0). \tag{24}$$

$\hat{\theta}$ is an estimate of θ at **Step 1~Step n**. Here, define a positive function $\bar{W}_1$

$$\bar{W}_1 = \frac{1}{2}e_1^2 + \frac{1}{2}\tilde{\theta}^T \Gamma^{-1}\tilde{\theta}, \quad (\tilde{\theta} = \theta - \hat{\theta},\ \Gamma = \Gamma^T > 0), \tag{25}$$

and take a time derivative of it.

$$\dot{\bar{W}}_1 = -k_1 e_1^2 + \tilde{g}_0 e_1 e_2 + \tilde{\theta}^T \Gamma^{-1}(-\dot{\hat{\theta}} + \tau_1), \tag{26}$$

$$\tau_1 = \Gamma \psi_1 e_1. \tag{27}$$

(Step i) $(2 \le i \le n-1)$ Define $e_i = z_i - \alpha_{i-1}$, $e_{i+1} = z_{i+1} - \alpha_i$, then we have

$$\dot{e}_i = \tilde{g}_0(e_{i+1} + \alpha_i) + \beta_i + w_i^T \theta - \eta_i \dot{\hat{\theta}}, \tag{28}$$

$$\beta_i = f_i - \frac{\partial \alpha_{i-1}}{\partial x_0} u_0 - \frac{\partial \alpha_{i-1}}{\partial \hat{\theta}_0} \tau_0 - \sum_{j=1}^{i-1} \frac{\partial \alpha_{i-1}}{\partial z_j}(\tilde{g}_0 z_{j+1} + f_j), \tag{29}$$

$$w_i = \psi_i - \frac{\partial \alpha_{i-1}}{\partial x_0}\phi_0 - \sum_{j=1}^{i-1} \frac{\partial \alpha_{i-1}}{\partial z_j}\psi_j, \tag{30}$$

$$\eta_i = \frac{\partial \alpha_{i-1}}{\partial \hat{\theta}}. \tag{31}$$

α_i is a virtual input to stabilize e_i and is determined as follows:

$$\alpha_i = \frac{1}{\tilde{g}_0}(-\tilde{g}_0 e_{i-1} - k_i e_i - \beta_i - w_i^T \hat{\theta} + \eta_i \tau_i + \alpha_i^*), \quad (k_i > 0), \tag{32}$$

$$\tau_i = \tau_{i-1} + \Gamma w_i e_i, \tag{33}$$

where α_i^* is to be determined later. A positive function W_i is defined by

$$\bar{W}_i = \bar{W}_{i-1} + \frac{1}{2}e_i^2. \tag{34}$$

Then, the time derivative of it is given as follows:

$$\dot{\bar{W}}_i = -\sum_{j=1}^{i} k_j e_j^2 + \tilde{g}_0 e_i e_{i+1} + \tilde{\theta}^T \Gamma^{-1}(-\dot{\hat{\theta}} + \tau_i) + \sum_{j=2}^{i} \eta_j(\tau_j - \dot{\hat{\theta}})e_j + \sum_{j=2}^{i} \alpha_j^* e_j. \tag{35}$$

(Step n) Define $e_n \equiv z_n - \alpha_{n-1}$, then we have

$$\dot{e}_n = u_1 + \beta_n + w_n^T \theta - \eta_n \dot{\hat{\theta}}, \tag{36}$$

where β_n, w_n, and η_n are defined by (29), (30), (31), respectively. The adaptive law of $\hat{\theta}$ and the auxiliary signals α_i^* are determined such as

$$\dot{\hat{\theta}} = \tau_n \equiv \tau_{n-1} + \Gamma w_n e_n = \Gamma \psi_1 e_1 + \sum_{i=2}^{n} \Gamma w_i e_i, \tag{37}$$

$$\alpha_i^* = \sum_{j=2}^{i-1} \eta_j \Gamma w_i e_j. \tag{38}$$

A positive function $W_1 = \bar{W}_n$ is defined by

$$W_1 \equiv \bar{W}_n \equiv \bar{W}_{n-1} + \frac{1}{2} e_n^2. \tag{39}$$

Then, the time derivative of it is derived as follows:

$$\begin{aligned}
\dot{W}_1 &= \tilde{g}_0 e_{n-1} e_n - \sum_{i=1}^{n} k_i e_i^2 + k_n e_n^2 + \left(u_1 + \beta_n + w_n^T \hat{\theta} - \eta_n \tau_n \right) e_n - \alpha_n^* e_n \\
&= - \sum_{i=1}^{n} k_i e_i^2 + k_n e_n^2 + \sum_{i=1}^{n} \Phi_{1i} e_i e_n + \Phi_2 e_n + e_n u_1 \\
&= - \sum_{i=1}^{n} \tilde{k}_i k_i e_i^2 - \sum_{i=1}^{n} (1 - \tilde{k}_i) k_i \left\{ e_i - \frac{\Phi_{1i}}{2(1 - \tilde{k}_i) k_i} e_n \right\}^2 \\
&\quad + \sum_{i=1}^{n} \frac{\Phi_{1i}^2}{4(1 - \tilde{k}_i) k_i} e_n^2 + k_n e_n^2 + \Phi_2 e_n + e_n u_1,
\end{aligned} \tag{40}$$

where $k_n > 0$, $0 < \tilde{k}_i < 1$ $(1 \le i \le 2)$, and Φ_{1i}, Φ_2 are determined properly such that (40) holds (Note that Φ_{1i}, Φ_2 satisfying (40) are not unique). Here we determine u_1 such as

$$u_1 = -\Phi_2 + v_1, \tag{41}$$

where v_1 is to be determined later. Then $\dot{W}_1$ is rewritten into

$$\begin{aligned}
\dot{W}_1 &= - \sum_{i=1}^{n} \tilde{k}_i k_i e_i^2 - \sum_{i=1}^{n} (1 - \tilde{k}_i) k_i \left\{ e_i - \frac{\Phi_{1i}}{2(1 - \tilde{k}_i) k_i} e_n \right\}^2 \\
&\quad + \sum_{i=1}^{n} \frac{\Phi_{1i}^2}{4(1 - \tilde{k}_i) k_i} e_n^2 + k_n e_n^2 + e_n v_1.
\end{aligned} \tag{42}$$

For the above relation, the following virtual system is introduced.

$$\begin{aligned}
\frac{d}{dt} \begin{bmatrix} e_1 \\ \vdots \\ e_{n-1} \\ e_n \end{bmatrix} &= \begin{bmatrix} -k_1 e_1 \\ \vdots \\ -k_{n-1} e_{n-1} \\ -k_n e_n + k_n e_n + \sum_{i=1}^{n} \Phi_{1i} e_i \end{bmatrix} + \begin{bmatrix} 0 \\ \vdots \\ 0 \\ 1 \end{bmatrix} v_1 \\
&\equiv \bar{f}(e) + \bar{g} v_1.
\end{aligned} \tag{43}$$

Similarly to **Step 0**, the stabilizing signal v_1 is derived as a solution of the corresponding optimal control problem. For that purpose, consider the following HJ equation and its solution V_1.

$$\mathcal{L}_{\bar{f}}V_1 - \frac{1}{4r_1}(\mathcal{L}_{\bar{g}}V_1)^2 + q_1 = 0, \tag{44}$$

$$V_1 = \sum_{i=1}^{n} \frac{1}{2}e_i^2. \tag{45}$$

The positive function q_1 and the positive weight r_1 are derived based on inverse optimality for the given solution V_1. The substitution of the solution V_1 (45) into HJ equation (44) yields

$$\begin{aligned}
&-\sum_{i=1}^{n} k_i e_i^2 i + k_n e_n^2 + \sum_{i=1}^{n} \Phi_{1i} e_i e_n - \frac{1}{4r_1}e_n^2 + q_1 \\
&= -\sum_{i=1}^{n} \tilde{k}_i k_i e_i^2 - \sum_{i=1}^{n}(1-\tilde{k}_i)k_i \left\{ e_i - \frac{\Phi_{1i}}{2(1-\tilde{k}_i)k_i} e_n \right\}^2 \\
&\quad + \sum_{i=1}^{n} \frac{\Phi_{1i}^2}{4(1-\tilde{k}_i)k_i} e_n^2 + k_n e_n^2 - \frac{1}{4r_1}e_n^2 + q_1 = 0.
\end{aligned} \tag{46}$$

Then q_1 and r_1 are given by

$$q_1 = \sum_{i=1}^{n} \tilde{k}_i k_i e_i^2 + \sum_{i=1}^{n}(1-\tilde{k}_i)k_i \left\{ e_i - \frac{\Phi_{1i}}{2(1-\tilde{k}_i)k_i} e_n \right\}^2, \tag{47}$$

$$r_1 = \frac{1}{4\left\{ k_n + \sum_{i=1}^{n} \frac{\Phi_{1i}^2}{4(1-\tilde{k}_i)k_i} \right\}}, \tag{48}$$

and v_1 is obtained as an solution of the corresponding optimal control problem.

$$v_1 = v_1^* \equiv -\frac{1}{2r_1}\mathcal{L}_g V_1 = -\frac{1}{2r_1}e_n = -2\left\{ k_n + \sum_{i=1}^{n} \frac{\Phi_{1i}^2}{4(1-\tilde{k}_i)k_i} \right\} e_n. \tag{49}$$

For the original process $(e_1, \cdots, e_n)$, $\dot{W}_1$ is evaluated as follows:

$$\dot{W}_1 = -q_1 + \frac{1}{4r_1}e_n^2 + e_n v_1 = -(q_1 + r_1 v_1^2) + r_1\left(v_1 + \frac{1}{2r_1}e_n\right)^2. \tag{50}$$

Then, v_1^* is an optimal control solution minimizing the cost functional $J_1(t)$.

$$J_1(t) = \int_0^t (q_1 + r_1 v_1^2)d\tau + W_1(t). \tag{51}$$

Finally, it is seen that $v_0 = v_0^*$ and $v_1 = v_1^*$ are optimal control solutions which minimize the following cost functional.

$$J(t) = \sum_{i=0}^{1} J_i(t) = \sum_{i=0}^{1} \left\{ \int_0^t (q_i + r_i v_i^2) d\tau + W_i(t) \right\}. \tag{52}$$

Furthermore, when $v_0 = v_0^*$, $v_1 = v_1^*$, it follows that

$$\dot{W} = -\sum_{i=0}^{1} (q_i + r_i v_i^2) \leq 0, \quad \left(W = \sum_{i=0}^{1} W_i \right). \tag{53}$$

Therefore, it is shown that all state variables are bounded, and that $e_1 \sim e_n \to 0$, $z_1 \sim z_n \to 0$, $x_0 \sim x_n \to 0$.
Theorem 2.1 *The proposed adaptive control systems are uniformly bounded, and $x_0 \sim x_n$ converge to zero asymptotically. Furthermore, $v_0 = v_0^*$ and $v_1 = v_1^*$ are optimal control solutions which minimize the cost functional J (52).*

3 Output Feedback Case

Next, we treat the output feedback case. Consider the chained system.

$$\begin{cases} \dot{x}_0 = u_0 + c_0 x_0, \\ \dot{x}_i = u_0 x_{i+1} + \phi_i(u_0, \bar{x}_i)^T \theta, \quad (1 \leq i \leq n-1), \\ \dot{x}_n = u_1 + \phi_n(u_0, x)^T \theta, \end{cases} \tag{54}$$

where only x_0 and x_1 (outputs) are available for measurement, but the other states $x_2 \sim x_n$ are not measurable. $x_0(0) \neq 0$, and θ is an unknown parameter. It is assumed that c_0 is known, and that

$$\phi_i(u_0, \bar{x}_i) = x_1 \cdot \bar{\phi}_i(u_0, \bar{x}_1). \tag{55}$$

$\phi_1(u_0, \bar{x}_1) \sim \phi_n(u_0, x)$ and $\bar{\phi}_1(u_0, \bar{x}_1) \sim \bar{\phi}_n(u_0, x)$ are smooth nonlinear functions. The control objective is to stabilize that system by utilizing control signals u_0 and u_1.
(Step 0) First, stabilize the subsystem (x_0) by utilizing u_0 based on optimal control strategy. For that purpose, introduce HJ equation (10), where its solution is given by V_0 (11), and the process is described as follows:

$$\dot{x}_0 = u_0 + c_0 x_0 \equiv \bar{f}_0 + \bar{g}_0 u_0. \tag{56}$$

The substitution of V_0 (11) into HJ equation (10) along the process (56) yields

$$c_0 x_0^2 - \frac{1}{4r_0} x_0^2 + q_0 = 0. \tag{57}$$

Then, q_0 and r (positive definite functions) are given by

$$q_0 = \left(\frac{1}{4r_0} - c_0 \right) x_0^2, \quad r_0 < \frac{1}{4c_0}, \tag{58}$$

and the control signal u_0^* is obtained as a solution for the corresponding optimal control problem.

$$u_0 = u_0^* \equiv -\frac{1}{2r_0}\mathcal{L}_{\bar{g}_0} V_0 = -\frac{1}{2r_0}x_0. \tag{59}$$

For the original process (x_0), $\dot{V}_0$ is evaluated as follows:

$$\dot{V}_0 = -q_0 + \frac{1}{4r_0}x_0^2 + x_0 u_0 = -(q_0 + r_0 u_0^2) + r_0\left(u_0 + \frac{1}{2r_0}x_0\right)^2. \tag{60}$$

Then, $u_0 = u_0^*$ is an optimal control solution which minimizes J_0.

$$J_0(t) = \int_0^t (q_0 + r_0 u_0^2) d\tau + V_0(t). \tag{61}$$

For simplicity of notation, x_0 is described as follows:

$$\dot{x}_0 = -\left(\frac{1}{2r_0} - c_0\right)x_0 = -(k_0 - c_0)x_0, \quad \left(k_0 \equiv \frac{1}{2r_0}\right). \tag{62}$$

(State scaling transformation) The same discontinuous state scaling transformation as (18) is introduced. Then we have

$$\begin{cases} \dot{z}_i = -k_0 z_{i+1} + a_{ii} z_i + \psi_i^T\theta, & (1 \le i \le n-1), \\ \dot{z}_n = u_1 + \psi_n^T\theta, \end{cases} \tag{63}$$

$$a_{ii} \equiv (n-i)(k_0 - c_0), \quad \psi_i \equiv \frac{\phi_i}{x_0^{n-i}} = z_1 x_0^{i-1}\bar{\phi}_i. \tag{64}$$

For $z = [z_1, \cdots, z_n]^T$, the state space form is obtained as follows:

$$\dot{z} = Az + b_1 u_1 + \psi^T\theta \tag{65}$$

$$y = Cz \tag{66}$$

$$A = \begin{bmatrix} a_{11} & -k_0 & 0 & \cdots & 0 \\ 0 & a_{22} & \ddots & \ddots & \vdots \\ \vdots & \ddots & \ddots & \ddots & 0 \\ \vdots & \ddots & \ddots & a_{n-1n-1} & -k_0 \\ 0 & \cdots & \cdots & \cdots & 0 \end{bmatrix}, \quad b_1 = \begin{bmatrix} 0 \\ \vdots \\ 0 \\ 1 \end{bmatrix}, \quad \psi^T = \begin{bmatrix} \psi_1^T \\ \vdots \\ \vdots \\ \psi_n^T \end{bmatrix},$$

$$C = \begin{bmatrix} 1 & 0 & \cdots & 0 \end{bmatrix}. \tag{67}$$

(Observer form) Since the states $z_2 \sim z_n$ are not available for measurement, an adaptive observer is constructed to estimate the unknown states.

$$\dot{\hat{z}} = A\hat{z} + b_1 u_1 + \psi^T\hat{\theta} - \zeta^T\dot{\hat{\theta}} + L(y - C\hat{z}), \tag{68}$$

$$\dot{\zeta}^T = (A - LC)\zeta^T - \psi^T, \tag{69}$$

where $L = [l_1, \cdots, l_n]^T$ is determined in order that $A - LC$ is stable. The state estimation errors are described as follows:

$$e \equiv z - \hat{z}, \quad \tilde{e} \equiv e + \zeta^T \tilde{\theta}, \tag{70}$$

$$\dot{\tilde{e}} = (A - LC)\tilde{e}. \tag{71}$$

Then, the overall system is written in the following form, where the states $z_1, \hat{z}_2 \sim \hat{z}_n$ are all available for measurement.

$$\dot{\tilde{e}} = (A - LC)\tilde{e}, \tag{72}$$

$$\dot{z}_1 = g_0 \hat{z}_2 + a_{11} z_1 + \psi_1^T \theta - k_0 \tilde{e}_2 + k_0 \zeta_2^T \tilde{\theta}, \tag{73}$$

$$\dot{\hat{z}}_i = a_{ii} \hat{z}_i - k_0 \hat{z}_{i+1} + l_i e_1 + \psi_i^T \hat{\theta} - \zeta_i^T \dot{\hat{\theta}}, \quad (2 \le i \le n-1), \tag{74}$$

$$\dot{\hat{z}}_n = u_1 + l_n e_1 + \psi_n^T \hat{\theta} - \zeta_n^T \dot{\hat{\theta}}, \tag{75}$$

$$g_0 \equiv -k_0. \tag{76}$$

The design procedure is continued for the new coordinate $[z_1, \hat{z}_2, \cdots, \hat{z}_n]$.
(Step 1) Define $\xi_1 = z_1$, $\xi_2 = \hat{z}_2 - \alpha_1$, then we have

$$\dot{\xi}_1 = g_0(\xi_2 + \alpha_1) + a_{11}\xi_1 + \psi_1^T \theta - k_0 \tilde{e}_2 + k_0 \zeta_2^T \tilde{\theta}. \tag{77}$$

α_1 is a virtual input and is determined so as to stabilize ξ_1.

$$\alpha_1 = \frac{1}{g_0}(-a_{11}\xi_1 - k_1\xi_1 - \psi_1^T \hat{\theta} + \nu_1), \quad (k_1 > 0), \tag{78}$$

where $\hat{\theta}$ is an estimate of θ, and ν_1 is stabilizing signal to be determined later. A positive definite function $\bar{W}_1$ is defined by

$$\bar{W}_1 = \frac{1}{2}\xi_1^2 + \frac{1}{2}\tilde{\theta}^T \Gamma^{-1} \tilde{\theta} + \frac{(n+1)}{2} \tilde{e}^T P \tilde{e}, \tag{79}$$

where $P (= P^T > 0)$ is determined such that $P(A - LC) + (A - LC)^T P = -2I$. Then, the time derivative of $\bar{W}_1$ is given by

$$\begin{aligned} \dot{\bar{W}}_1 &= -k_1\xi_1^2 + g_0\xi_1\xi_2 - (n+1)\|\tilde{e}\|^2 - k_0\tilde{e}_2\xi_1 + \nu_1\xi_1 + \tilde{\theta}^T \Gamma^{-1}(-\dot{\hat{\theta}} + \tau_1) \\ &\le -k_1\xi_1^2 + g_0\xi_1\xi_2 - (n+1)\|\tilde{e}\|^2 + \tilde{e}_2^2 + \frac{k_0^2\xi_1^2}{4} + \nu_1\xi_1 + \tilde{\theta}^T \Gamma^{-1}(-\dot{\hat{\theta}} + \tau_1), \end{aligned} \tag{80}$$

$$\tau_1 = \Gamma(\psi_1 + k_0\zeta_2)\xi_1. \tag{81}$$

Here, ν_1 is determined as

$$\nu_1 = -\frac{k_0^2}{4}\xi_1, \tag{82}$$

and the following relation is derived.

$$\dot{\bar{W}}_1 \le -k_1\xi_1^2 + g_0\xi_1\xi_2 - n\|\tilde{e}\|^2 + \tilde{\theta}^T\Gamma^{-1}(-\dot{\hat{\theta}} + \tau_1). \tag{83}$$

(Step i) $(2 \le i \le n-1)$ Define $\xi_i = \hat{z}_i - \alpha_{i-1}$, $\xi_{i+1} = \hat{z}_{i+1} - \alpha_i$, then we have

$$\dot{\xi}_i = g_0(\xi_{i+1} + \alpha_i) + \beta_i + w_i^T\hat{\theta} + h_i^T\tilde{\theta} + \eta_i\dot{\hat{\theta}} + \frac{\partial\alpha_{i-1}}{\partial z_1}k_0\tilde{e}_2, \tag{84}$$

$$\begin{aligned}\beta_i &= a_{ii}\hat{z}_i + l_ie_1 + \frac{\partial\alpha_{i-1}}{\partial x_0}(k_0 - c_0)x_0 - \frac{\partial\alpha_{i-1}}{\partial z_1}(a_{11}z_1 + g_0\hat{z}_2)\\ &\quad -\sum_{j=2}^{i-1}\frac{\partial\alpha_{i-1}}{\partial\hat{z}_j}(a_{jj}\hat{z}_j + g_0\hat{z}_{j+1} + l_je_1),\end{aligned} \tag{85}$$

$$w_i = \psi_i - \frac{\partial\alpha_{i-1}}{\partial z_1}\psi_1 - \sum_{j=2}^{i-1}\frac{\partial\alpha_{i-1}}{\partial\hat{z}_j}\psi_j, \tag{86}$$

$$h_i = -\frac{\partial\alpha_{i-1}}{\partial z_1}(\psi_1 + k_0\zeta_2), \tag{87}$$

$$\eta_i = -\zeta_i^T + \sum_{j=2}^{i-1}\frac{\partial\alpha_{i-1}}{\partial\hat{z}_j}\zeta_j^T - \frac{\partial\alpha_{i-1}}{\partial\hat{\theta}}. \tag{88}$$

α_i is a virtual input to stabilize ξ_i and is determined as follows:

$$\alpha_i = \frac{1}{g_0}(-g_0\xi_{i-1} - k_i\xi_i - \beta_i - w_i^T\hat{\theta} - \eta_i\tau_i + \nu_i + \alpha_i^*), \quad (k_i > 0), \tag{89}$$

$$\tau_i = \tau_{i-1} + \Gamma h_i\xi_i, \tag{90}$$

where α_i^* is to be determined later. A positive definite function $\bar{W}_i$ is defined by

$$\bar{W}_i = \bar{W}_{i-1} + \frac{1}{2}\xi_i^2. \tag{91}$$

Then, the time derivative of it is evaluated as follows:

$$\begin{aligned}\dot{\bar{W}}_i &\le -\sum_{j=1}^{i}k_j\xi_j^2 + g_0\xi_i\xi_{i+1} - (n+2-i)\|\tilde{e}\|^2 + \tilde{\theta}^T\Gamma^{-1}(-\dot{\hat{\theta}} + \tau_i)\\ &\quad +\sum_{j=2}^{i}\eta_j(\dot{\hat{\theta}} - \tau_j)\xi_j + \sum_{j=2}^{i}\alpha_j^*\xi_j + \xi_i\frac{\partial\alpha_{i-1}}{\partial z_1}k_0\tilde{e}_2 + \xi_i\nu_i\\ &\le -\sum_{j=1}^{i}k_j\xi_j^2 + g_0\xi_i\xi_{i+1} - (n+2-i)\|\tilde{e}\|^2 + \tilde{\theta}^T\Gamma^{-1}(-\dot{\hat{\theta}} + \tau_i)\\ &\quad +\sum_{j=2}^{i}\eta_j(\dot{\hat{\theta}} - \tau_j)\xi_j + \sum_{j=2}^{i}\alpha_j^*\xi_j + \tilde{e}_2^2 + \frac{1}{4}\left(\frac{\partial\alpha_{i-1}}{\partial z_1}k_0\right)^2\xi_i^2 + \xi_i\nu_i.\end{aligned} \tag{92}$$

Here, determine ν_i such as

$$\nu_i = -\frac{1}{4}\left(\frac{\partial \alpha_{i-1}}{\partial z_1}k_0\right)^2 \xi_i, \tag{93}$$

and the following relation holds.

$$\begin{aligned}\dot{W}_i \leq & -\sum_{j=1}^{i} k_j \xi_j^2 + g_0 \xi_i \xi_{i+1} - (n+1-i)\|\tilde{e}\|^2 + \tilde{\theta}^T \Gamma^{-1}(-\dot{\hat{\theta}} + \tau_i) \\ & + \sum_{j=2}^{i} \eta_j (\dot{\hat{\theta}} - \tau_j)\xi_j + \sum_{j=2}^{i} \alpha_j^* \xi_j. \end{aligned} \tag{94}$$

(Step n) Define $\xi_n = \hat{z}_n - \alpha_{n-1}$, then we have

$$\dot{\xi}_n = u_1 + \beta_n + w_n^T \theta + h_n^T \tilde{\theta} + \eta_n \dot{\hat{\theta}} + \frac{\partial \alpha_{n-1}}{\partial z_1} k_0 \tilde{e}_2, \tag{95}$$

where β_n, w_n, h_n and η_n are defined by (85), (86), (87), (88), respectively. Here we determine the adaptive law of $\hat{\theta}$ and the auxiliary signals α_i^* such as

$$\dot{\hat{\theta}} = \tau_n \equiv \tau_{n-1} + \Gamma h_n \xi_n = \Gamma(\psi_1 + k_0 \zeta_2)\xi_1 + \sum_{i=2}^{n} \Gamma h_i \xi_i, \tag{96}$$

$$\alpha_i^* = -\sum_{j=2}^{i-1} \eta_j \Gamma h_i \xi_j. \tag{97}$$

A positive definite function $W_1 = \bar{W}_n$ is defined by

$$W_1 = \bar{W}_n = \bar{W}_{n-1} + \frac{1}{2}\xi_n^2. \tag{98}$$

Then, the time derivative of it is evaluated as follows:

$$\begin{aligned}\dot{W}_1 \leq & -\sum_{i=1}^{n} k_i \xi_i^2 + k_n \xi_n^2 + g_0 \xi_{n-1}\xi_n - \|\tilde{e}\|^2 + (u_1 + \beta_n + w_n^T \hat{\theta} + \eta_n \tau_n)\xi_n \\ & - \alpha_n^* \xi_n + \frac{1}{4}\left(\frac{\partial \alpha_{n-1}}{\partial z_1}k_0\right)^2 \xi_n^2 \\ = & -\sum_{i=1}^{n} k_i \xi_i^2 + k_n \xi_n^2 - \|\tilde{e}\|^2 + \sum_{i=1}^{n} \Phi_{1i}\xi_i \xi_n + \Phi_2 \xi_n + u_1 \xi_n \\ = & -\sum_{i=1}^{n} \tilde{k}_i k_i \xi_i^2 - \sum_{i=1}^{n}(1-\tilde{k}_i)k_i \left\{\xi_i - \frac{\Phi_{1i}}{2(1-\tilde{k}_i)k_i}\xi_n\right\}^2 - \|\tilde{e}\|^2 \\ & + \sum_{i=1}^{n} \frac{\Phi_{1i}^2}{4(1-\tilde{k}_i)k_i}\xi_n^2 + k_n \xi_n^2 + \Phi_2 \xi_n + \xi_n u_1, \end{aligned} \tag{99}$$

where $k_n > 0$, $0 < \tilde{k}_i < 1$ $(1 \leq i \leq 2)$, and Φ_{1i}, Φ_2 are determined properly such that (99) holds. Here determine u_1 such as

$$u_1 = -\Phi_2 + v_1, \tag{100}$$

where v_1 is to be determined later. Then, $\dot{W}_1$ is rewritten into

$$\begin{aligned}\dot{W}_1 \leq &-\sum_{i=1}^{n} \tilde{k}_i k_i \xi_i^2 - \sum_{i=1}^{n}(1-\tilde{k}_i)k_i \left\{\xi_i - \frac{\Phi_{1i}}{2(1-\tilde{k}_i)k_i}\xi_n\right\}^2 - \|\tilde{e}\|^2 \\ &+\sum_{i=1}^{n} \frac{\Phi_{1i}^2}{4(1-\tilde{k}_i)k_i}\xi_n^2 + k_n\xi_n^2 + k_n\xi_n^2 + \xi_n v_1.\end{aligned} \tag{101}$$

For the above relation, the following virtual system is introduced.

$$\begin{aligned}\frac{d}{dt}\begin{bmatrix}\xi_1 \\ \vdots \\ \xi_{n-1} \\ \xi_n\end{bmatrix} &= \begin{bmatrix}-k_1\xi_1 \\ \vdots \\ -k_{n-1}\xi_{n-1} \\ -k_n\xi_n + k_n\xi_n + \sum_{i=1}^{n}\Phi_{1i}\xi_i\end{bmatrix} + \begin{bmatrix}0 \\ \vdots \\ 0 \\ 1\end{bmatrix} v_1 \\ &\equiv \bar{f}(e) + \bar{g}v_1.\end{aligned} \tag{102}$$

Similarly to **Step 0**, the stabilizing signal v_1 is derived as a solution of the corresponding optimal control problem. For that purpose, consider the following HJ equation (44) and its solution V_1.

$$V_1 = \sum_{i=1}^{n} \frac{1}{2}\xi_i^2 \tag{103}$$

The substitution of V_1 (103) into HJ equation (44) along (102) yields

$$\begin{aligned}&-\sum_{i=1}^{n} k_i\xi_i^2 + k_n\xi_n^2 + \sum_{i=1}^{n}\Phi_{1i}\xi_i\xi_n - \frac{1}{4r_1}\xi_n^2 + q_1 \\ &\quad = -\sum_{i=1}^{n} \tilde{k}_i k_i \xi_i^2 - \sum_{i=1}^{n}(1-\tilde{k}_i)k_i \left\{\xi_i - \frac{\Phi_{1i}}{2(1-\tilde{k}_i)k_i}\xi_n\right\}^2 \\ &\quad +\sum_{i=1}^{n} \frac{\Phi_{1i}^2}{4(1-\tilde{k}_i)k_i}\xi_n^2 + k_n\xi_n^2 - \frac{1}{4r_1}\xi_n^2 + q_1 = 0.\end{aligned} \tag{104}$$

Then q_1 and r_1 (positive definite functions) are given by

$$q_1 = \sum_{i=1}^{n} \tilde{k}_i k_i \xi_i^2 + \sum_{i=1}^{n}(1-\tilde{k}_i)k_i \left\{\xi_i - \frac{\Phi_{1i}}{2(1-\tilde{k}_i)k_i}\xi_n\right\}^2, \tag{105}$$

$$r_1 = \frac{1}{4\left\{k_n + \sum_{i=1}^{n}\frac{\Phi_{1i}^2}{4(1-\tilde{k}_i)k_i}\right\}}, \tag{106}$$

and v_1 is obtained as a solution of the corresponding optimal control problem.

$$v_1 = v_1^* \equiv -\frac{1}{2r_1}\mathcal{L}_g V_1 = -\frac{1}{2r_1}\xi_n = -2\left\{k_n + \sum_{i=1}^{n}\frac{\Phi_{1i}^2}{4(1-\tilde{k}_i)k_i}\right\}\xi_n. \quad (107)$$

For the original process ($\xi_1 \sim \xi_n$), $\dot{W}_1$ is evaluated as follows:

$$\begin{aligned}\dot{W}_1 &\leq -q_1 + \frac{1}{4r_1}\xi_n^2 + \xi_n v_1 - \|\tilde{e}\|^2 \\ &= -(q_1 + r_1 v_1^2) + r_1\left(v_1 + \frac{1}{2r_1}\xi_n\right)^2 - \|\tilde{e}\|^2. \end{aligned} \quad (108)$$

Then, $v_1 = v_1^*$ is a sub-optimal control solution which minimizes the upper bound of the following cost functional.

$$J_1(t) = \int_0^t (q_1 + r_1 v_1^2)d\tau + W_1(t). \quad (109)$$

Finally, it is shown that $v_0 = v_0^*$ and $v_1 = v_1^*$ are sub-optimal control solutions which minimize the upper bound of the following cost functional J.

$$J(t) = \sum_{i=0}^{1} J_i(t) = \sum_{i=0}^{1}\left\{\int_0^t (q_i + r_i v_i^2)d\tau + W_i(t)\right\}. \quad (110)$$

Furthermore, when $v_0 = v_0^*$, $v_1 = v_1^*$, it follows that

$$\dot{W} \leq -\sum_{i=0}^{1}(q_i + r_i v_i^2) \leq 0, \quad \left(W = \sum_{i=0}^{1} W_i\right). \quad (111)$$

Therefore, all state variables are bounded, and $\xi_1 \sim \xi_n, \hat{z}_2 \sim \hat{z}_n, z_1 \sim z_n, x_0 \sim x_n \to 0$.

Theorem 3.1 *The proposed adaptive control systems are uniformly bounded, and $x_0 \sim x_n$ converge to zero asymptotically. Furthermore, $v_0 = v_0^*$ and $v_1 = v_1^*$ are sub-optimal control solutions which minimize the upper bound of the cost functional J (110).*

4 Numerical Example

The numerical simulation study is performed. Consider the chained system.

$$\dot{x}_0 = d_0 u_0, \quad \dot{x}_1 = u_0 x_2, \quad \dot{x}_2 = u_1, \quad (d_0 = 1 - 0.5\epsilon^2), \quad (112)$$

where x_0 and x_1 are available for measurement, but x_2 is unmeasurable. d_0 is an unknown parameter. This problem originally comes from the control of a mobile robot, which is described as folows:

$$\dot{x}_c = v \cdot \cos(\theta + \epsilon), \quad \dot{y}_c = v \cdot \sin(\theta + \epsilon,) \quad \dot{\theta} = \omega, \quad (113)$$

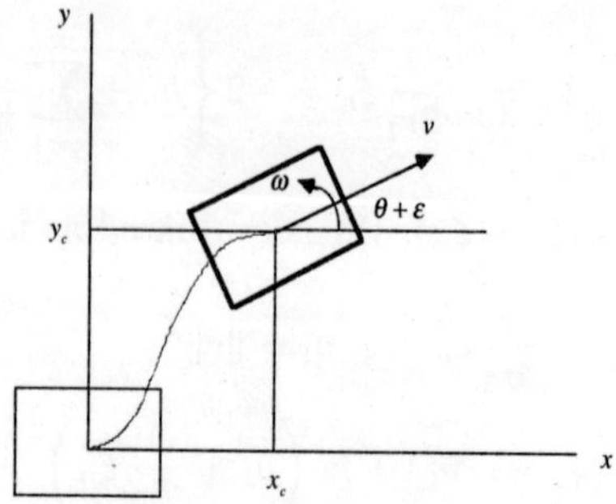

Fig. 1. Mobile Robot

where ϵ is an unknown measurement error. The control objective is to set the mobile to the origin (0, 0) with the actual angle $\theta + \epsilon \to 0$. This system is transformed into the chained forms by setting

$$x_0 = x_c, \quad x_1 = y_c, \quad x_2 = \theta + \epsilon, \quad u_0 = v, \quad u_1 = \omega, \tag{114}$$

and by deducing the second-order approximation. For numerical simulation, we consider two cases, where Case 1 has a similar structure to the conventional method, but the control gain is different from the conventional one. On the contrary, Case 2 has quite different structure from the conventional control scheme, where highly nonlinear terms are involved.

(Case 1)

$\Phi_{11} = 0, \ \Phi_{12} = 0,$

$$\Phi_2 = \beta_2 - \frac{\partial \alpha_1}{\partial z_1} k_0 z_1 \hat{d}_0 - \left(\zeta_2 + \frac{\partial \alpha_1}{\partial \hat{d}_0}\right) \tau_2 - k_0 \xi_1 + \frac{1}{4}\left(\frac{\partial \alpha_1}{\partial z_1} k_0\right)^2 \xi_2.$$

(Case 2)

$$\Phi_{11} = -k_0, \ \Phi_{12} = \frac{1}{4}\left(\frac{\partial \alpha_1}{\partial z_1} k_0\right)^2, \ \Phi_2 = \beta_2 - \frac{\partial \alpha_1}{\partial z_1} k_0 z_1 \hat{d}_0 - \left(\zeta_2 + \frac{\partial \alpha_1}{\partial \hat{d}_0}\right) \tau_2.$$

Simulation conditions are determined such as $x_0(0) = x_1(0) = x_2(0) = 1$, $\hat{z}_2(0) = 0$, $k_0 = k_1 = k_2 = g = 1$, $\tilde{k}_1 = \tilde{k}_2 = 0.5$, $l_1 = 10$, $l_2 = -25$. Simulation results are shown in Fig.2 and Fig.3.

5 Conclusion

Design methodologies of adaptive control for uncertain nonholonomic systems are presented based on the notion of inverse optimality. It is seen that the resulting control strategies are optimal or sub-optimal to certain meaningful cost functionals. The design methods by both state and output feedbacks are shown. Robust controller design for parameter variations, and extension to wide class of nonholonomic systems are left in our future research.

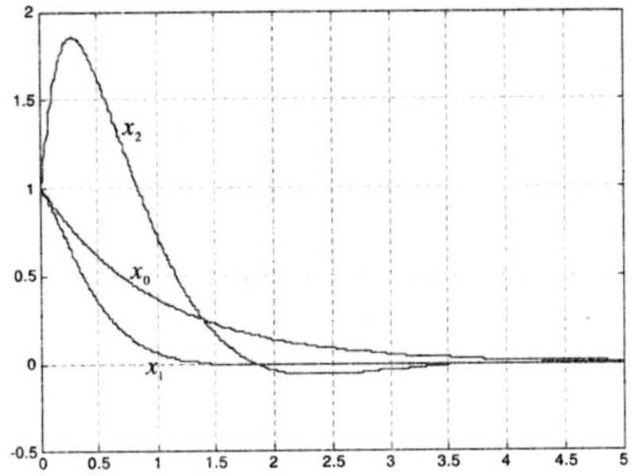

Fig. 2. Simulation Result 1 (x_0, x_1, x_2) (Case 1).

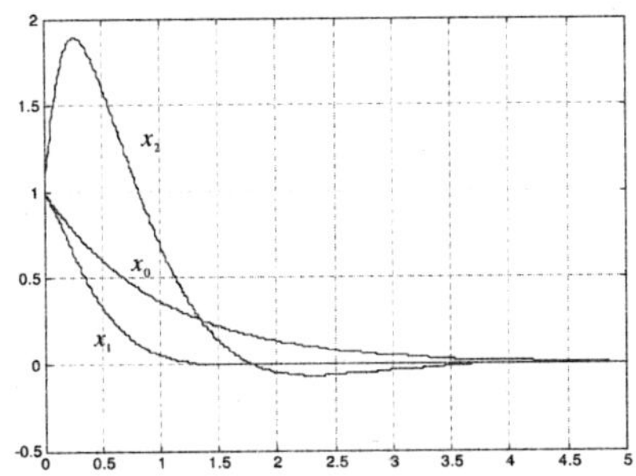

Fig. 3. Simulation Result 1 (x_0, x_1, x_2) (Case 2).

References

1. Jiang, Z. P. (2000), Robust exponential regulation of nonholonomic systems with uncertainties, *Automatica*, Vol.36, No.2, 189-209.
2. Ge, S. S., Wang, Z., Lee, T. H. (2003), Adaptive stabilization of uncertain nonholonomic systems by state and output feedback, *Automatica*, Vol.39, No.8, 1451-1460.
3. Krstić, M., Deng, H. (1998), *Stabilization of Nonlinear Uncertain Systems*, Springer.
4. Miyasato, Y. (2002), Redesign of adaptive control systems based on the notion of optimality, *Transaction of SICE*, Vol.38, No.9, 765–774.
5. Astolfi, A. (1996), Discontinuous control of nonholonomic systems, *Systems and Control Letters*, Vol.27, 37-45,.

Design of a Data-Based Self-Tuning PID Controller

Toru Yamamoto and Kenji Takao

Garaduate school of Education, Hiroshima University, 1-1-1, Kagamiyama, Higashi-hiroshima, Hiroshima, 739-8524, Japan `yama@hiroshima-u.ac.jp`, `takao17@hiroshima-u.ac.jp`

1 Introduction

In recent years, many complicated control algorithms such as adaptive control theory or robust control theory have been proposed and implemented. However, in industrial processes, PID controllers[1, 2, 3] have been widely employed for about 80% or more of control loops. The reasons are summarized as follows. (1) the control structure is quit simple; (2) the physical meaning of control parameters is clear; and (3) the operators' know-how can be easily utilized in designing controllers. Therefore, it is still attractive to design PID controllers. However, since most process systems have nonlinearities, it is difficult to obtain good control performances for such systems simply using the fixed PID parameters. Therefore, PID parameters tuning methods using neural networks(NN)[4] and genetic algorithms(GA)[5] have been proposed until now. According to these methods, the learning cost is considerably large, and these PID parameters cannot be adequately adjusted due to the nonlinear properties. Therefore, it is quite difficult to obtain good control performances using these conventional schemes.

By the way, development of computers enables us to memorize, fast retrieve and read out a large number of data. By these advantages, the following method has been proposed: Whenever new data is obtained, the data is stored. Next, similar neighbors to the information requests, called '*queries*', are selected from the stored data. Furthermore, the local model is constructed using these neighbors. This data-based(DB) modeling method, is called *Just-In-Time (JIT)* method[6, 7] , *Lazy Learning* method[8, 9] or Model-on-Demand(MoD)[10], and these scheme have lots of attention in last decade.

In this paper, a design scheme of PID controllers based on the DB modeling method is discussed. A few PID controllers have been already proposed based on the JIT method[11] and the MoD method[12] which belong to the DB modeling methods. The JIT method is used as the purpose of supplementing the feedback controller with a PID structure. However, the tracking property is not guaranteed enough due to the nonlinearities in the case where reference signals are changed, because the controller does not includes any integral action in the whole control system. On the other hand, the latter method has a PID control structure. PID parameters are tuned by operators' skills, and they are stored in the data-base in advance. And also, a suitable set of PID parameters is generated using the stored data. However, the good control performance cannot be necessarily obtained in the case where nonlinearities are included in the controlled object and/or system parameters are changed, because PID parameters are not tuned in an on-line manner corresponding to characteristics of the controlled object.

In this paper, a design scheme of PID controllers based on the DB modeling method is newly proposed. According to the proposed method, PID parameters which are obtained using the DB modeling method are adequately tuned in proportion to control errors, and modified PID parameters are stored in the data-base. Therefore, more suitable PID parameters corresponding to characteristics of the controlled object are newly stored. Moreover, an algorithm to avoid the excessive increase of the stored data, is further discussed. This algorithm yields the reduction of memories and computational costs. Finally, the effectiveness of the newly proposed control scheme is examined on a simulation example.

2 PID Controller Design Based on Data-Based Modeling Method

2.1 DB modeling method

First, the following discrete-time nonlinear system is considered:

$$y(t) = f(\phi(t-1)), \tag{1}$$

where $y(t)$ denotes the system output and $f(\cdot)$ denotes the nonlinear function. Moreover, $\phi(t-1)$ is called '*information vector*', which is defied by the following equation:

$$\phi(t) := [\ y(t-1), \cdots, y(t-n_y), u(t-1), \cdots u(t-n_u)\ ], \tag{2}$$

where $u(t)$ denotes the system input. Also, n_y and n_u respectively denote the orders of the system output and the system input, respectively. According to the DB modeling method, the data is stored in the form of the information vector ϕ expressed in Eq.(2). Moreover, $\phi(t)$ is required in calculating the estimate of the output $y(t+1)$ called '*query*'. That is, after some similar neighbors to the query are selected from the data-base, the predictive value of the system can be obtained using these neighbors.

2.2 Controller design based on DB modeling method

In this paper, the following control law with a PID structure is considered:

$$\Delta u(t) = \frac{k_c T_s}{T_I} e(t) - k_c \left(\Delta + \frac{T_D}{T_s} \Delta^2 \right) y(t) \tag{3}$$

$$= K_I e(t) - K_P \Delta y(t) - K_D \Delta^2 y(t), \tag{4}$$

where $e(t)$ denotes the control error signal defined by

$$e(t) := r(t) - y(t). \tag{5}$$

$r(t)$ denotes the reference signal. Also, k_c, T_I and T_D respectively denote the proportional gain, the reset time and the derivative time, and T_s denotes the sampling interval. Here, K_P, K_I and K_D included in Eq.(4) are derived by the relations $K_P = k_c$, $K_I = k_c T_s / T_I$ and $K_D = k_c T_D / T_s$. Δ denotes the differencing operator defined by $\Delta := 1 - z^{-1}$. Here, it is quite difficult to obtain a good control performance due to nonlinearities, if PID parameters(K_P, K_I, K_D) in Eq.(4) are fixed. Therefore, a new control scheme is proposed, which can adjust PID parameters in an on-line manner corresponding to characteristics of the system. Thus, instead of Eq.(4), the following PID control law with variable PID parameters is employed:

$$\Delta u(t) = K_I(t) e(t) - K_P(t) \Delta y(t) - K_D(t) \Delta^2 y(t). \tag{6}$$

Now, Eq.(6) can be rewritten as the following relations:

$$u(t) = g(\phi'(t)) \tag{7}$$

$$\phi'(t) := [\mathbf{K}(t), r(t), y(t), y(t-1), y(t-2), u(t-1)] \tag{8}$$

$$\mathbf{K}(t) := [K_P(t), K_I(t), K_D(t)], \tag{9}$$

where $g(\cdot)$ denotes a linear function. By substituting Eq.(7) and Eq.(8) into Eq.(1) and Eq.(2), the following equation can be derived:

$$y(t+1) = h(\tilde{\phi}(t)) \tag{10}$$

$$\tilde{\phi}(t) := [\ y(t), \cdots, y(t-n_y+1), \mathbf{K}(t), r(t), u(t-1), \cdots, u(t-n_u+1)\ ], \tag{11}$$

where $n_y \geq 3$, $n_u \geq 2$, and $h(\cdot)$ denotes a nonlinear function. Therefore, $\mathbf{K}(t)$ is given by the following equations:

$$\mathbf{K}(t) = F(\bar{\phi}(t)) \tag{12}$$

$$\bar{\phi}(t) := [\ y(t+1), y(t), \cdots, y(t-n_y+1), r(t), u(t-1), \cdots, u(t-n_u+1)\ ], \tag{13}$$

where $F(\cdot)$ denotes a nonlinear function. Since the future output $y(t+1)$ included in Eq.(13) cannot be obtained at t, $y(t+1)$ is replaced by $r(t+1)$.

Because the control system so that can realize $y(t+1) \rightarrow r(t+1)$, is designed in this paper. Therefore, $\bar{\phi}(t)$ included in Eq.(13) is newly rewritten as follows:

$$\begin{aligned} \bar{\phi}(t) := [\ & r(t+1), r(t), y(t), \cdots, y(t-n_y+1), \\ & u(t-1), \cdots, u(t-n_u+1)\ ]. \end{aligned} \tag{14}$$

After the above preparation, a new PID control scheme is designed based on the DB modeling method. The controller design algorithm is summarized as follows.

[STEP 1] Generate initial data-base

The DB modeling method cannot work if the past data is not saved at all. Therefore, PID parameters are firstly calculated using Zieglar & Nichols method[2] or Chien, Hrones & Reswick(CHR) method[3] based on historical data of the controlled object in order to generate the initial data-base. That is, $\mathbf{\Phi}(j)$ indicated in the following equation is generated as the initial data-base:

$$\mathbf{\Phi}(j) := [\ \bar{\phi}(j)\ ,\ \mathbf{K}(j)\ ], \quad j = 1, 2, \cdots, N(0) \tag{15}$$

where $\bar{\phi}(j)$ and $\mathbf{K}(j)$ are given by Eq.(14) and Eq.(9). Moreover, $N(0)$ denotes the number of information vectors stored in the initial data-base. Note that all PID parameters included in the initial information vectors are equal, that is, $\mathbf{K}(1) = \mathbf{K}(2) = \cdots = \mathbf{K}(N(0))$ in the initial stage.

[STEP 2] Calculate distance and select neighbors

Distances between the query $\bar{\phi}(t)$ and the information vectors $\bar{\phi}(i)(i \neq k)$ are calculated using the following $\mathcal{L}_1$-norm with some weights:

$$d(\bar{\phi}(t), \bar{\phi}(j)) = \sum_{l=1}^{n_y+n_u+1} \left| \frac{\bar{\phi}_l(t) - \bar{\phi}_l(j)}{\max\limits_m \bar{\phi}_l(m) - \min\limits_m \bar{\phi}_l(m)} \right|, \tag{16}$$

$$(\ j = 1, 2, \cdots, N(t)\ )$$

where $N(t)$ denotes the number of information vectors stored in the data-base when the query $\bar{\phi}(t)$ is given. Furthermore, $\bar{\phi}_l(j)$ denotes the l-th element of the j-th information vector. Similarly, $\bar{\phi}_l(t)$ denotes the l-th element of the query at t. Moreover, $\max\bar{\phi}_l(m)$ denotes the maximum element among the l-th element of all information vectors($\bar{\phi}(j)$, $j = 1, 2, \cdots, N(t)$) stored in the data-base. Similarly, $\min\bar{\phi}_l(m)$ denotes the minimum element. Here, k pieces with the smallest distances are chosen from all information vectors.

[STEP 3] Construct local model

Next, using k neighbors selected in STEP 2, the local model is constructed based on the following Linearly Weighted Average(LWA)[13]:

$$\mathbf{K}^{old}(t) = \sum_{i=1}^{k} w_i \mathbf{K}(i), \qquad \sum_{i=1}^{k} w_i = 1, \tag{17}$$

where w_i denotes the weight corresponding to the i-th information vector $\bar{\phi}(i)$ in the selected neighbors, and is calculated by:

$$w_i = \sum_{l=1}^{n_u+n_y+1} \left(1 - \frac{[\bar{\phi}_l(t) - \bar{\phi}_l(i)]^2}{[\max\limits_m \bar{\phi}_l(m) - \min\limits_m \bar{\phi}_l(m)]^2} \right). \tag{18}$$

[STEP 4] Data adjustment

In the case where information corresponding to the current state of the controlled object is not effectively saved in the data-base, a suitable set of PID parameters cannot be effectively calculated. That is, it is necessary to adjust PID parameters so that the control error decreases. Therefore, PID parameters obtained in STEP 3 are updated corresponding to the control error, and these new PID parameters are stored in the data-base. The following steepest descent method is utilized in order to modify PID parameters:

$$\mathbf{K}^{new}(t) = \mathbf{K}^{old}(t) - \eta \frac{\partial J(t+1)}{\partial \mathbf{K}(t)} \tag{19}$$

$$\eta := [\eta_P, \eta_I, \eta_D], \tag{20}$$

where η denotes the learning rate, and the following $J(t+1)$ denotes the error criterion:

$$J(t+1) := \frac{1}{2}\varepsilon(t+1)^2 \tag{21}$$

$$\varepsilon(t) := y_r(t) - y(t). \tag{22}$$

$y_r(t)$ denotes the output of the reference model which is given by:

$$y_r(t) = \frac{z^{-1}T(1)}{T(z^{-1})} r(t) \tag{23}$$

$$T(z^{-1}) := 1 + t_1 z^{-1} + t_2 z^{-2}. \tag{24}$$

Here, $T(z^{-1})$ is designed based on the reference literature[14]. Moreover, each partial differential of Eq.(19) is developed as follows:

$$\left.\begin{aligned}\frac{\partial J(t+1)}{\partial K_P(t)} &= \frac{\partial J(t+1)}{\partial \varepsilon(t+1)}\frac{\partial \varepsilon(t+1)}{\partial y(t+1)}\frac{\partial y(t+1)}{\partial u(t)}\frac{\partial u(t)}{\partial K_P(t)}\\ &= \varepsilon(t+1)(y(t)-y(t-1))\frac{\partial y(t+1)}{\partial u(t)}\\ \frac{\partial J(t+1)}{\partial K_I(t)} &= \frac{\partial J(t+1)}{\partial \varepsilon(t+1)}\frac{\partial \varepsilon(t+1)}{\partial y(t+1)}\frac{\partial y(t+1)}{\partial u(t)}\frac{\partial u(t)}{\partial K_I(t)}\\ &= -\varepsilon(t+1)e(t)\frac{\partial y(t+1)}{\partial u(t)}\\ \frac{\partial J(t+1)}{\partial K_D(t)} &= \frac{\partial J}{\partial \varepsilon(t+1)}\frac{\partial \varepsilon(t+1)}{\partial y(t+1)}\frac{\partial y(t+1)}{\partial u(t)}\frac{\partial u(t)}{\partial K_D(t)}\\ &= \varepsilon(t+1)(y(t)-2y(t-1)+y(t-2))\frac{\partial y(t+1)}{\partial u(t)}.\end{aligned}\right\} \quad (25)$$

Note that *a priori* information with respect to the system Jacobian $\partial y(t+1)/\partial u(t)$ is required in order to calculate Eq.(25). Here, using the relation $x = |x|\text{sign}(x)$, the system Jacobian can be obtained by the following equation:

$$\frac{\partial y(t+1)}{\partial u(t)} = \left|\frac{\partial y(t+1)}{\partial u(t)}\right| \text{sign}\left(\frac{\partial y(t+1)}{\partial u(t)}\right), \quad (26)$$

where $\text{sign}(x) = 1(x > 0), -1(x < 0)$. Now, if the sign of the system Jacobian is known in advance, by including $|\partial y(t+1)/\partial u(t)|$ in η, the usage of the system Jacobian can make easy[15]. Therefore, it is assumed that the sign of the system Jacobian is known in this paper.

[STEP 5] Remove redundant data

In implementing to real systems, the newly proposed scheme has a constraint that the calculation from STEP 2 to STEP 4 must be finished within the sampling time. Here, storing the redundant data in the data-base needs excessive computational time. Therefore, an algorithm to avoid the excessive increase of the stored data, is further discussed. The procedure is carried out in the following two steps.

First, the information vectors $\mathbf{\Phi}(\bar{i})$ which satisfy the following first condition, are extracted from the data-base:

[First condition]

$$d(\bar{\phi}(t), \bar{\phi}(i)) \leq \alpha_1, \quad i = 1, 2, \cdots, N(t) - k \quad (27)$$

where $\mathbf{\Phi}(\bar{i})$ is defined by

$$\mathbf{\Phi}(\bar{i}) := \left[\bar{\phi}(\bar{i}) \ \mathbf{K}(\bar{i})\right]. \quad \bar{i} = 1, 2, \cdots \quad (28)$$

Moreover, the information vectors $\mathbf{\Phi}(\hat{\imath})$ which satisfy the following second condition, are further chosen from the extracted $\mathbf{\Phi}(\bar{\imath})$:

[Second condition]

$$\sum_{l=1}^{3}\left\{\frac{\mathbf{K}_l(\bar{\imath})-\mathbf{K}_l^{new}(t)}{\mathbf{K}_l^{new}(t)}\right\}^2 \le \alpha_2, \tag{29}$$

where $\mathbf{\Phi}(\hat{\imath})$ is defined by

$$\mathbf{\Phi}(\hat{\imath}) := \left[\bar{\phi}(\hat{\imath}),\ \mathbf{K}(\hat{\imath})\right]. \quad \hat{\imath} = 1, 2, \cdots \tag{30}$$

If there exist plural $\mathbf{\Phi}(\hat{\imath})$, the information vector with the smallest value in the second condition among all $\mathbf{\Phi}(\hat{\imath})$, is only removed. By the above procedure, the redundant data can be removed from the data-base.

3 Simulation Example

In order to evaluate the effectiveness of the newly proposed scheme, a simulation example for a nonlinear system is considered. As a controlled object, the following Hammerstein model[16] is discussed:

[System 1]

$$\left.\begin{aligned} y(t) &= 0.6y(t-1) - 0.1y(t-2) + 1.2x(t-1) - 0.1x(t-2) + \xi(t) \\ x(t) &= 1.5u(t) - 1.5u^2(t) + 0.5u^3(t) \end{aligned}\right\} \tag{31}$$

[System 2]

$$\left.\begin{aligned} y(t) &= 0.6y(t-1) - 0.1y(t-2) + 1.2x(t-1) - 0.1x(t-2) + \xi(t) \\ x(t) &= 1.0u(t) - 1.0u^2(t) + 1.0u^3(t) \end{aligned}\right\} \tag{32}$$

where $\xi(t)$ denotes the white Gaussian noise with zero mean and variance 0.01^2. Static properties of the System 1 and System2 are shown in Fig.1. From Fig.1, it is apparent that gains of the System 2 are larger than ones of the System 1 at $y \ge 1.0$.

Here, the reference signal $r(t)$ is given as follows:

$$r(t) = \begin{cases} 0.5(0 \le t < 50) \\ 1.0(50 \le t < 100) \\ 2.0(100 \le t < 150) \\ 1.5(150 \le t \le 200). \end{cases} \tag{33}$$

The information vector $\bar{\phi}$ are given as follows:

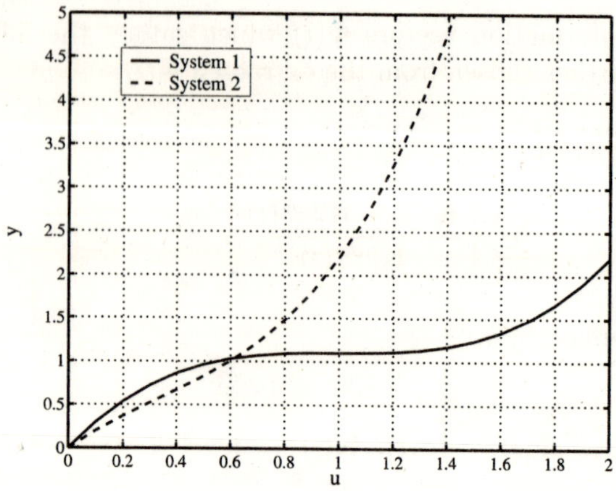

Fig. 1. Static properties of the System 1 and System 2.

$$\bar{\phi}(t) := [\ r(t+1), r(t),\\ y(t), y(t-1), y(t-2), u(t-1)\ ]. \tag{34}$$

Desired characteristic polynomial $T(z^{-1})$ included in the reference model was designed as follows:

$$T(z^{-1}) = 1 - 0.271z^{-1} + 0.0183z^{-2}, \tag{35}$$

where $T(z^{-1})$ was designed based on the reference literature[14]. Furthermore, user-specified parameters included in the proposed method are determined as shown in Table 1.

Table 1. User-specified parameters included in the proposed method.

Orders of the information vector	$n_y = 3$
	$n_u = 2$
Number of neighbors	$k = 6$
Learning rates	$\eta_P = 0.8$
	$\eta_I = 0.8$
	$\eta_D = 0.2$
Coefficients to inhibit the data	$\alpha_1 = 0.5$
	$\alpha_2 = 0.1$
Initial number of data	$N(0) = 6$

For the purpose of the comparison with the conventional scheme, the fixed PID control scheme widely used in industrial processes is employed, whose PID parameters are tuned by CHR method[3]. Then, PID parameters are as follows:

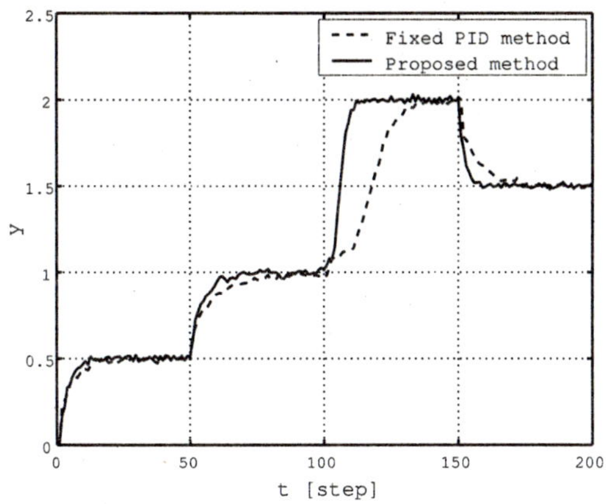

Fig. 2. Control results using the proposed method(solid line) and the fixed PID control(broken line) for System1.

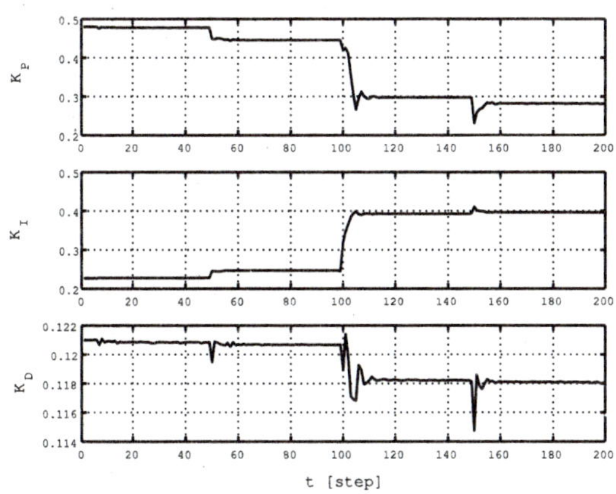

Fig. 3. Trajectories of PID parameters corresponding to Fig.2.

$$K_P = 0.486, \quad K_I = 0.227, \quad K_D = 0.122. \tag{36}$$

Moreover, a PID controller using a NN, called NN-PID controller, is applied for the purpose of the comparison. In this method, a NN is used in order to complement the fixed PID controller.

Firstly, control results for System 1 (eq.(31)) are shown in Fig.2 using the proposed method and the fixed PID controller. And trajectories of PID parameters in the proposed method are shown in Fig.3.

From Fig.2, owing to nonlinearities of the controlled object, the control result using the fixed PID controller is not good. On the other hand, from Fig.2 and Fig.3, the good control result can be obtained using the proposed method because PID parameters are adequately adjusted.

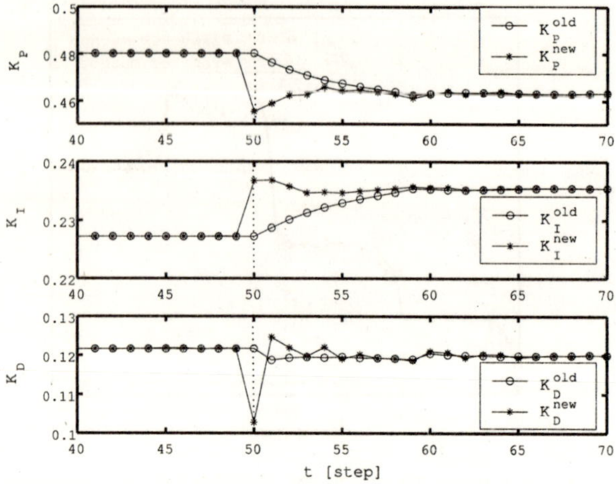

Fig. 4. Detailed behavior of PID parameters adjustment around $t = 100$[step].

The adjustment of PID parameters is further discussed. Fig.4 shows the behabior of updating scheme in [STEP 4], in the case where the reference signals is changed from 0.5 thru 1.0 around 50[step]. From Fig.4, it is clear that a new set of PID parameters is gradually stored in the data-base, and PID parameters are suitably adjusted according to the change of the reference signal.

Moreover, $*$ in Fig.5 shows the point in which a new information vector is stored in the data-base. That is, the removal of the redundant data in [STEP 5] adequately works, and the new data, *i.e*, the information vector, is stored instead of the redundant data. Furthermore, Fig.5 illustrates that a new data is stored in only the transient state where the reference signals are changed, and the replacement of the redundant data by the new data is hardly worked in the steady state where the reference signal is constant. Then, the number of data stored in the data-base was 49. Using the algorithm ([STEP 5]) to remove needless data, the number of data stored in the data-base can be effectively reduced from 206 to 49.

In addition, the error ε_{ep} given as a following equation is 0.0417 using the proposed method.

$$\varepsilon_{ep}(epoc) := \frac{1}{N}\sum_{t=1}^{N}\left\{\frac{\varepsilon(t)}{r(t)}\right\}^2, \tag{37}$$

where N denotes the number of steps per 1[epoc], and N is set as 200 in this case. Furthermore, the number of iteration was set as 1, because PID parameters can be adjusted in an on-line manner by the proposed method.

Moreover, the PID controller using the neural network, called NN-PID controller, was applied to this system (eq.(31)), for the purpose of comparison,

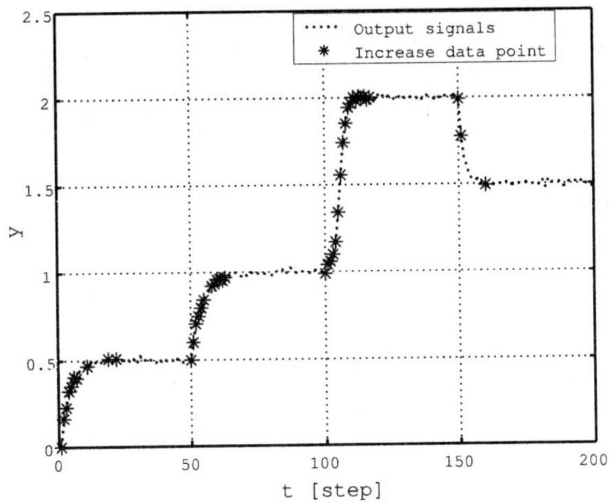

Fig. 5. Replacement behavior of the redundant data by the new data, where * shows the point in which a new information vector is stored in the data-base.

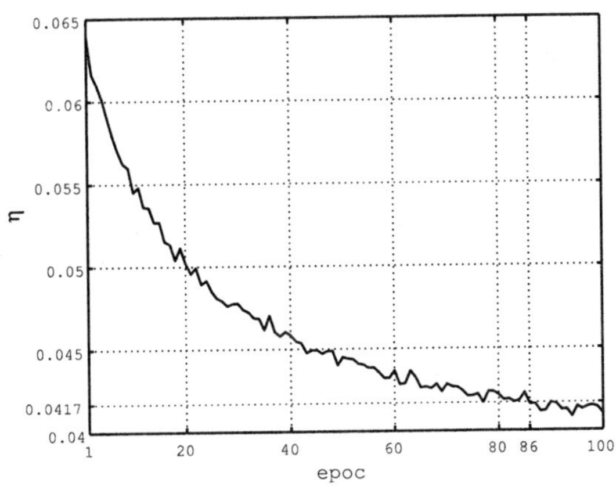

Fig. 6. Error behaviors using the controller fused the fixed PID with the NN-PID for Hammerstein model.

where the NN was utilized in order to supplement the fixed PID controller. Error behaviors of ε_{ep} expressed in Eq.(37) are shown in Fig.6, and control results are shown in Fig.7. From Fig.6, the necessary number of learning iterations is 86[epoc] until the control result using the NN-PID controller can be obtained the same control performances as the proposed method, that is, until $\epsilon_{ep} \leq 0.0417$ is satisfied. Therefore, the effectiveness of the proposed method was examined for nonlinear systems.

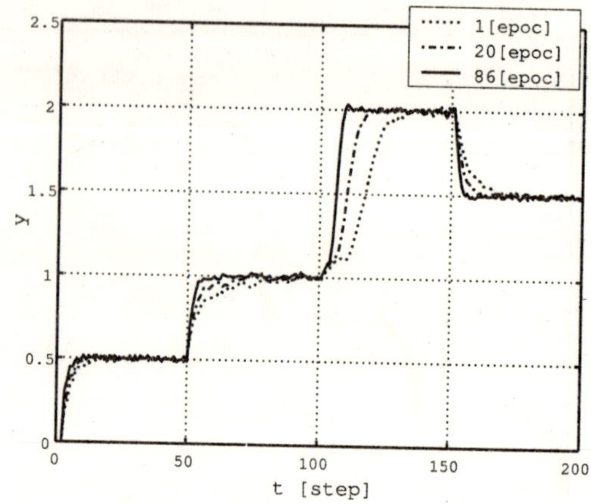

Fig. 7. Control result using the controller fused the fixed PID with the NN-PID for Hammerstein model.

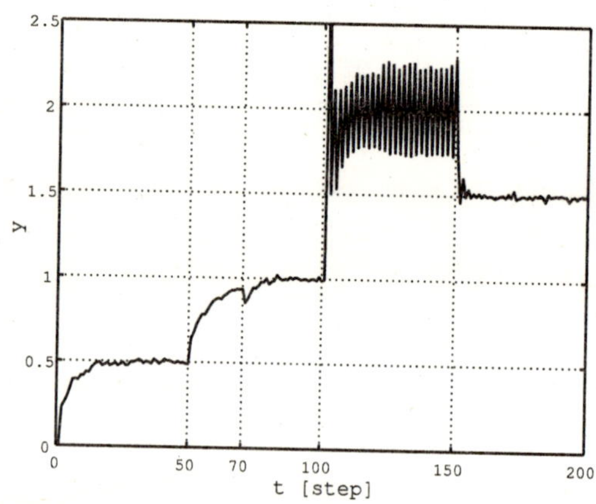

Fig. 8. Control result using the fixed PID controller in the case where the system is changed from System1 to System2.

Next, the case where system parameters change is considered. It assumes that the system changes from Eq.(31) to Eq.(32) at $t = 70$. Firstly, the control result is shown in Fig.8 using the fixed PID controller, where PID parameters are used the same parameters as shown in Eq.(36). Although the gain of the controlled object becomes high after $t = 70$, PID parameters are not changed according to the system gain. Therefore, the control performance is not good. On the other hand, the proposed control scheme is employed for this case. The control result and trajectories of PID parameters are shown in Fig.9 and Fig.10. From Fig.9 and Fig.10, it is clear that a good control performance

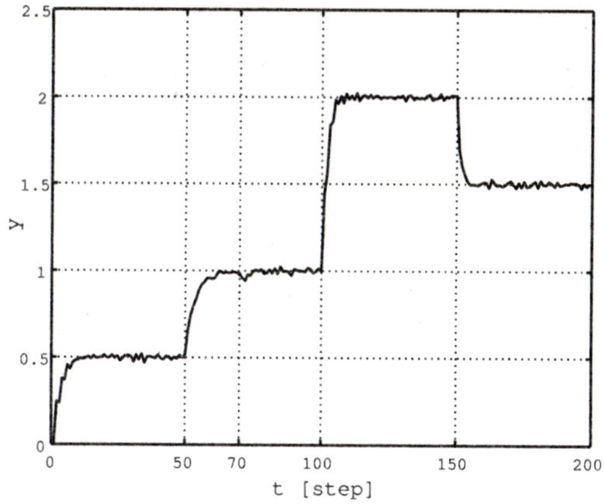

Fig. 9. Control result using the proposed method in the case where the system parameters are changed.

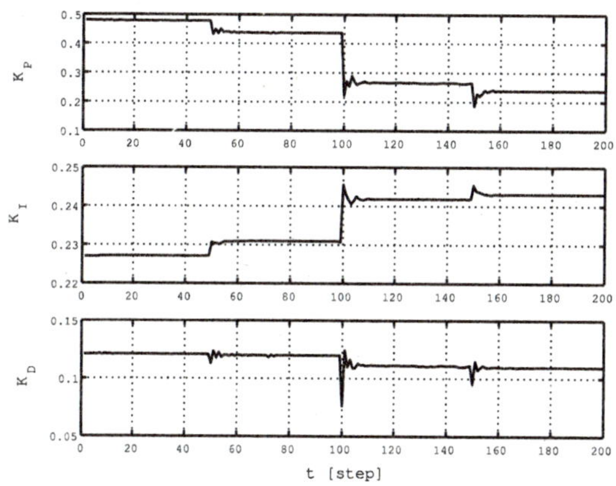

Fig. 10. Trajectories of PID parameters corresponding to Fig.9.

can be obtained, because PID parameters are adjusted adequately using the proposed method even if system parameters are changed.

4 Conclusions

In this paper, a new design scheme of PID controllers using the DB modeling method has been proposed. Many PID controller design schemes using NNs and GAs have been proposed for nonlinear systems up to now. In employing these scheme for real systems, however, it is a serious problem that the learning cost becomes considerably large. On the other hand, according to the proposed method, such computational burdens can be effectively reduced using the algorithm for removing the redundant data. In addition, the effectiveness of the proposed method have been verified by a numerical simulation example.

The application of the newly proposed scheme for real systems and the extension to multivariable cases are currently under consideration.

References

1. Åström, K.J., Hägglund, T. (1988), *Automatic Tuning of PID Controllers*, Instrument Society of America.
2. Zieglar, J.G., Nichols, N.B. (1942), Optimum settings for automatic controllers, *Trans. ASME*, Vol.64, No.8, 759-768.
3. Chien, K.L., Hrones, J.A., Reswick, J.B. (1972), On the Automatic Control of Generalized Passive Systems, *Trans. ASME*, Vol.74, 175-185.
4. Omatu, O., Marzuki, K., Rubiyah, Y. (1995), *Neuro-Control and Its Applications*, Springer-Verlag, London.
5. Porter, B., Jones, A.H. (1992), Genetic tuning of digital PID controllers, *Electronics Letter*, Vol.28, 843-844.
6. Stenman, A., Gustafsson, F., Ljung, L. (1996), Just in time models for dynamical systems, *35th IEEE Conference on Decision and Control*, 1115-1120.
7. Zheng, Q., Kimura, H. (2001), A New Just-In-Time Modeling Method and Its Applications to Rolling Set-up Modeling (in Japanese), *Trans. on SICE*, Vol.37, No.7, 640-646.
8. Zhang, J., Yim, Y., Yang, J. (1997), Intelligent selection of instances for prediction functions in lazy learning algorithms, *Artificial Intelligence Review*, Vol.11, 175-191.
9. Bontempi, G., Birattari, M., Bersini, H. (1999), Lazy learning for local modeling and control design, *International Journal of Control*, Vol.72, No.7-8, 643-658.
10. Stenman, A. (1999), *Model on demand*, Algorithms, analysis and applications, PhD thesis, Department of Electrical Engineering Linköping University.
11. Zheng, Q., Kimura, H. (2001), Just-in-time PID Control, *The 44th Japan Joint Automatic Control Conference*, Tokyo, 336-339.
12. Ohta, J., Yamamoto, S. (2004), Detabase-Driven Tuning of PID Controllers (in Japanese), *Trans. on SICE*, Vol.40, No.6, 664-669.
13. Atkeson, C.G., Moore, A.W., Schaal, S. (1997), Locally weighted learning for control, *Artificial Intelligence Review*, Vol.11, 75-114.
14. Yamamoto, T., Shah, S.L. (1998), Design and Experimental Evaluation of a Multivariable Self-Tuning PID Controller, *Proc. of IEEE Conference on Control Applications*, Trieste, 1230-1234.
15. Omatu, S., Yamamoto, T. (1996), *Self-Turning Control (in Japanese)*, SICE Learned Book.
16. Zi-Qiang, L. (1994), On identification of the controlled plants described by the Hammerstein system, *IEEE Trans. Automat. Contr.*, Vol.AC-39, 569-573.

Design and Application of Simple 2-Delay Input Adaptive Control System Based on Relative Degree Model

Naoki Mizuno[1], Kazuyuki Fujiwara[1] and Akira Satou[2]

[1] Nagoya Institute of Technology, Dept. of Mechanical Engineering, Gokiso-cho, Showa-ku, Nagoya, 466-8555 Japan `nmizuno@nitech.ac.jp`
[2] Yokogawa Electric Corporation, 2-9-32 Nakamachi, Musashino, 180-8750, Japan

1 Introduction

In the MRACS, it must be assumed that the plant is a minimum phase system, since the adaptive controller is designed based on cancellation of all plant zeros. This assumption restricts the application of the MRACS to real plants, especially in the case of discrete time control, because the discrete time form of the minimum phase continuous time plant often becomes a non-minimum phase system, depending on the dynamic behaviour of the plant and sampling period. Åström, Hagander and Sternby [1] have concluded that the all continuous-time systems, whose relative degree greater than one, sampled at a fast sampling with zero order hold device gives the discrete time model having zeros outside or on the unit circle of the z-plane.

For this serious problem, the conventional discrete time adaptive control algorithm has been modified to handle non-minimum phase system and different types of discrete time adaptive control algorithms have been proposed by several authors. However, many algorithms (for example, adaptive pole placement, explicit criterion minimization etc.) are more complicated than the conventional MRACS and are difficult to apply the control system which requires the fast sampling rate like a digital servo control systems.

In order to circumvent these problems, Fujii and Mizuno had proposed a very simple adaptive control method using a discrete time autoregressive (AR) plant model with dead time [2] ,[3]. Moreover, other types of model reference adaptive control system using a new discrete time model described by delta operator was proposed [4]. A key principle of these methods is that the fixed idea that a discrete-time model of the continuous-time system is an ARMA model described by z-operator has been changed. Based on this idea, Mizuno and Fujii [5] proposed another new discrete-time model, called limiting-zero model, and a design method of adaptive control system based on the model.

On the other hand, Miyasato [6] has proposed a design method of MRAC (which means the perfect tracking at each sampling instances) for non-minimum phase systems by utilizing 2-delay input. This method has been extended to improve the input characteristics [7] - [10]. In these papers, it is shown that under several assumptions, the stable model reference adaptive control systems can be constructed even for non-minimum phase systems.

However, from the practical point of view, simpler control algorithms are needed for high speed servo system which requires very high sampling rate.

In this manuscript, we propose a new simple discrete-time model, called 2-delay relative degree model, which is a simplified version of 2-delay limiting-zero model [11]. If a 2-delay relative degree model is adopted as the discrete-time model of the continuous-time system, the adaptive controller with only one adjustable parameter can be designed, since the poles and zeros of the 2-delay relative degree model are known beforehand. Therefore, the model reference control is achieved with simple adaptive controller even for continuous-time system having the relative degree greater than one at a fast sampling.

First, a short review of the relation between the continuous-time system and its normal (1-delay) discrete-time form described by z-operator is given. Next, we give the brief description of the 2-delay input model and zero assignment by feed forward compensation [6], [12]. Then, we describe the limiting zeros of the 2-delay input model and the motivation of recommending the 2-delay relative degree model as a discrete-time model of the continuous-time system. In the main part of this paper, we propose a new design method of the MRACS based on the 2-delay relative degree model. Moreover, in order to investigate the property of the proposed method, computer simulations are carried out for some types of continuous-time systems. These results give some guide line of applying the proposed method to real systems. Finally, we give a simulation result for a detailed model of the electro-hydraulic servo system [13] using the proposed method.

2 Continuous-Time System and Discrete-Time System

Consider a single-input, single-output continuous-time system described by

$$A_c(s)y(t) = B_c(s)u(t-\tau) \tag{1}$$

or

$$y(t) = G(s)u(t-\tau) \tag{2}$$

$$G_{T/2}(z) = z^{-d}c^T(zI-A)^{-1}b \tag{3}$$

where $G(s)$ is a strictly proper rational transfer function and τ is a time delay. $y(t)$ and $u(t)$ denote the continuous-time system input and output, respectively. $A_c(s)$ and $B_c(s)$ are the polynomials in the differential operator, $s = d/dt$, of order n and m as follows.

$$A_c(s) = s^n + a_{Cn-1}s^{n-1} + .. + a_{C1}s + a_{C0} \tag{4}$$

$$B_c(s) = b_{Cm}s^m + b_{Cm-1}s^{m-1} + .. + b_{C1}s + b_{C0} \quad , \quad b_{Cm} \neq 0, \quad m < n \tag{5}$$

2.1 1-Delay Discrete-Time Model of the Continuous-Time System and its Limiting-Zeros

Assume that the continuous-time plant interconnected a zero-order hold device is sampled at interval T. Then, the exact discrete-time model of the above continuous-time system is obtained by standard z-transformation as follows.

$$A_d(z)y(i) = B_d(z)u(i) \tag{6}$$

where z is a shift operator, such as $zy(i) = y(i+1)$. $y(i)$ and $u(i)$ denote the discrete-time system input and output at sampling instance i, $A_d(z)$, $B_d(z)$ are polynomials in the unit shift operator z of order n and $n-1$ respectively.

$$A_d(z) = z^n + a_{Dn-1}z^{n-1} + ... + a_{D1}z + a_{D0} \tag{7}$$

$$B_d(z) = b_{Dn-1}z^{n-1} + ... + b_{D1}z + b_{D0} \tag{8}$$

The state space representation of the above system is also given by the following form.

$$x(i+1) = A_T x(i) + b_T u(i-d) \tag{9}$$

$$y(i) = c_T^T x(i) \tag{10}$$

where $x(i)$ is a discrete state, $A \in R^{n\times n}$, $b, c \in R^n$ are the system matrrix and vectors of the discrete-time model with sampring period T. d is a time delay in the discrete-time representation ($d = [\tau/(T/2)]$). As is well known, when a continuous time system is sampled at sampling period T, the continuous-time poles p_i are transformed to $e^{p_i T}$. On the other hand, a relation between the continuous-time zeros and the discrete-time zeros is more complicated. Moreover, from Eqs.(6) - (10), the discrete-time model has excess zeros compared with the continuous-time ones. However, a simple relationship between the zeros of $B_c(s)$ and $B_d(z)$ for the rapid sampling was given by Åström *et al.* [1]. The result is as follows.

[Limiting-Zeros]

If $B_c(s)$ has the zeros $z_1, ..., z_m$ then the zeros of $B_d(z)$ converge to the zeros (limiting-zeros) of the following polynomials as T tends to 0.

$B_n(z)$: the zeros of this polynomial are $e^{z_1 T}, ..., e^{z_m T}$

$B_L^r(z)$: the zeros of this polynomial only depend on the relative degree of the form :

$$B_L^r(z) = b_1^r z^{r-1} + b_2^r z^{r-2} + ... + b_r^r, r = n^* - m \tag{11}$$

$$b_p^r = \sum_{j=1}^{p} (-1)^{p-j} j^r (r+1)!/[(r+1-p+j)!(p-j)!] \tag{12}$$

For $r = (n-m)$ greater than one, the polynomial $B_L^r(z)$ has zeros outside (or on) the unit circle. Because of the above property of the conventional 1-delay discrete-time model of the continuous-time system, the MRACS are difficult to apply for practical control problems.

2.2 2-Delay Discrete-Time Model of the Continuous-Time System and Zero Assignment

In this section, we give a brief description of the 2-delay input discrete-time model of the continuous-time system and zero assignment by 2-delay input control [6] . By utilizing the 2-delay sampling method, the discrete-time model of the single-input, single-output continuous time system becomes to the 2-inputs, single-output discrete-time system..

Let *(A,b,c,d)* ($A \in R^{n\times n}$, $b, c \in R^n$, d(integer) > 0) be a discrete-time model of (A.1) based on the usual sampling method with the sampling period $T/2$, that is,

$$G_{T/2}(z) = z^{-d} c^T (zI - A)^{-1} b \tag{13}$$

where d is a time delay in the discrete-time representation ($d = [\tau/(T/2)]$), and $G_{T/2}(z)$ is the pulse transfer function of the controlled system with sampling period $T/2$. For this controlled system, if the degree n and the time delay d are known, then the 2-delay sampling representation of the controlled system can be seen as the 2-inputs $[u_1(i),\ u_2(i)]$, single-output $[y(i)]$ system described as follows:

$$A(z^{-1})y(i) = z^{-d}(B_1(z^{-1})u_1(i) + B_2(z^{-1})u_2(i)) \tag{14}$$

$$A(z^{-1}) = 1 + \sum_{k=1}^{n} a_k z^{-k} = z^{-n} \det(zI - A^2) \tag{15}$$

$$B_1(z^{-1}) = \sum_{k=1}^{n} b_{1k} z^{-(k-1)} = z^{-(n-1)} c^T adj(zI - A^2)Ab \tag{16}$$

$$B_2(z^{-1}) = \sum_{k=1}^{n} b_{2k} z^{-(k-1)} = z^{-(n-1)} c^T adj(zI - A^2) b \tag{17}$$

The next assumption is introduced for the model.

[Assumption 1]

b_{11} or b_{21} is non-zero and $B_1(z^{-1})$ and $B_2(z^{-1})$ are relatively prime.

If the above assumption holds and when the controlled system is known, the equivalent zeros of the controlled system can be arbitrary assigned by the following 2-delay inputs control.

$$u_1(i) = G_1(z^{-1}) v(i) \tag{18}$$

$$u_2(i) = G_2(z^{-1}) v(i) \tag{19}$$

where

$$G_k(z^{-1}) = g_{k1} + \cdots + g_{kn-1} z^{-(n-2)} \ (k = 1, 2) \tag{20}$$

are the polynomial solutions of the following identity (Diophantine identity).

$$G_{10}(z^{-1}) B_1(z^{-1}) + G_{20}(z^{-1}) B_2(z^{-1}) = B^*(z^{-1}) \tag{21}$$

$$G_1(z^{-1}) \equiv G_{10}(z^{-1}) + B_2(z^{-1}) T(z^{-1}) \tag{22}$$

$$G_2(z^{-1}) \equiv G_{20}(z^{-1}) - B_1(z^{-1}) T(z^{-1}) \tag{23}$$

where, $B^*(z^{-1})$

$$B^*(z^{-1}) = b_0^* + b_1^* z^{-1} \cdots + b_{n_{b*}}^* z^{-n_{b*}} \tag{24}$$

is any Hurwits polynomia and $T(z^{-1})$ is an arbitraly polynomial. From the assumption 1, $G_{k0}(z^{-1})$ $(k = 1, 2)$ $G_{k0}(z^{-1})$ $(k = 1, 2)$ satisfying (21), always exist. Moreover, by using the 2-delay inputs (18) and (19), all equivalent zeros from the signal $v(i)$ to the output $y(i)$ are assigned to the zeros of the polynomial $B^*(z^{-1})$.

However, in case of adaptive control, all parameters in polynomials $B_1(z^{-1})$, $B_2(z^{-1})$ should be estimated and the Diophantine identity should also be solved based on the estimated parameters to assign equivalent zeros. In this case, the numerical difficulty may occur when the estimated polynomials have common factors.

If the polynomials $B_1(z^{-1})$, $B_2(z^{-1})$ of the 2-delay input model have some special properties, the procedure of assigning equivalent zeros may become more easy to implement. From this point of view, we investigate the limiting zeros of the 2-delay input model of the continuous-time system.

2.3 Limiting-Zeros of the 2-Delay Discrete-Time Model of the Continuous-Time System

Based on the limiting-zeros of the 1-delay discrete-time model of the continuous-time system, we have obtained the following result about limiting-zeros of the 2-delay discrete-time model.

[Limiting-Zeros of 2-delay input model]

If $B_c(s)$ has the zeros $z_1, ..., z_m$ then the zeros of $B_1(z^{-1})$, $B_2(z^{-1})$ converge to the zeros (2-delay limiting-zeros) of the following polynomials as T tends to 0 respectively.

$B_{in}(z^{-1})(i = 1, 2)$: the zeros of this m th order polynomial are $e^{z_1 T'}, ..., e^{z_m T'}$ (T': function of T)

$B_{iL}^{r}(z^{-1})(i = 1, 2)$: the zeros of this r th order polynomials only depend on the relative degree.

$$B_{1L}^{r}(z^{-1}) = \frac{1}{r!}\left(\frac{T}{2}\right)^{r}\sum_{k=0}^{r-1} b_{1k} z^{-k} \tag{25}$$

$$B_{2L}^{r}(z^{-1}) = \frac{1}{r!}\left(\frac{T}{2}\right)^{r}\sum_{k=0}^{r-1} b_{2k} z^{-k} \tag{26}$$

$$b_{1k} = \sum_{l=0}^{k}\{(2l+2)^r - (2l+1)^r\}\frac{(-1)^{k-l} r!}{(k-l)!(r+l-k)!} \tag{27}$$

$$b_{2k} = \sum_{l=0}^{k}\{(2l+1)^r - (2l)^r\}\frac{(-1)^{k-l} r!}{(k-l)!(r+l-k)!} \tag{28}$$

The fact that the limiting-zeros of the system only depend the relative degree of the continuous-time system is the key property of introducing the 2-delay relative degree model.

If the convergence speed of the excess zeros to the limiting-zeros is faster than the convergence speed of the other poles and zeros to corresponding discrete-time ones, then $B_1(z^{-1})$, $B_2(z^{-1})$ can be expressed for the rapid sampling as [11]

$$B_i(z^{-1}) = B_{in}(z^{-1})B_{iL}(z^{-1})(i = 1, 2) \tag{29}$$

where $B_{in}(z^{-1})$ has the unknown zeros which converge to $e^{z_1 T'}, ..., e^{z_m T'}$ (z_i denote the continuous-time zeros) as the sampling period tents to 0, and $B_{iL}(z^{-1})$ has only the known 2-delay limiting-zeros (sometimes unstable). Moreover, if the sampling period is sufficiently short compared with the time constant of the system, all poles and zeros of the model become -1. This leads to the possibility of stable pole-zero cancellation in the discrete-time model of the continuous-time system. This is the motivation of proposing the relative degree model of the plant.

3 Controller Design Using 2-Dely Relative Degree Model

The idea mentioned above leads to the following new representation of the discrete-time model.

The control problem considered here is to determine a suitable control input $u(t)$ such that the discrete-time asymptotic model-following is achieved for any continuous-time system with stable continuous-time zeros.

Assume that the relative degree of the continuous-time system is known and the sampling interval is short compared with the time constant of the plant, then the plant can be described as follows.

$$A(z^{-1})y(i) = z^{-d}(B_1(z^{-1})u_1(i) + B_2(z^{-1})u_2(i)) \tag{30}$$

$$A_e(z^{-1}) = A(z^{-1}) - (1 - z^{-1})^n \tag{31}$$

$$B_{ie}(z^{-1}) = B_i(z^{-1}) - K(1 - z^{-1})^m B_{in}(z^{-1})B_{iL}(z^{-1})(i = 1, 2) \tag{32}$$

where the polynomials $A_e(z^{-1})$ and $B_{ie}(z^{-1})(i = 1, 2)$ are residual polynomials which converge to zero as the sampling period T tends to 0. Based on this property, we will ignore the polynomial $A_e(z^{-1})$ and $B_{ie}(z^{-1})$ then model the controlled system by the approximate model called "2-delay relative degree" of the form:

$$A_r(z^{-1})y(i) = Kz^{-1}\left(B_{1L}(z^{-1})u_1(i) + B_{2L}(z^{-1})u_2(i)\right) \tag{33}$$

$$A_r(z^{-1}) = (1 - z^{-1})^r \tag{34}$$

If all continuous-time zeros and the same numbers of the poles lie in left half of the s-plane, this model contains stable pole/zero cancellation pairs and has stable/unstable zeros of $B_{iL}(z^{-1})$. Note that the 2-delay relative degree model of the plant is equivalent to the exact 2-delay discrete-time model of the following system.

$$y(t) = \frac{1}{s^r}u(t) \tag{35}$$

In almost all cases, the 2-delay relative degree model of the system has unstable zeros of $B_{iL}(z^{-1})$, the equivalent zeros of the controlled system should be assigned to stable ones using the same feed forward compensators as eqs.(18) and (19) but with $G_i(z^{-1})(i = 1, 2)$ of the solution of the following polynomial identity similar to Eq.(21).

$$G_{10}(z^{-1})\,B_{1L}(z^{-1}) + G_{20}(z^{-1})\,B_{2L}(z^{-1}) = B^*(z^{-1}) \tag{36}$$

The above identity can be solved without numerical difficulty, because the polynomial $B_{iL}(z^{-1})$ is known and has simple parameters as shown in Eqs. (24)-(28).

In the proposed control scheme, the control low is composed of two components as follows.

$$u_1(i) = \left(v(i) + G_1(z^{-1})\bar{v}(i)\right)\frac{1}{K} \tag{37}$$

$$u_2(i) = \left(v(i) + G_2(z^{-1})\bar{v}(i)\right)\frac{1}{K} \tag{38}$$

In above equations, $v(i)$ is the feedback component of the input generated by

$$L(z^{-1})v(i) = D(z^{-1})\left(y_m(i) - y(i)\right) \tag{39}$$

where $y_m(i)$ is a desired output to be followed and

$$L(z^{-1}) = 1 + \cdots + l_{d-1}z^{-(d-1)} \tag{40}$$

$$D(z^{-1}) = d_0 + d_1 z^{-1} + .. + d_{r-1}z^{-(r-1)} \tag{41}$$

are the solutions of the following Diophantine equation.

$$A^*(z^{-1}) = A_r(z^{-1})L(z^{-1}) + z^{-d}B_L(z^{-1})D(z^{-1}) \tag{42}$$

and $A^*(z^{-1})$

$$A^*(z^{-1}) = 1 + a_1^* z^{-1} + .. + a_r^* z^{-r} \tag{43}$$

is any Hurwits polynomial and $B_L(z^{-1})$ is

$$B_L(z^{-1}) = B_{1L}(z^{-1}) + B_{2L}(z^{-1}) \tag{44}$$

On the other hand, $\bar{v}(i)$ is the feedforward component which compensates the tracking performance.

If the polynomial $B^*(z^{-1})$ is chosen as a Hurwits one, the perfect model following property can be achieved by synthesising $\bar{v}(i)$ as

$$B^*(z^{-1})\bar{v}(i) = z^d A_r(z^{-1})y_m(i) \tag{45}$$

If the property of the desired output $y_m(i)$ is modeled by

$$\Omega(z^{-1})y_m(i) = 0 \tag{46}$$

Then the input property can be improved by setting the arbitraly polynomial $T(z^{-1})$ to the solution of the following polynomial equation.

$$G_{10}(z^{-1}) - G_{20}(z^{-1}) = \Omega(z^{-1})L'(z^{-1}) - B_L(z^{-1})T(z^{-1}) \tag{47}$$

In the above controller synthesis, all parameters in the design equations except the gain K are known. If the only one parameter K of the relative degree model of the plant is unknown, first, we estimate the unknown parameter then construct the adaptive controller based on the estimated parameter. For simplicity, we describe the controlled system in the following vector form.

[Controlled system]

$$y'(i) = Ku'(i-1) \tag{48}$$

where

$$y'(i) = A_r(z^{-1})y(i) \tag{49}$$

$$u'(i) = B_{1L}(z^{-1})u_1(i) + B_{2L}(z^{-1})u_2(i) \tag{50}$$

We consider the adaptive identifier and the adaptive control law as follows.

[Adaptive identifier]

$$\hat{y}(i) = \hat{\theta}^{\mathrm{T}}(i-1)\xi(i-1) \tag{51}$$

$\hat{\theta}(i-1)$ is a current estimate of K at the time instant $t = (i-1)T$.

$$\hat{K}(i) = \hat{K}(i-1) + \frac{P(i-1)u'(i-1)}{1 + u'(i-1)P(i-1)u'(i-1)}\varepsilon(i) \tag{52}$$

$$\varepsilon(i) = y'(i) - \hat{y}'(i) \tag{53}$$

$$P(i) = \frac{1}{\lambda_1(i)}\left(P(i-1) - \frac{\lambda_2(i)P(i-1)u'(i-1)u'(i-1)P(i-1)}{\lambda_1(i) + \lambda_2(i)u'(i-1)P(i-1)u'(i-1)}\right) \tag{54}$$

where $P(0) = P(0)^T > 0, 0 < \lambda_1(i) \leq 1\ ,\ 0 \leq \lambda_2(i) < 2$.

[Adaptive control law]

$$u_1(i) = \left(v(i) + G_1(z^{-1})\bar{v}(i)\right)\frac{1}{\hat{K}(i)} \tag{55}$$

$$u_2(i) = \left(v(i) + G_2(z^{-1})\bar{v}(i)\right)\frac{1}{\hat{K}(i)} \tag{56}$$

Figure 1 shows the structure of the adaptive control system based on 2-delay relative degree model.

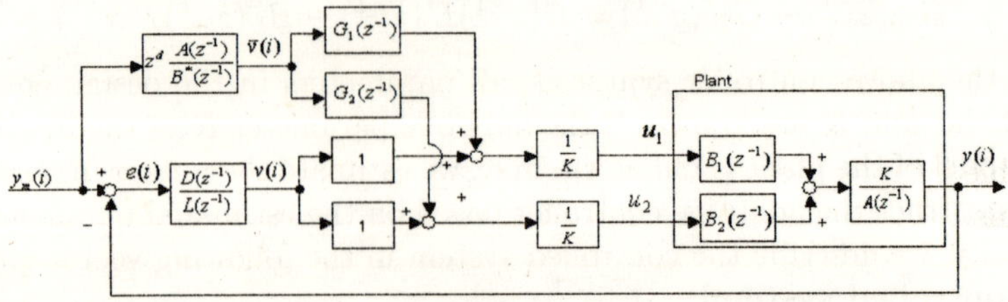

Fig. 1. Structure of proposed adaptive control system

4 Validity of 2-delay Relative Degree Model

We recommend the 2-delay relative degree model as a plant model. However, the validity of the 2-delay relative degree model of the plant should be investigated.

In order to investigate the effectiveness of the proposed method, first, computer simulations are performed for different types of continuous-time plants having the relative degree greater than one.

For example, we have applied the proposed method to some types of plant with same relative degree and evaluated the control performance using the relative degree model. There are three different plants with relative degree 2.

$$A: \ y(t) = \frac{s+0.6}{(s+0.8)^3}u(t) \tag{57}$$

$$B: \ y(t) = \frac{s+0.1}{s^2(s+0.8)}u(t) \tag{58}$$

$$C: \ y(t) = \frac{(s+0.6)^2}{(s+0.8)^4}u(t) \tag{59}$$

For these three types of plats, the same relative degree models are obtained as follows.

$$(1-2z^{-1}+z^{-2})y(i) = z^{-1}K\frac{T^2}{8}(3+z^{-1})u_1(i) + z^{-1}K\frac{T^2}{8}(1+3z^{-1})u_2(i) \tag{60}$$

Based on this model, the model following control system is designed. The design parameters of the control system are set at

$$A^*(z^{-1}) = (1-a^*z^{-1})^2, a^* = 0.98 \tag{61}$$

$$B^*(z^{-1}) = 1 \tag{62}$$

The control performances for each model are measured by the cost function with the desired output $y_m(t) = \sin(2\pi t) + \sin(3\pi t) + \sin(5\pi t)$.

$$IAE = \int_0^{T_0} |e(t)|\,dt \tag{63}$$

Table 1 shows the results for three types of plants with different sampling intervals.

Table 1. Performance of control system based on relative degree model

Controlled System	System A	System B	System C
IAE for T/2 = 2.5[msec]	0.1924	0.4590	0.5029
IAE for T/2 = 5[msec]	0.5604	1.4532	1.7120

These results show that the practical performance can be obtained by choosing the appropriate short sampling interval. However, the achieved performance of the control system varies depending on the characteristics of the controlled system. Thus, some guideline for selecting the sampling interval is needed.

To clarify the relation between the time constant of the controlled system and the appropriate sampling interval, simulation based evaluation is carried out for the following plant.

$$y(t) = \frac{\tau}{(\tau s + 1)s} u(t) \tag{64}$$

This plant is a simplified model of typical servo system with time constant τ. For this plant, the same 2-delay input control system as described above is applied with different sampling interval T and different frequency of the desired output f. Figure 2 shows the performance index for different sampling condition T/τ and frequency f. In the left figure, we can see that the worst case occurred when $T * f \approx 0.0005$ for different T/τ. The right figure shows the relation between T/τ and $IAE * f$ for worst case. Based on this property, we can select the appropriate sampling interval which achieves the desired performance.

From these numerical analysis for some types of plants, it is confirmed that rather wide class of plant can be well modelled and controlled by the 2-delay relative degree model based controller.

5 Application of Proposed Method

Next, as a practical situation in the application, we apply this method to a detailed model of an electro-hydraulic servo system as shown in Fig. 3. The control objective is the position control of the mechanical system. The model

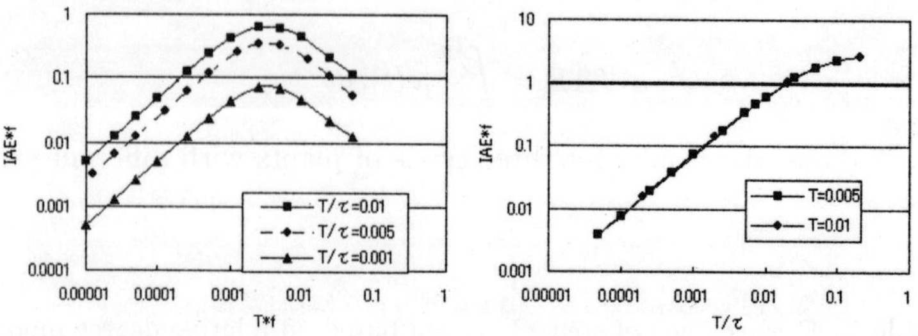

(a) Control performance vs. $T * f$ for different T/τ

(b) Control performance vs. T/τ for worst case

Fig. 2. Performance of control system for different conditions

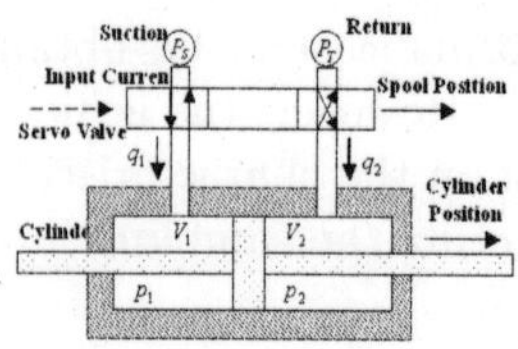

Fig. 3. Schematic diagram of the electro-hydraulic servo system

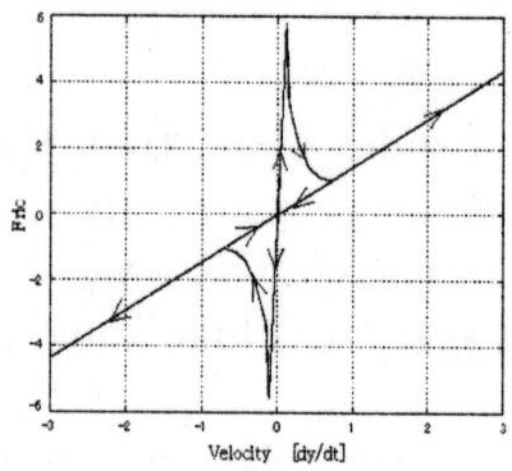

Fig. 4. Non-linear friction property of servo valve

which contains some realistic non-linearities including non-linear friction as shown in Fig. 4, is originally proposed by the adaptive and learning control research group in the society of instrument and control engineers (SICE) for benchmarking many types of adaptive control techniques [13] (See Appendix).

In this case, it can be assumed that the continuous-time characteristics of this system is simply described by Eq. (64) .

For this servo system, the adaptive controller can be designed based on the relative degree model described by Eq. (60). Only typical simulation results for different cylinder load are shown (Figs. 6-7).

In these figures, the upper trace shows the output, the middle and the lower show input and estimated parameter respectively.

From these figures, it can be seen that the proposed adaptive controller works well for different cylinder load by adjusting only one parameter.

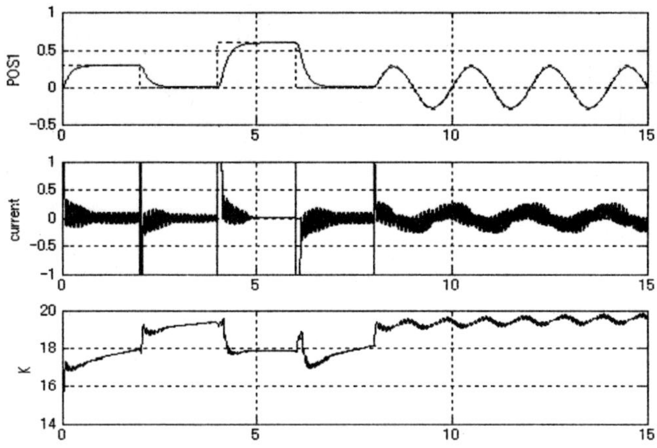

Fig. 5. Simulation result for light load ($M_t = 1$)

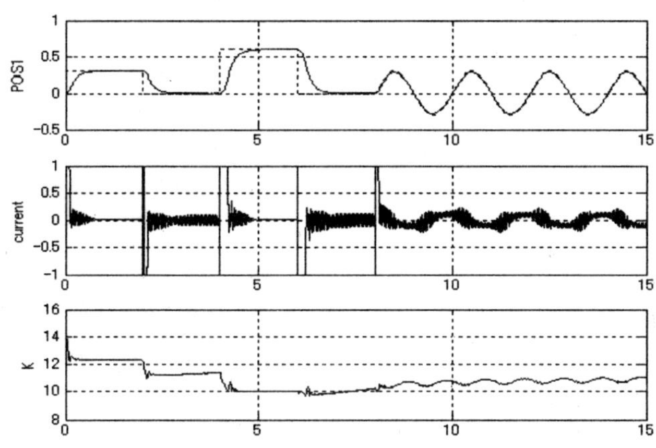

Fig. 6. Simulation result for middle load ($M_t = 2$)

6 Conclusions

We have presented the discrete-time adaptive control system using 2-delay limiting-zero model, especially to handle the continuous-time system having the relative degree greater than one. The effectiveness of the algorithm was confirmed from the simulation studies for simple controlled system and the detailed model of electro-hydraulic servo system. In the later case, the pro-

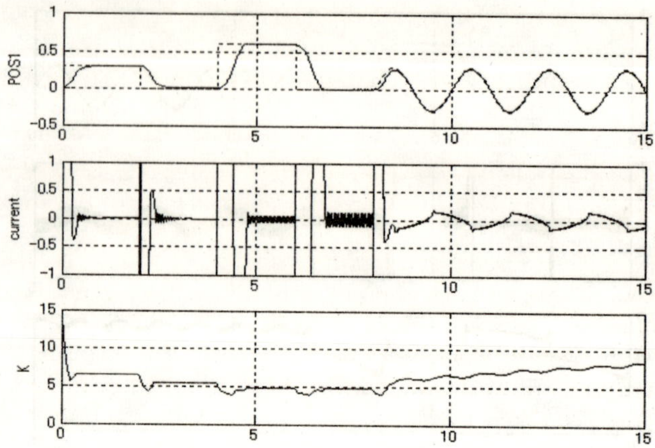

Fig. 7. Simulation result for heavy load ($M_t = 5$)

posed method has a very simple structure with only one estimated parameter. So the adaptive controller is easily implemented with a microprocessor.

References

1. Åström, K.J., P.Hagander and J. Sternby. (1984). Zeros of Sampled Systems. *Automatica* 20-1, pp.31- 38.
2. Fujii, S. and N.Mizuno. (1981). A Discrete Model Reference Adaptive Control Using an Autoregressive Model with Dead Time of the Plant. *Preprints of 8th IFAC Congress,* VII,pp.120-125.
3. Fujii, S., N.Mizuno, Y. Nomura. (1984). A Discrete Multivariable Model Reference Adaptive Control Algorithm with Application to a Real Plant. *Preprints of 9th IFAC Congress*, VII,pp. 99-104.
4. Goodwin, G.C., R. Lozano Leal, D.Q.Mayne and R.H.Middleton. (1986). Rapprochement between Continuous and Discrete Model Reference Adaptive Control. *Automatica,* 22-2, pp.199-207.
5. Mizuno, N and S. Fujii (1989). Discrete-Time Adaptive Control for Continuous-Time Systems Using Limiting-Zero Model and Its Application, *Preprints of IFAC Symposium on Adaptive Systems in Control and Signal Processing,* Vol.1, 237-242.
6. Miyasato, Y. (1992), Model reference adaptive control for non minimum phase systems by 2-delay feedback. *Proceedings of the IFAC Symposium on Adaptive Systems in Control and Signal Processing, Grenoble, France,* pp.197-202.
7. Mizuno, N. and Y. Fujiwara, (1996). Extended 2-delay model reference adaptive controllers with reduced input amplitude, *Preprints of 13th IFAC World Congress,* **Vol. K**, 7-12.
8. Mizuno, N. (1997). An Adaptive Tracking Control System with 2-Delay Feedforward Compensarors, *Proceedings of ECC '97,* WE-M-C2, pp. 1-6

9. Mizuno, N. and E. Sato. (1998). Extended 2-delay model reference adaptive control algorithms with application to a real system, *Preprints of IEEE CCA'98*, pp.1115-1119.
10. Mizuno, N. and E.Sato. (1999). Design And Evaluation Of Adaptive Tracking Controllers Using 2-Delay Input, *Preprints of 14th IFAC World Congress,* Vol.I, pp.221-226.
11. Mizuno, N. and A.Sato. (2005). Discrete-Time Adaptive Control for Continuous-Time Systems Using 2-Dely Limiting-Zero Model, *Preprints of 16th IFAC World Congress,* (to appier).
12. Mita, T., Y.Chida, Y.Kaku and H.Numasato. (1990), Two delay digital control and its applications-avoiding the problems on unstable limiting zeros. *IEEE Trans. Automat. Contr.*, AC-35, pp.962-970.
13. Masuda, S, N. Mizuno, T. Yamamoto, I. Mizumoto and T. Fukao (2001).On an Electro-Hydraulic Servo Benchmark Problem for Adaptive Control Techniques, *J. of the So. of Instrum. And and Control Eng.*,40-10, pp.6758-773.

A Model of Electro-hydraulic Servo Cylinder

The dynamics of the hydraulic servo system is separated into four parts.

1) Spool Dynamics of the Servo Valve

$$x_s = \frac{\omega_s^2}{s^2 + 2\zeta\omega_s + \omega_s^2} i \tag{A.1}$$

where x_s denotes the spool position of the ideal servo valve, and i is the command current input to the servo valve .

2) Flow Characteristics of the Oil

For $x_s \geq 0$:

$$q_1 = sgn(P_s - P_1) q_r \bar{x}_s \sqrt{\frac{|P_s - P_1|}{P_R}} \tag{A.2}$$

$$q_2 = -sgn(P_2 - P_T) \tag{A.3}$$

For $x_s < 0$:

$$q_1 = -sgn(P_1 - P_T) q_r \bar{x}_s \sqrt{\frac{|P_1 - P_T|}{P_R}} \tag{A.4}$$

$$q_2 = sgn(P_S - P_2) q_r \bar{x}_s \sqrt{\frac{|P_S - P_2|}{P_R}} \tag{A.5}$$

where q_1 , q_2 are the oil flow to the chamber 1 and chamber 2 respectively. P_1 and P_2 denote the oil pressure in each chamber. q_r is the nominal flow of the servo valve. P_S and P_T are the source oil pressure and the return oil pressure respectively. P_R denotes the nominal pressure reduction of the cylinder.

Moreover, $\bar{x}_s$ is the real opening of the servo valve, which can be modeled using the wrap of the valve Δ as follows.

$$\bar{x}_s = \begin{cases} 0, & |x_s| < \Delta \\ |x_s| - \Delta, & \Delta \leq |x_s| < 1 + \Delta \\ 1, & 1 + \Delta \leq |x_s| \end{cases} \tag{A.6}$$

3) Pressure Dynamics of the Cylinder

$$\frac{dP_1}{dt} = \frac{K_\phi}{V_{10} + A_1 y}\left(q_1 - A_1\frac{dy}{dt} - C_i(P_1 - P_2) - C_o P_1\right) \tag{A.7}$$

$$\frac{dP_2}{dt} = \frac{K_\phi}{V_{20} - A_2 y}\left(q_2 + A_2\frac{dy}{dt} + C_i(P_1 - P_2) - C_o P_2\right) \tag{A.8}$$

where K_ϕ denotes the volume compliance coefficient of the oil. C_i and C_o are the inner and outer leak coefficient of the cylinder respectively. V_{10} and V_{20} denote the volumes of the chamber 1 and 2. A_1 and A_2 are the cross areas of chamber 1 and 2.

4) Cylinder Dynamics

$$M_t = \frac{d^2 y}{dt^2} = A_1 P_1 - A_2 P_2 - Fric\left(y, \frac{dy}{dt}\right) \tag{A.9}$$

where M_t is the total mass of the cylinder and load, and $Fric$ denotes the non-linear friction described by Fig. 4.

The parameters in the model are given by the following Table 2. All values in this table are normalized by the real specification of the servo system.

Table 2. Parameters in the model

Name and Symbol	Value
Wrap:Δ	0.0027
Normal flow of the servo valve:q_r	9.8754
Source pressure:P_S	22.73
Return pressure:P_T	0.5413
Normal pressure reduction:P_R	9.7431
Volume compliance coefficient of the oil: K_ϕ	866.05
Inner leak coefficient: C_i	0.0261
Outer leak coefficient:C_o	0.0106
Natural resonance frequency: ω_s	$2\pi\times58$
Dumping factor: ζ	0.65

Nonlinear Multivariable Decoupling Control Using Neural Networks

Tianyou Chai[1], Lianfei Zhai[1] and Heng Yue[2]

[1] Key Laboratory of Process Industry Automation, Ministry of Education, China
[2] Research Center of Automation, Northeastern University, Shenyang, 110004, P. R. China tychai@mail.neu.edu.cn

1 Introduction

In many complex industrial processes, there are usually strong couples between control loops. These coupling effects often result in undesired performance for control systems. As such, it is important to investigate the realization of decoupling control, a topic that has received an increased attention in control engineering practice. Decoupling control was initially developed for deterministic linear systems. Many typical decoupling control approaches for linear systems have been proposed. Kavangh[6] realized decoupling control through a pre-decoupling compensator that transforms the controlled transfer function matrix into a diagonal matrix or diagonal dominance. Falb and Wolovich[5] achieved decoupling of state equation through state feedbacks. The decoupling method of inverse-Nyquist array was put forward by Rosenbrock[9] and the decoupling method of Bristol-Shinskey was presented by Shinskey[10].

For linear systems with unknown parameters, many adaptive decoupling control algorithms have been proposed. For example, multivariable adaptive decoupling controllers were presented by McDermott and Mellichamp[8], and Lang, et al.[7], which combined the decoupling design with a self-tuning control structure. Also, pole placement adaptive decoupling control algorithm was developed by Wittenmark, et al.[12]. In the approaches of [1], [2] and [3], the coupling effects among control loops were viewed as measurable disturbances so that they can be eliminated through feedforward compensation.

To deal with coupling in nonlinear systems, neural networks and fuzzy methods were adopted. Relative approaches can be found in [13], [14] and [15]. The above decoupling control methods not only require the controlled processes with accurate mathematic model, but also are difficult to be implemented in DCS.

In recent years, distributed control systems (DCS) have become more and more popular in industrial process control because of their high reliability. In practice, control engineers divide multivariable industrial processes into many

SISO systems without considering the coupling effects among each loop so as to use the standard control module software of DCS to realize the required control tasks. If the coupling effects are very strong, they would generally result in bad performance of the control systems or even lead to an unstable operation. Therefore, it is very important to develop effective decoupling controllers that are applicable in DCS.

This paper is organized as follows. A statement of the problem is given in Section 2. Combining generalized minimum variance control based decoupling design with a feedforward compensator for unmodeled dynamics, a decoupling control law for nonlinear multivariable systems is proposed in Section 3. In Section 4, a nonlinear multivariable decoupling control algorithm using neural network is presented, which consists of the proposed control law, the recursive least square (RLS) identify algorithm, and neural networks. In Section 5, simulation studies are described to illustrate the effectiveness and robust performance of the proposed decoupling control algorithm. The industrial experiments of proposed decoupling control algorithm in a ball mill pulverizing system of 200MW power units are given in Section 6, and the conclusions are drawn in Section 7.

2 Statement of the Problem

The industrial process to be controlled is assumed to be a n-input-n-output nonlinear system described by

$$\mathbf{y}(t+k) = \mathbf{f}[\mathbf{y}(t), \cdots, \mathbf{y}(t-n_a+1), \mathbf{u}(t), \cdots, \mathbf{u}(t-n_b)] \tag{1}$$

where $\mathbf{y} = [y_1, \cdots, y_n]^T \in R^n$ and $\mathbf{u} = [u_1, \cdots, u_n]^T \in R^n$ are system output and input vectors, respectively; n_a and n_b are system orders; $\mathbf{f}(\cdot) = [f_1(\cdot), \cdots, f_n(\cdot)]^T \in R^n$ is a vector-valued nonlinear function that is assumed unknown and continuously differentiable. Qualitatively, the purpose of decoupling control is to determine an input control law so that all the signals in the closed loop system remain bounded, whilst the output vector $\mathbf{y}$ is confined in some pre-specified range, and the influence of the couples is suppressed as small as possible.

Since the industrial process usually works around its operating point, on its operating point (Suppose it is the origin.), equation (1) can be divided into an approximate linear model (with lower order) plus a higher order nonlinear term through Taylor's expansion, as expressed in the following:

$$\begin{aligned}\mathbf{y}(t+k) &= -\bar{\mathbf{A}}\mathbf{y}(t) + \mathbf{B}\mathbf{u}(t) + \mathbf{v}[\mathbf{y}(t), \cdots, \mathbf{y}(t-n_a+1), \mathbf{u}(t), \cdots, \mathbf{t}(t-n_b)] \\ &= \Theta \cdot \mathbf{x}(t) + \mathbf{v}[\mathbf{x}(t)] \end{aligned} \tag{2}$$

where $\bar{\mathbf{A}}$ and $\mathbf{B}$ are polynomial matrices in terms of the unit back shift operator z^{-1} with $\bar{\mathbf{A}}$ being diagonal. Both $\bar{\mathbf{A}}$ and $\mathbf{B}$ can be expressed in the following format.

$$\bar{\mathbf{A}} = \mathbf{A}_1 + \mathbf{A}_2 z^{-1} + \cdots + \mathbf{A}_{n_a} z^{-n_a+1}$$

$$\mathbf{B} = \mathbf{B}_0 + \mathbf{B}_1 z^{-1} + \cdots + \mathbf{B}_{n_b} z^{-n_b}$$

The data vector and the coefficient vector are defined as

$$\mathbf{x}^T(t) = [-\mathbf{y}^T(t), \cdots, -\mathbf{y}^T(t-n_a+1), \mathbf{u}^T(t), \cdots, \mathbf{u}^T(t-n_b)]$$

$$\Theta = [\mathbf{A}_1, \cdots, \mathbf{A}_{n_a}; \mathbf{B}_0, \cdots, \mathbf{B}_{n_b}] = [\theta_1, \cdots, \theta_n]^T$$

The nonlinear term represents the unmodeled dynamics, whilst it can be presented by equation

$$\begin{aligned} \mathbf{v}[\mathbf{x}(t)] &= \mathbf{y}(t+k) - \Theta \cdot \mathbf{x}(t) \\ &= \mathbf{f}(\cdot) + \bar{\mathbf{A}}(z^{-1})\mathbf{y}(t) - \mathbf{B}(z^{-1})\mathbf{u}(t) \end{aligned} \tag{3}$$

which is the function about high order terms of $\mathbf{x}(t)$.

3 The Control Law

To achieve decoupling design, equation (2) can be transformed into

$$\mathbf{A}\mathbf{y}(t+k) = \bar{\mathbf{B}}\mathbf{u}(t) + \bar{\bar{\mathbf{B}}}\mathbf{u}(t) + \mathbf{v}[\mathbf{x}(t)] \tag{4}$$

where $\mathbf{A} = \mathbf{I} + z^{-1}\bar{\mathbf{A}}$, $\mathbf{B} = \bar{\mathbf{B}} + \bar{\bar{\mathbf{B}}}$, $\mathbf{I}$ denotes the identity matrix, $\bar{\mathbf{B}}$ is a diagonal polynomial matrix, $\bar{\bar{\mathbf{B}}}$ is a polynomial matrix with zero diagonal elements.

Combining the performance index [2] with a nonlinear feedforward compensator of $\mathbf{v}[\mathbf{x}(t)]$, we introduce the following performance index:

$$\mathbf{J} = \|\mathbf{P}\mathbf{y}(t+k) - \mathbf{R}\mathbf{w}(t) + \mathbf{Q}\mathbf{u}(t) + \mathbf{S}\mathbf{u}(t) + \mathbf{K}\mathbf{v}[\mathbf{x}(t)]\|^2 \tag{5}$$

where $\mathbf{P}$, $\mathbf{Q}$, $\mathbf{R}$ and $\mathbf{K}$ are the diagonal weighting polynomial matrices, $\mathbf{S}$ is a weighting polynomial matrix with zero diagonal elements, $\mathbf{w} = [w_1, \cdots, w_n]^T \in R^n$ denotes the reference signal.

Introduce the Diophantine equation

$$\mathbf{P} = \mathbf{F}\mathbf{A} + z^{-k}\mathbf{G} \tag{6}$$

where $\mathbf{F}$ and $\mathbf{G}$ are the diagonal polynomial matrices , the orders of them are $n_g = \max\{n_a - 1, n_p - 1\}$ and $n_f = k - 1$ respectively, n_p denotes the order of $\mathbf{P}$. Multiplying equation (4) by $\mathbf{F}$ and using equation (6), the following equation can be obtained

$$\mathbf{P}\mathbf{y}(t+k) = \mathbf{G}\mathbf{y}(t) + \mathbf{F}\bar{\mathbf{B}}\mathbf{u}(t) + \mathbf{F}\bar{\bar{\mathbf{B}}}\mathbf{u}(t) + \mathbf{F}\mathbf{v}[\mathbf{x}(t)] \tag{7}$$

It can be proved that the optimal controller minimizing (5) is described by

$$[\mathbf{F}\bar{\mathbf{B}} + \mathbf{Q}]\mathbf{u}(t) = \mathbf{R}\mathbf{w}(t) - \mathbf{G}\mathbf{y}(t) - [\mathbf{F}\bar{\bar{\mathbf{B}}} + \mathbf{S}]\mathbf{u}(t) - [\mathbf{F} + \mathbf{K}]\mathbf{v}[\mathbf{x}(t)] \tag{8}$$

Multiplying equation (8) by $\bar{\mathbf{B}}$, multiplying equation (4) by $[\mathbf{F}\bar{\mathbf{B}}+\mathbf{Q}]$ and using equation (6), the following equation can be received

$$[\mathbf{P}\bar{\mathbf{B}}+\mathbf{Q}\mathbf{A}]\mathbf{y}(t+k)=\bar{\mathbf{B}}\mathbf{R}\mathbf{w}(t)+[\mathbf{Q}\bar{\bar{\mathbf{B}}}-\bar{\mathbf{B}}\mathbf{S}]\mathbf{u}(t)+[\mathbf{Q}-\bar{\mathbf{B}}\mathbf{K}]\mathbf{v}[\mathbf{x}(t)] \quad (9)$$

Multiplying equation (4) by $\mathbf{G}$, multiplying equation (8) by $\mathbf{A}$ and using equation (6) lead

$$\{\mathbf{P}\mathbf{B}+\mathbf{A}[\mathbf{Q}+\mathbf{S}]\}\mathbf{u}(t)=\mathbf{A}\mathbf{R}\mathbf{w}(t)-[\mathbf{A}\mathbf{K}+\mathbf{P}]\mathbf{v}[\mathbf{x}(t)] \quad (10)$$

Equation (10) can be regarded as the closed loop equation of the control system. From equations (9) and (10), it can be seen that if $\mathbf{P},\mathbf{Q},\mathbf{R},\mathbf{K}$ and $\mathbf{S}$ are chosen as follows:

$$\mathbf{P}\bar{\mathbf{B}}+\mathbf{Q}\mathbf{A}=\bar{\mathbf{B}}\mathbf{R} \quad (11)$$

$$\mathbf{Q}\bar{\bar{\mathbf{B}}}=\bar{\mathbf{B}}\mathbf{S} \quad (12)$$

$$\mathbf{Q}=\bar{\mathbf{B}}\mathbf{K} \quad (13)$$

$$|\mathbf{P}\mathbf{B}+\mathbf{A}[\mathbf{Q}+\mathbf{S}]|\neq 0, \qquad |z|>1 \quad (14)$$

then the closed loop system will be stable whilst a decoupling control effect and the tracking errors can be eliminated. (11)-(14) can be exactly satisfied in a steady state. In addition, the least square solution can be obtained for a transient state[8].

Furthermore, equation (8) can be expressed by

$$\bar{\mathbf{H}}\mathbf{u}(t)=\mathbf{R}\mathbf{w}(t)-\mathbf{G}\mathbf{y}(t)-\bar{\bar{\mathbf{H}}}\mathbf{u}(t)-[\mathbf{F}+\mathbf{K}]\mathbf{v}[\mathbf{x}(t)] \quad (15)$$

where $\bar{\mathbf{H}}=\mathbf{F}\bar{\mathbf{B}}+\mathbf{Q}$ is a diagonal polynomial matrix, $\bar{\bar{\mathbf{H}}}=\mathbf{F}\bar{\bar{\mathbf{B}}}+\mathbf{S}$ is a polynomial matrix with zero diagonal elements. The significance of equation (15) can be seen from the expression of controller in every single channel

$$\begin{aligned} u_i(t)=\frac{1}{\bar{h}_{ii}}\cdot\{r_{ii}w_i(t)-g_{ii}y_i(t)-[\bar{\bar{h}}_{i1}u_1(t)+\cdots+\bar{\bar{h}}_{i(i-1)}u_{i-1}(t) \\ +\bar{\bar{h}}_{i(i+1)}u_{i+1}(t)+\cdots+\bar{\bar{h}}_{in}u_n(t)]-(f_{ii}+k_{ii})v_i[\mathbf{x}(t)]\} \end{aligned} \quad (16)$$

where r_{ij}, g_{ij}, f_{ij}, k_{ij}, $\bar{h}_{ij}$ and $\bar{\bar{h}}_{ij}$ are the elements in the ith row and the jth column of matrices $\mathbf{R},\mathbf{G},\mathbf{F},\mathbf{K},\bar{\mathbf{H}}$ and $\bar{\bar{\mathbf{H}}}$ respectively. From (16), it can be known that the ith control input $u_i(t)$ only relies on the output $y_i(t)$, the reference input $w_i(t)$ and the unmodeled dynamics $v_i[\mathbf{x}(t)]$ of the ith control loop and the other control inputs $u_1(t),\cdots,u_{i-1}(t),u_{i+1},\cdots,u_n(t)$. In this way, it can be easily implemented on a DCS. In DCS, control modules are based on the single loop for controller design. When the unmodeled dynamics are so small to be omitted, it will not be considered in the control law. When the compensating terms of decoupling and suppressing for unmodeled dynamics are not used, this control law becomes a conventional controller, which is easy to be implemented on DCS using standard control modules. Since the other control inputs and the estimation of $\mathbf{v}[\mathbf{x}(t)]$ are easily to be obtained, when the compensating signal is introduced into such single loop controller, the decoupling control is easy to be realized on DCS.

4 Nonlinear Multivariable Decoupling Control Algorithm

4.1 Establishing the model for controller design

The process (2) can be written into the expression in every single channel, as expressed in the following:

$$y_i(t+k) = \mathbf{x}^T(t)\theta_i + v_i[\mathbf{x}(t)], \quad i = 1, \cdots, n \tag{17}$$

Offline identify the parameters of the process model using the following recursive least square (RLS) algorithm.

$$\hat{\theta}_i(t+1) = \hat{\theta}_i(t) + \mathbf{K}_i(t+1)[y_i(t+k) - \mathbf{x}^T(t)\hat{\theta}_i(t)] \tag{18}$$

$$\mathbf{K}_i(t+1) = \frac{\mathbf{P}_i(t)\mathbf{x}(t)}{1 + \mathbf{x}^T(t)\mathbf{P}_i(t)\mathbf{x}(t)} \tag{19}$$

$$\mathbf{P}_i(t+1) = \mathbf{P}_i(t) - \frac{\mathbf{P}_i(t)\mathbf{x}(t)\mathbf{x}^T(t)\mathbf{P}_i(t)}{1 + \mathbf{x}^T(t)\mathbf{P}_i(t)\mathbf{x}(t)} \tag{20}$$

After identify the model parameters offline, the process model for controller design can be obtained.

4.2 Estimation of the unmodeled dynamics using neural networks

When the unmodeled dynamics $\mathbf{v}[\mathbf{x}(t)]$ is very small, it can be omitted out of the controller design. Otherwise, compensate it using a feedforwad compensator. As for the estimation of the unmodeled dynamics, a Back-Propagation (BP) neural network is employed, as shown in Fig. 1, which has one hidden layer and the linear output. It has $n(n_a + n_b + 1)$ inputs and n outputs

$$\hat{\mathbf{v}}(t) = \mathbf{NN}[\mathbf{x}(t)] \tag{21}$$

where **NN** denotes the format of the neural network.

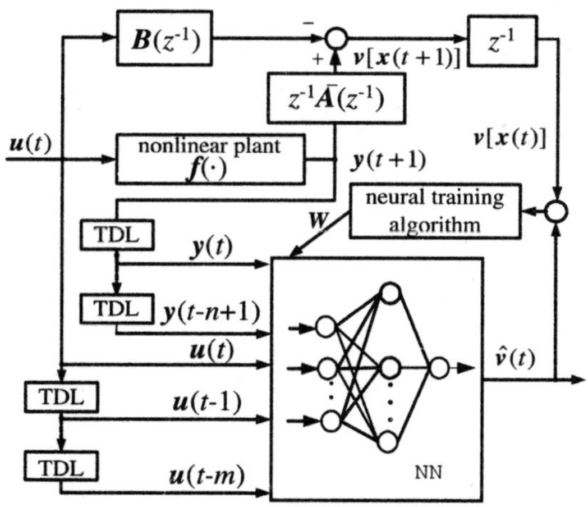

Fig. 1. Estimating $\mathbf{v}[\mathbf{x}(t)]$ from neural network

Suppose the hidden layer of the neural network has μ nodes. Adjust the connection weighting matrices W_{ij} and V_{ij} by the following formulas. The error function E is defined as:

$$E = \frac{1}{2}\sum_{l=1}^{n}\{v_l[x(t)] - \hat{v}_l(t)\}^2 \tag{22}$$

Adjust the connection weighting matrices using the following procedure:

$$W_{ji}(t+1) = W_{ji}(t) - \eta \cdot \frac{\partial E}{\partial W_{ji}(t)} + \alpha \cdot \Delta W_{ji}(t) \tag{23}$$

$$V_{lj}(t+1) = V_{lj}(t) - \eta \cdot \frac{\partial E}{\partial V_{lj}(t)} + \alpha \cdot \Delta V_{lj}(t) \tag{24}$$

$$\frac{\partial E}{\partial W_{ji}(t)} = -[z_j(t)(1 - z_j(t))d_i(t)]\sum_{l=1}^{n}\{(v_l[x(t)] - \hat{v}_l(t))\hat{v}_i(t)W_{ji}(t)\} \tag{25}$$

$$\frac{\partial E}{\partial V_{lj}(t)} = -\{v_l[x(t)] - \hat{v}_l(t)\}z_j(t) \tag{26}$$

$$i = 1, \cdots, n(n_a + n_b + 1); j = 1, \cdots, \mu; l = 1, \cdots, n$$

where W_{ji} is the weighting matrix from the ith element of the input layer to the jth element of the hidden layer; V_{lj} denotes the weighting matrix from the jth element of the hidden layer to the lth element of the output layer; η is the learning rate and α is the momentum factor; ∂W_{ji} and ∂V_{lj} are the value of the previous weight changes; z_j is the jth node output of the hidden layer; d_i denotes the ith node input of the neural network.

4.3 Nonlinear multivariable decoupling control algorithm

The nonlinear multivariable decoupling control algorithm can be divided into two parts: 1) offline controller design and 2) adaptive compensating algorithm for the unmodeled dynamics. The algorithm of offline controller design can be summarized as follows. *Step1.* Choose the system orders n_a and n_b; *Step2.* Identify the parameters of (4) using the RLS algorithm (18)-(20); *Step3.* Properly choose $\mathbf{P}, \mathbf{R}, \mathbf{Q}, \mathbf{S}$ and $\mathbf{K}$ so as to satisfy (11)-(14); *Step4.* Calculate $\mathbf{F}$ and $\mathbf{G}$ by using (6); *Step5.* Establish neural network **NN** to approximate the unmodeled dynamics $\mathbf{v}[\mathbf{x}(t)]$, train it in batch and adjust weighting matrices by using (23)-(26).

The decoupling control algorithm with an adaptive compensator for the unmodeled dynamics can be summarized as follows. *Step1.* Read input-output data to construct vector $\mathbf{x}(t)$; *Step2.* Estimate $\mathbf{v}[\mathbf{x}(t)]$ using the neural network **NN** online (Here, the element $\mathbf{u}(t)$ in $\mathbf{x}(t)$ is substituted by $\mathbf{u}(t-1)$ because $\mathbf{u}(t)$ is unknown); *Step3.* Calculate the control input $\mathbf{u}(t)$ using equation (8), the chosen weighting polynomial matrices $\{\mathbf{P}, \mathbf{Q}, \mathbf{R}, \mathbf{K}$ and $\mathbf{S}\}$ and calculated matrices $\{\mathbf{F}$ and $\mathbf{G}\}$; *Step4.* Impose $\mathbf{u}(t)$ on the process and get the new output $\mathbf{y}(t)$ of the process; *Step5.* Obtain the tutorial signal from equation (3); *Step6.* Train the neural networks **NN** once using thetraining algorithm (23)-(26); *Step7.* Let $t = t + 1$, and return to *Step1.*

5 Simulations

To illustrate the effective and robust performance of the proposed decoupling control algorithm, simulation studies are described in this section.

Example 5.1 The process to be controlled is described by the following equations.

$$\begin{aligned} y_1(t+1) &= 0.4\sin[y_1(t)]/[0.5 + y_2(t-1)] \\ &\quad - 0.45y_1(t-1) + 1.1u_1(t) + 0.8u_2(t) + \xi_1(t+1) \qquad (27) \\ y_2(t+1) &= \cos[y_2^2(t)] - 0.3y_2(t-1) + 1.25u_2(t) \\ &\quad + 0.29\cos[u_1^2(t)] + 0.1u_2(t-1) + 0.32u_1(t-1) + \xi_2(t+1) \qquad (28) \end{aligned}$$

where, $\xi_1(t+1)$ and $\xi_2(t+1)$ are random disturbances. The process model for controller design can be expressed as $\mathbf{Ay}(t+1) = \mathbf{Bu}(t)$. Select $n_a = 2$ and $n_b = 1$. Inspire the process by a small noise series in the neighbourhood of its equilibrium, get the input-output data of the process, then identify the parameters of (4) offline by using the RLS algorithm (18)-(20). The estimations of **A** and **B** can be obtained, which are presented by

$$\hat{\mathbf{A}} = \begin{pmatrix} 1 - 0.3102z^{-1} + 0.445z^{-2} & 0 \\ 0 & 1 + 1.0809z^{-1} + 0.3048z^{-2} \end{pmatrix}$$

$$\hat{\mathbf{B}} = \begin{pmatrix} 1.1077 & 0.7982 \\ 0.3155z^{-1} & 1.2481 + 0.1016z^{-2} \end{pmatrix}$$

A BP neural network $\mathbf{NN}_1$ is used, where the input and output vectors are expressed by

$$\begin{aligned} \mathbf{x}(t) &= [y_1(t-1), y_1(t), y_2(t-1), y_2(t), u_1(t-1), u_1(t), u_2(t-1), u_2(t)] \\ \mathbf{v}[\mathbf{x}(t)] &= [v_1(t), v_2(t)]^T = \mathbf{y}(t+1) - \Theta\mathbf{x}(t) \end{aligned}$$

and only one hidden layer is employed with 20 nodes. Adjust the connection weighting matrices of the neural network $\mathbf{NN}_1$ by using (23)-(26), with $\alpha = 0.5$ and $\eta = 0.8$. Matrices $\mathbf{P}, \mathbf{Q}$ and $\mathbf{R}$ have been chosen as follows $\mathbf{P} = \mathbf{R} = \mathbf{I}$, $Q = diag[\lambda_1(1 - z^{-1}), \lambda_2(1 - z^{-1})]$, where $\lambda_1 = 0.0308$ and $\lambda_2 = 0.1152$.According to (8), the control input $\mathbf{u}(t)$ implemented on the system is

$$\begin{aligned} &\begin{pmatrix} 1.1385 - 0.0308z^{-1} & 0 \\ 0 & 1.3633 - 0.0136z^{-1} \end{pmatrix} \mathbf{u}(t) \\ &= \mathbf{w}(t) - \begin{pmatrix} 0 & 0.7982 \\ 0.3155z^{-1} & 0 \end{pmatrix} \mathbf{u}(t) \\ &- \begin{pmatrix} 0.3102 - 0.445z^{-1} & 0 \\ 0 & -1.0809 - 0.3048z^{-1} \end{pmatrix} \mathbf{y}(t) - \hat{\mathbf{v}}(t) \qquad (29) \end{aligned}$$

The simulation result is shown in Fig.2. From Fig.2, it can be seen that the performance of proposed decoupling controller is very good for the most of time.

Example 5.2 In order to study the robustness of the controller, the parameters of the process are changed but the controller is not changed. The simulation is described as follow. After the parameters in the process (27) and (28) varied, the new process can be described by

$$\begin{aligned} y_1(t+1) &= 0.6\sin[y_1(t)]/[0.6+y_2(t-1)] \\ &\quad - 0.41y_1(t-1) + 1.3u_1(t) + 0.5u_2(t) + \xi_1(t+1) \qquad (30) \\ y_2(t+1) &= \cos[y_2^2(t)] - 0.5y_2(t-1) + 1.5u_2(t) \\ &\quad + 0.4545\cos[u_1^2(t)] + 0.3u_2(t-1) + 0.2u_1(t-1) + \xi_2(t+1) \qquad (31) \end{aligned}$$

Control the new process by using the control input (29), employ another BP neural network $\mathbf{NN}_2$ to estimate the unmodeled dynamics, which has the same structure as $\mathbf{NN}_1$. The simulation result is shown in Fig.3. It can be seen from Fig.3, although the parameters of the process model are varied, the system outputs can track the reference inputs. It shows that the proposed algorithm has the strong robustness for the parameters variation.

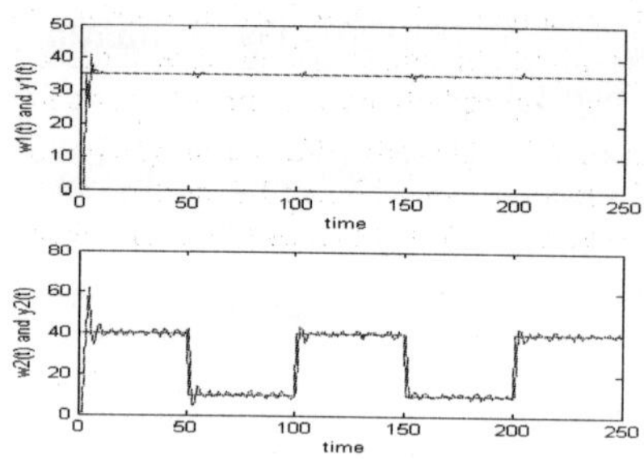

Fig.2 Setpoint responses

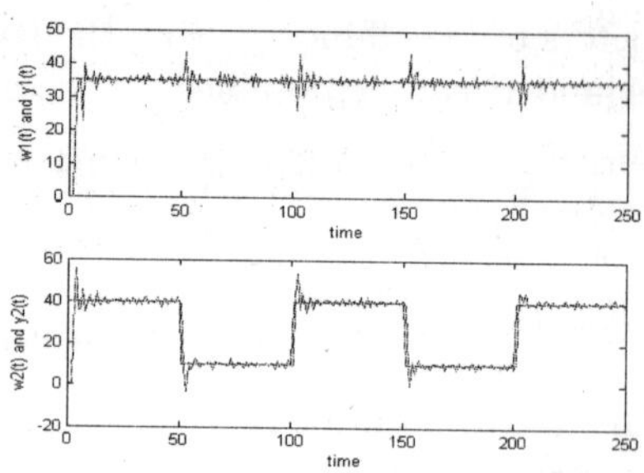

Fig.3 Setpoint responses of the process with varied parameters

Example 5.3 In order to study the adaptive ability of controller to the variation of the process structure, the structure of process is changed but the controller is not changed. The simulation experiment is described as follow. After the structure of the process (27) and (28) varied, the new process can be presented by

$$\begin{aligned} y_1(t+1) &= 0.4\sin[y_1(t)]/[0.5+y_2(t-1)] \\ &\quad - 0.45\sin[y_1(t-1)] + 1.1u_1(t) + 0.8u_2(t) + \xi_1(t+1) \qquad (32) \\ y_2(t+1) &= \cos[y_2^2(t)] - 0.3y_2(t-1) + 1.25u_2(t) \\ &\quad + 0.29\cos[u_1^2(t)] + 0.1u_2(t-1) + 0.32u_1(t-1) + \xi_2(t+1) \qquad (33) \end{aligned}$$

Implement the control input of (29) on the new process, whilst employ another BP neural network $\mathbf{NN}_3$ to estimate the unmodeled dynamics, which has the same structure as to estimate the unmodeled dynamics, which has the same structure as $\mathbf{NN}_1$. The simulation result is shown in Fig.4. From Fig.4, it can be seen that the system outputs still track the reference inputs, through the structure of the process model is varied. It shows that the proposed algorithm also has the strong robustness for the structure variation.

From Fig.3 and Fig.4, it shows that the system outputs can track the reference inputs when the structure and the parameters of the process model are changed. The proposed nonlinear multivariable decoupling control using neural network has strong robust performance.

6 Application on Ball Mill Pulverizing Systems

The proposed intelligent decoupling control system has been applied in ball mill coal-pulverizing systems of 200MW units in a power plant. Simulations and industrial experiments are studied in this section.

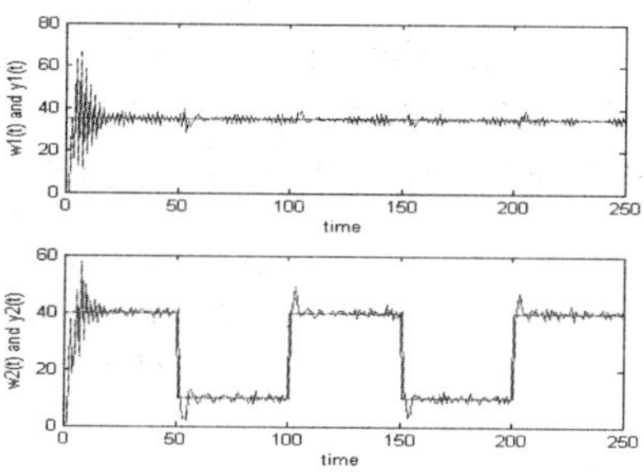

Fig.4 Setpoint responses of the process with varied structure

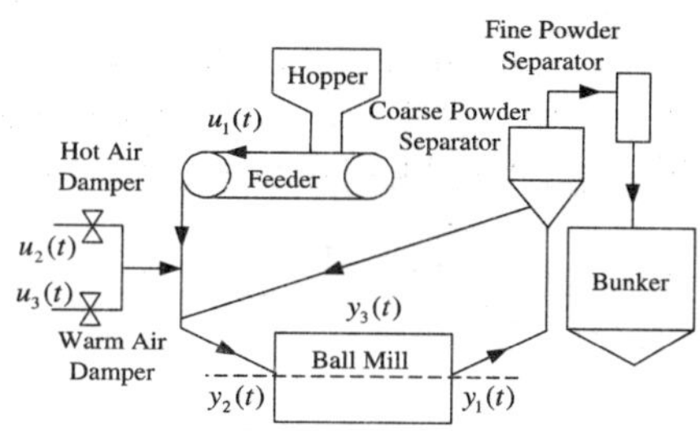

Fig.5 Flow chart of the ball mill pulverizing system

6.1 Process description

Ball mill pulverizing systems are important heat-power equipments in power plants. They are widely used at 200MW power units in China, where thousands of such power generation systems are installed. Ball mill pulverizing systems are used to pulverize raw coal into fine powder and then dry them so that coal powder can be sent into the boilers for burning. The flow chart of ball mill pulverizing system is shown in Fig. 5. The raw coal is sent from hopper to ball mill by coal feeder. In the ball mill, it is pulverized to fine powder by knocking and grinding of iron ball when the ball mill rotates. At the same time, the coal powder is dried and brought out of the ball mill by drying air, which is produced through a mixture of hot air and warm air by adjusting the hot air and warm air damper position. Then it is sent into the separator, where the coarse powder and fine powder are separated. Coarse powder is returned into the ball mill for regrinding and fine powder is sent into the bunker and then to boiler for burning.

The three inputs of this process are the feeder speed, the hot air damper position and the warm air damper position. The three outputs of the plant are the outlet temperature of the mill, the inlet pressure of the mill and the differential pressure between the inlet and outlet of the mill, which represents the mill load. The three outputs must be kept in certain regions respectively so that system works continuously and smoothly.

Since this system has integrated complexities such as strong coupling, serious nonlinearity, large time delay, big variation of boundary condition, the conventional SISO control system cannot be used automatically. As such, the ball mill pulverizing system relies on human operations, leading to various accidents such as mill-blockage, over-temperature and mis-emission of pulverized coal powder. To prevent accidents, human operators are employed to control this system at an uneconomical operating point where a large volume of energy has been wasted. It has therefore become important yet difficult issue on how to realize decoupling control of such systems in China.

6.2 Process model

Tao, et al.[11] established a nonlinear dynamic first principle model, which consists of the energy balance equation in the ball mill as follows:

$$\begin{aligned}\frac{d[(C_{gq}W_{gq}+C_mW_m)T_m]}{dt} &= C_{gz}G_{gz}T_{gz}+C_{lk}G_{lf}T_{lk}+\frac{B_{gm}C_{gm}T_c}{3.6}+Q_o\\ &-\frac{C_mB_mT_m}{3.6}-C_tf(G_{gz}+G_{lf})T_m-Q_c\end{aligned} \tag{34}$$

and the mass balance equation of the form:

$$\frac{dW_m}{dt}=\frac{B_{gm}-B_m}{3.6} \tag{35}$$

Moreover, the equation of inlet pressure P_{in} is given by:

$$\frac{dP_{in}}{dt}=\frac{RT}{V}(G_{gz}+G_{lf}+\frac{B_{gm}\triangle W}{3.6})-\sqrt{\frac{P_{in}-\triangle P+P_0}{R_1}} \tag{36}$$

with

$$\frac{d\triangle P}{dt}=3(1+0.8\mu)\frac{\omega_{thr}^2}{2V_1}(G_i-G_o) \tag{37}$$

$$G_r=\sqrt{\frac{\triangle P_rK_r^2}{f_r(273+T_r)}} \tag{38}$$

$$G_w=\sqrt{\frac{\triangle P_wK_w^2}{f_w(273+T_w)}} \tag{39}$$

being used as the physical models for the differential pressure $\triangle P$, the hot-air and warm-air mass flow rates. The coal feeding rate is given by:

$$B_{gm} = B_{gmax} K_{gm} \tag{40}$$

with input variables $u_1(t) = K_{gm}(t)$, $u_2(t) = K_r(t)$, $u_3(t) = K_w(t)$, and the output variables $y_1(t) = T_m(t)$, $y_2(t) = P_{in}(t)$, $y_3(t) = \triangle P(t)$ with related symbols being defined as follows: $C_{gq}, C_m, C_{gz}, C_{lk}, C_{tf}$ and C_{gm} are the specific heats of steel balls, coal powder, drier air, cold air, ventilation and raw coal; $T_{gz}, T_{lk}, T_m, T_c, T_r$ and T_w are the temperatures of drier air, cold air, mill outlet, raw coal, hot air and warm air; G_{gz} is the mass flow rate of drier air, which embraces that of the hot air G_r and the warm air G_w; G_{lf} is the mass flow rate of air leak; G_i is the mass flow rate of air at inlet of ball mill; G_o is the mass flow rate of air out of ball mill; W_{gq} is the ball mass inside the mill; W_m is the mill load; B_m is powder yield of the mill; B_{gm} is the coal feeding rate; Q_o is the quantity of heat generated from friction and knocking, and Q_c is the quantity of heat consumed by evaporation of water in raw coal; $\triangle W$ is quantity of water evaporation during milling; P_0 is the sum of outlet pressure of mill exhauster and zero position pressure; R_1 is the sum of resistances of coarse powder separator, fine powder separator and powder elevation; V is the mill volume; R is the ideal gas constant; T is the thermodynamic temperature of the inlet tube; V_1 is the gas volume inside the mill; ω_{thr} is the air speed at the mill throat; μ is the concentration of the coal powder inside the mill; $\triangle P_r$ and $\triangle P_w$ are the differential pressures between the inlet and outlet of hot air damper and warm air damper; f_r and f_w are resistance coefficients of hot air damper and warm air damper; K_r, K_w, K_{gm} are the feeder speed, the hot air damper position and the warm air damper position; B_{gmax} is the maximum coal feeding rate.

6.3 Decoupling control system design

By practising in the power plant, the warm air damper position u_3 is usually closed because of the shortage of drying capability. On the other hand, the mill differential pressure y_3 has complex and uncertain relations with the mill load W_m, it is impossible to be kept constantly. Thus u_3 and y_3 are excluded from our decoupling control system design, and the process is simplified into a 2-input-2-output system.

From the analysis of process dynamics and industrial experiments, it can be known that u_2 (the hot air damper position) has strong impact on y_1 (the outlet temperature), but u_1 (the feeder speed) has relatively slight impact on y_2 (the inlet pressure). So the decoupling control of the channel from u_1 to y_1 an be realized through above approach based on the following obtained linearized channel model.

$$A_1 y_1(t+1) = B_{11} u_1(t) + B_{12} u_2(t) + v_1(t) \tag{41}$$

with the weighting items chosen as $P_1 = R_1 = 1$, $K_1 = 0$, $Q_1 = \lambda_1(1 - z^{-1})$. The control input of this loop becomes

$$[F_1B_{11} + Q_1]u_1(t) = w_1 - G_1y_1(t) - F_1B_{12}u_2(t) - F_1\hat{v}_1(t) \tag{42}$$

As for the channel from u_2 to y_2, a linear controller is employed for an improved reliability.

6.4 Simulations

Simulation experiments using the first principle model above mentioned have been made before industrial implementation.

Example 6.1 In one operating point, the approximate linear model of the first channel is obtained as $A_1 = 1 - 1.3346z^{-1} + 0.2692z^{-2}$, $B_{11} = -0.03 - 0.0192z^{-2}$, $B_{12} = 0.11 + 0.0567z^{-2}$. Let $\lambda_1 = -0.1$, a BP neural network with one hidden layer and 20 hidden neurons is adopted. The simulation results are shown in Fig. 6 and Fig. 7. The tracing error of y_1 is reduced gradually with online training of the neural network, and y_2 is kept generally steady. The performances of two controllers are satisfactory.

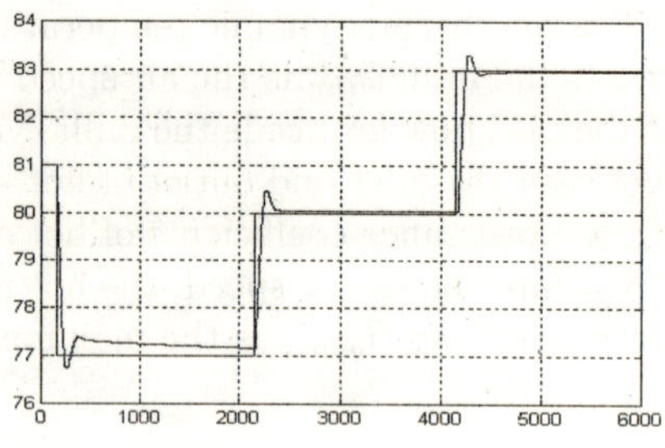

Fig. 6. The setpoint and response of the outlet temperature

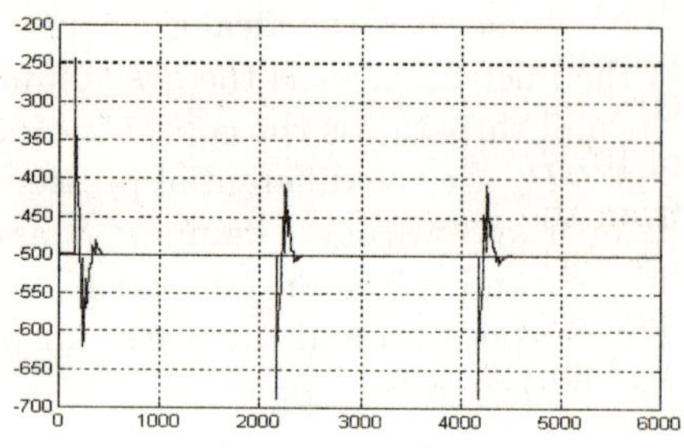

Fig. 7. The setpoint and response of the inlet pressure

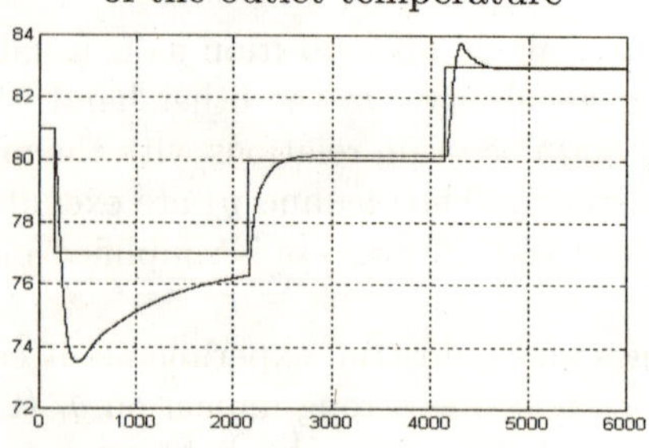

Fig. 8. The outlet temperature follows its setpoint gradually

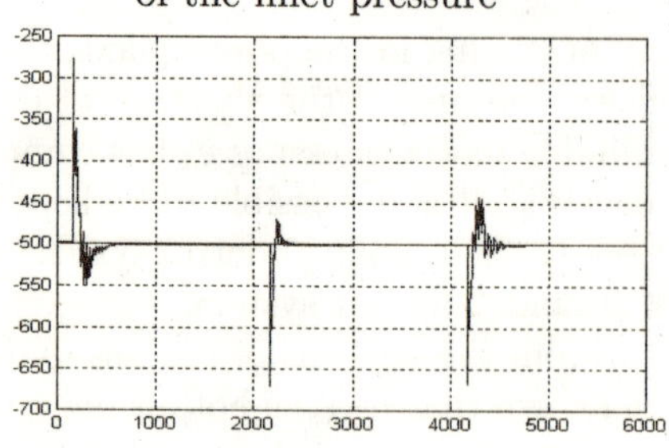

Fig. 9. The inlet pressure keeps around its setpoint

Example 6.2 To show the adaptive ability of the control system, the following inaccurate linear model of the first channel is used $\hat{A}_1 = 1 - 1.043z^{-1} +$

$0.575z^{-2}, \hat{B}_{11} = -0.045 - 0.0028z^{-2}, \hat{B}_{12} = 0.59 + 0.0139z^{-2}$. Compared with the model of Example 6.1, the unmodeled dynamics of system in this example is very large. With the new linear model, the simulation results are shown in Fig.8 and Fig.9. Although there is large tracing error of y_1 at the beginning, the neural network can gradually estimate the unmodeled dynamics online and compensate its influence.

6.5 Industrial experiment

To illustrate the effectiveness of the feedforward compensation for the unmodeled dynamics $\mathbf{v}[\mathbf{x}(t)]$, a comparative experiment between the proposed controller and a controller without the feedforward compensator is performed, where the warm air is taken as a disturbance and the responses of the otherloops are reviewed. When the control system operates in a steady state, the warm air damper position u_3 is increased by 10% (i.e., by shifting the damper under a temporary manual control) so as to introduce the warm air disturbance. Fig. 10 and Fig.11 show the experiment results of the disturbance attenuation, from where it can be clearly seen that the outlet temperature y_1 and the inlet pressure y_2 can return to their original setpoints. The measurement ranges of the four variables (i.e., the outlet temperature y_1, the inlet pressure y_2 , the feeder speed u_1 and the hot air damper position u_2) are $60^oC \sim 80^oC$ of $2^oC/grid$, $-1.0kPa \sim 0.0kPa$ of $0.1kPa/grid$ and $0 \sim 100\%$ of $10\%/grid$, respectively. The time scale is 3 minutes per grid. It can be seen from the experiment results that the fluctuations of the outlet temperature are within 2^oC and the transient response time is about 6 minutes when the intelligent decoupling control is adopted. Moreover, the fluctuation of the outlet temperature is about 4^oC and the transient response time is more than 15 minutes when the controller without the feedforward compensator is used. Apparently, the performance of the closed loop control system has been remarkably improved through the feedforward compensator.

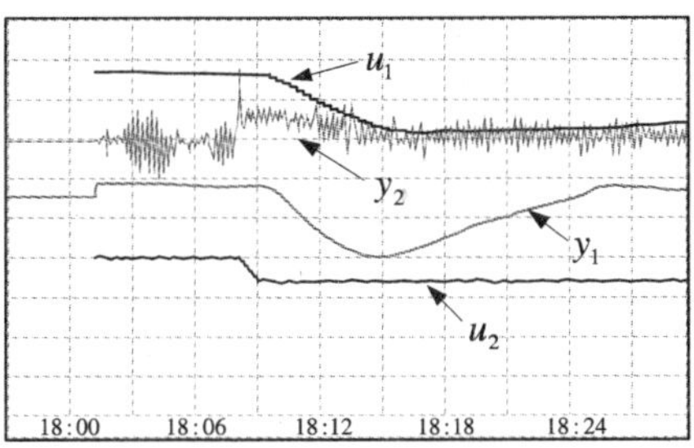

Fig. 10. Experiment result of the proposed decoupling controller

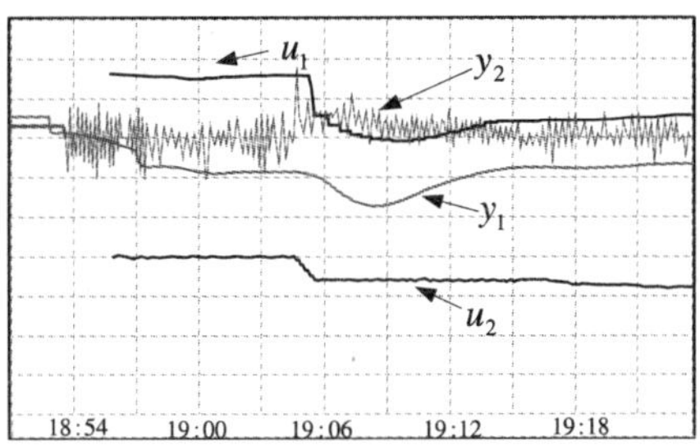

Fig. 11. Experiment result of controller without feedforward compensator

7 Conclusion

This paper proposes a nonlinear multivariable decoupling control law which combines generalized minimum variance control based decoupling design with feedforward compensator for unmodeled dynamics. Using the recursive least square (RLS) algorithm to identify the linear term offline and introducing a neural network to estimate the unmodeled dynamics online, a multivariable decoupling control algorithm is presented. When the unmodeled dynamics is so small to be omitted, this control becomes generalized minimum variance control based decoupling design. The proposed control algorithm can be easily implemented on DCS. Both simulations for unknown nonlinearity process with time varying parameters and structure, and the industrial experiments in 200MW power units, show the effectiveness and robust performance of the proposed decoupling control algorithm to deal with the decoupling control problem of the complex industrial processes.

ACKNOWLEDGEMENT

This paper is supported by the Chinese National Basic Research Program (2002CB312201), the Chinese National Hi-Tech Development Program (2004A A412030) and Key Laboratory of Process Industry Automation, Ministry of Education, China.

References

1. Chai, T.Y. (1988), A self-tuning decoupling controller for a class of multivariable systems and global convergence analysis, *IEEE Transactions on Automatic Control*, Vol. 33, 767-771.
2. Chai, T.Y. (1990), Direct adaptive decoupling control for general stochastic multivariable systems, *International Journal of Control*, Vol. 51, 885-909.
3. Chai, T.Y., Wang, G. (1992), Globally convergent multivariable adaptive decoupling controller and its application to a binary distillation column, *International Journal of Control*, Vol. 55, 415-429.
4. Chai, T.Y., Yue, H., Bai, Y. (1999), Intelligent control of coal-pulverizing systems with ball-tube mill, *Proc. 14th IFAC World Congress*, Beijing, Vol. O, 97-102.
5. Falb, P.L., Wolovich, W.A. (1967), Decoupling in the design and synthesis of multivariable control systems, *IEEE Transactions on Automatic Control*, Vol. 12, 651-659.
6. Kavangh, R.J. (1957), Noninteracting controls in linear multivariable systems, *Transactions of the American Institute of Electronic Engineers, Part II, Applications and Industry*, Vol. 76, 95-99.
7. Lang, S.J., Gu, X.Y., Chai, T.Y. (1986), A multivariable generalized self-tuning controller with decoupling design, *IEEE Transactions on Automatic Control*, Vol. 31, 474-477.
8. McDermott, P.E., Mellichamp, D.A. (1986), A decoupling pole-placement self-tuning controller for a class of multivariable process, *Optimal Control Applications and Methods*, Vol. 7, 55-79.

9. Rosenbrock, H.H. (1969), Design of multivariable control systems using inverse Nyquist array, *IEE Proc. Control Theory and Applications*, Vol. 116, 1929-1936.
10. Shinskey, F.G. (1979), *Process Control Systems*, 2nd ed. McGraw-Hill Book Company.
11. Tao, W.H., Chai, T.Y., Yue, H. (2004), Research of dynamic mathematical model and simulation of ball coal mill system, *Journal of System Simulation*, Vol. 16, 778-780. (in Chinese)
12. Wittenmark, B., Middleton, R., Goodwin, G. (1987), Adaptive Decoupling of Multivariable Systems, *International Journal of Control*, Vol. 46, 1993-2009.
13. Wu, L.M., Chai, T.Y. (1995), Decoupling of a class of nonlinear discrete time systems using neural networks, *Proc. of the 1995 American Control Conference*, Vol. 6, 4275-4279.
14. Yue, H., Chai, T.Y. (1998), Adaptive decoupling control of multivariable nonlinear non-minimum phase systems using neural networks, *Proc. of 1998 American Control Conference*, Vol. 1-6, 513-514.
15. Zhu, K.Y., Qin, X.F., Chai, T.Y. (1999), A new decoupling design of self-tuning multivariable generalized predictive control, *International Journal of Adaptive Control*, Vol. 13, 183-196.

NCF Representation and Closed-Loop Nonparametric Estimation

Tong Zhou

Department of Automation, Tsinghua University, Beijing 100084, P.R.China
tzhou@mail.tsinghua.edu.cn

1 Introduction

In industrial applications, safety and economy considerations often require that system identification is performed in closed-loop, in addition to the case in which the plant is unstable and closed-loop identification is necessary[6, 13, 3, 7]. On the other hand, NCF based models have recently begun to attract popularity, mainly due to their appreciable properties in describing the dynamics of unstable plants, and close relations with the gap metric and the ν-gap metric that are widely considered to be appropriate in control oriented model uncertainty quantification [11, 5, 4, 12, 8].

Although model sets with NCF perturbations (NCFPMS) are well utilized in robust control theories, it is not very easy to identify them from experimental data. An essential issue here seems to establish mathematically appreciable relations between NCFs and physically measurable variables [8, 5, 2, 10]. To solve this identification problem, it is suggested in [14, 15] to divide it into two steps. That is, at first, frequency response estimates are obtained from experimental data for NCFs. And then, a parametric NCFPMS is identified from these nonparametric estimates.

This paper summaries the major results on the most important step, that is, the nonparametric estimation step, in this two-step approach towards NCF-PMS estimation. A parameterization is given for all the NRCFs of plants internally stabilizable by a known controller. On the basis of this parameterization, a procedure is proposed for estimating the frequency response of plant NRCFs, through constrained data fitting when time domain closed-loop experimental data is available and likelihood maximization when frequency domain closed-loop experimental data is available. A method is given for the selection of the NRCFs of the auxiliary plant, in order to guarantee that the time domain estimator is asymptotically unbiased and complex normally distributed in case that the probing signals are periodic, or the frequency domain estimator is correct. On the other hand, for general probing signals, the asymptotic statistical properties of the estimate is completely determined by a normalized random matrix with all its columns having complex normal distributions.

The notations used in this paper are fairly standard. Their detailed definitions are therefore not included, but can be found, for example, in [4, 8, 5, 14].

2 NCF Parameterization

Typically, a closed-loop system used in identification has a structure like that depicted in Figure 1. Here, r_1 and r_2 are probing signals which are designed to obtain some dynamical characteristics of the plant; d_1 and d_2 are respectively measurement errors of the plant input and output, while e represents the composite influence of exterior disturbances on the closed-loop system. On the other hand, P and C stand respectively for the plant and the controller. In this paper, we only consider linear time invariant plants and controllers. It is assumed that both the plant and the controller are of m inputs and m outputs. Moreover, we also assume that the controller is known and of finite McMillan degree.

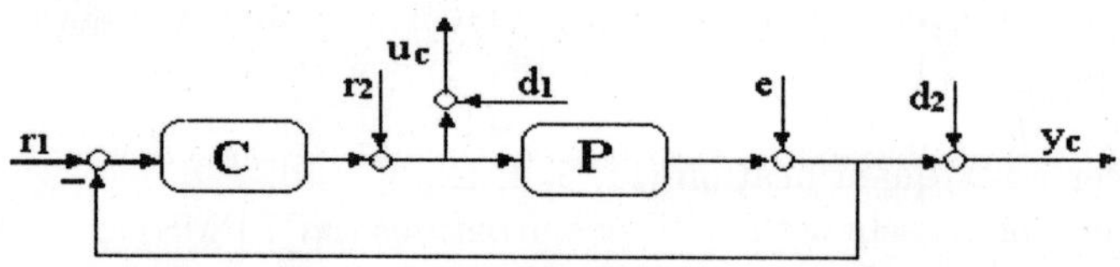

Fig.1. Closed-Loop Identification Experiment

To derive an estimate for the frequency response of the plant NRCFs, the following lemma is introduced, which gives a parameterization for all the NRCFs of plants that compose an internally stable closed-loop system with a known controller. Note that when the controller C is known, successful closed-loop identification experiments imply that the plant has been internally stabilized by the controller. This means that the aforementioned parameterization is closely related to the nonparametric estimation problem. In fact, in the proposed estimation procedure, this parameterization plays an essential role which becomes clear in Sections 3 and 4.

To make statements concise, some symbols are at first defined.

Assume that $N_{rc}(z)$ and $D_{rc}(z)$, $D_{lc}(z)$ and $N_{lc}(z)$, are respectively NRCFs and NLCFs of the controller $C(z)$. As $C(z)$ is of finite degree, these factors always exist. Let $P_x(z)$ denote a transfer function matrix (TFM) which is internally stabilizable by $C(z)$ and has $N_{rx}(z)$ and $D_{rx}(z)$ as its NRCFs over $\mathcal{H}_\infty$. Denote $N_{rc}^{\sim}(z)D_{rx}(z) - D_{rc}^{\sim}(z)N_{rx}(z)$ by $H(z)$, and let $D_{lh}(z)$ and $N_{lh}(z)$ represent its left coprime factors with $D_{lh}(z)$ inner. Assume that $\Phi(z)$ is an outer TFM such that there exists an outer TFM $M(z)$ satisfying $M^{\sim}(z)M(z) = \Phi^{\sim}(z)\Phi(z) + H^{\sim}(z)H(z) - I$.

As $D_{lh}(z)$ is square and inner, it can be simply proved that the Smith-McMillan form of $D_{lh}(z)$ can be expressed as $U_h(z)\Sigma_h(z)V_h(z)$ with $\Sigma_h(z) =$

$diag\left\{ \left. \dfrac{\alpha_{ih}(z)}{\alpha_{ih}(z^{-1})} \right|_{i=1}^{m} \right\}$, in which $\alpha_{ih}(z)$ is a monic polynomial. Moreover, let z_{iq} denote a zero of $\alpha_{ih}(z)$ with multiplicity k_{iq}, then, it is obvious that $|z_{iq}| \le 1$, $q = 1, 2, \cdots, p_i$, $i = 1, 2, \cdots, m$.

Using these factorizations, a parameterization is developed for all the NRCFs of TFMs internally stabilizable by controller $C(z)$.

Lemma 1. Assume that $\Phi(z)$ is an outer TFM such that the desirable TFM $M(z)$ exists. Denote the ith row of $U_h^{-1}(z)$ by $\xi_{ih}(z)$. Then, $N_r(z)$ and $D_r(z)$ are NRCFs of a TFM internally stabilized by $C(z)$, if and only if there exists a stable TFM $L(z)$, such that $L^{\sim}(z)L(z) = I$ and

$$\begin{bmatrix} N_r(z) \\ D_r(z) \end{bmatrix} = \left(\begin{bmatrix} N_{rx}(z) \\ D_{rx}(z) \end{bmatrix} + \begin{bmatrix} D_{rc}(z) \\ -N_{rc}(z) \end{bmatrix} [D_{lh}^{-1}(z)L(z)M(z) + H(z)] \right) \Phi^{-1}(z) \quad (1)$$

$$\left. \frac{d^l[\xi_{ih}(z)L(z)]}{dz^l} \right|_{z=z_{iq}} = - \left. \frac{d^l[\xi_{ih}(z)N_{lh}(z)M^{-1}(z)]}{dz^l} \right|_{z=z_{iq}} \quad (2)$$

in which $0 \le l \le k_{iq} - 1$, $1 \le q \le p_i$, and $1 \le i \le m$. Moreover, when a plant is internally stabilizable by $C(z)$, then, for every pair of its NRCFs, there exist an outer $\Phi(z)$ and an inner $L(z)$ such that Equations (1) and (2) are satisfied.

In the interpolation conditions of Equation (2), derivatives of the unknown inner TFM $L(z)$ are involved. This makes it not very convenient to directly apply the NRCF parameterization to the formulated nonparametric estimation problem. A simpler form is required. Noting that $L(z)$ is stable, it can be claimed that $\sup_{|z|<1} \bar{\sigma}(L(z)) \le 1$ from the maximum modulus principle [1]. For $i = 1, 2, \cdots, m$; $q = 1, 2, \cdots, p_i$, define a $k_{iq} \times k_{iq}$ dimensional matrix A_{iq} to be a Jordan block with its diagonal elements being z_{iq}, and define $m \times k_{iq}$ dimensional matrices C_{iq+} and C_{iq-} as

$$C_{iq+}^T = - \left. \begin{bmatrix} \eta_{ih}(z) \\ \frac{d\eta_{ih}(z)}{dz} \\ \vdots \\ \frac{1}{(k_{iq}-1)!} \frac{d^{k_{iq}-1}\eta_{ih}(z)}{dz^{k_{iq}-1}} \end{bmatrix} \right|_{z=z_{iq}}, \quad C_{iq-}^T = \left. \begin{bmatrix} \xi_{ih}(z) \\ \frac{d\xi_{ih}(z)}{dz} \\ \vdots \\ \frac{1}{(k_{iq}-1)!} \frac{d^{k_{iq}-1}\xi_{ih}(z)}{dz^{k_{iq}-1}} \end{bmatrix} \right|_{z=z_{iq}}$$

in which $\eta_{ih}(z) = \xi_{ih}(z)N_{lh}(z)M^{-1}(z)$. Moreover, construct matrices A, C_+ and C_- as $A = diag\{diag\{A_{iq}|_{q=1}^{p_i}\}_{i=1}^{m}\}$, $C_- = [C_{1-} \;\; C_{2-} \;\; \cdots \;\; C_{m-}]$, $C_+ = [C_{1+} \;\; C_{2+} \;\; \cdots \;\; C_{m+}]$, in which $C_{i-} = [C_{i1-} \;\; C_{i2-} \;\; \cdots \;\; C_{ip_i-}]$, $C_{i+} = [C_{i1+} \;\; C_{i2+} \;\; \cdots \;\; C_{ip_i+}]$. Then, it can be proved that [1] the interpolation conditions in Equation (2) are satisfied if and only if $\frac{1}{2\pi j} \oint_{\partial\mathcal{D}} L^T(z)C_-(zI - A)^{-1}dz = C_+$. Here, $\partial\mathcal{D} = \{\, z \mid |z| = 1, \;\; z \in \mathcal{C} \,\}$. On the other hand, the spectral radius of matrix A is obviously not greater than 1 from its construction. Therefore, the Lyapunov equation $S - A^H S A = C_-^H C_- - C_+^H C_+$ has a unique solution [1]. Denote it by $S^{\#}$. When matrix $S^{\#}$ is invertible, define matrix D and matrix valued functions (MVF) $\Theta_{il}(z)$, $i, l = 1, 2$, as

$$D = I + \begin{bmatrix} C_+ \\ C_- \end{bmatrix} (S^{\#})^{-1}(I - aA)^{-1}[C_+^H \; -C_-^H] \tag{3}$$

$$\begin{bmatrix} \Theta_{11}(z) & \Theta_{12}(z) \\ \Theta_{21}(z) & \Theta_{22}(z) \end{bmatrix} = D + \begin{bmatrix} C_+ \\ C_- \end{bmatrix} [A^H S^{\#}(zI - A)]^{-1}[C_+^H \; -C_-^H]D \tag{4}$$

in which a is an arbitrary complex number with magnitude being 1. Then, from the two-sided Nudelman interpolation theories [1], we have the following results.

Lemma 2. There exists a MVF $L(z)$ in $\mathcal{H}_\infty$ such that $\sup_{|z|<1} \bar{\sigma}(L(z)) \leq 1$, $\frac{1}{2\pi j}\oint_{\partial\mathcal{D}} L^T(z)C_-(zI - A)^{-1}dz = C_+$, if and only if matrix $S^{\#}$ is positive semi-definite. Moreover, when $S^{\#} \geq 0$ and $det(S^{\#}) = 0$, there exists a unique desirable MVF; when $S^{\#} > 0$, all the desirable MVFs can be parameterized as $L^T(z) = [\Theta_{11}(z)G(z) + \Theta_{12}(z)][\Theta_{21}(z)G(z) + \Theta_{22}(z)]^{-1}$, in which $G(z)$ is an arbitrary MVF satisfying $\sup_{|z|<1} \bar{\sigma}(G(z)) \leq 1$.

When $S^{\#} \geq 0$ and $det(S^{\#}) = 0$, $N_r(z)$ and $D_r(z)$ are uniquely determined by $N_{rx}(z)$, $D_{rx}(z)$, etc., and are independent of experimental data. This is clearly not very interesting in system identification. Therefore, only the case $S^{\#} > 0$ will be discussed in this paper.

Note that $N_{rx}(z)$, $D_{rx}(z)$, $N_{rc}(z)$ and $D_{rc}(z)$ are all real and causal. It is easy to see that $\xi_{ih}(z)$ is real, $i = 1, 2, \cdots, m$. Moreover, if z_{iq} is a zero of $\alpha_{ih}(z)$ with $\Im\{z_{iq}\} \neq 0$, then, $\bar{z}_{iq}$ must also be a zero of $\alpha_{ih}(z)$ with the same multiplicity. On these observations, it can be proved that $a = \pm 1$ is a necessary and sufficient condition for $\Theta_{il}(z)$, $i, l = 1, 2$, to be a TFM. Therefore, the interpolation function $L(z)$ satisfying Equation (2) is a TFM if and only if $G(z)$ is a TFM and a equals 1 or -1.

Summarizing the above results, the following conclusions are reached, which give a complete yet simple parameterization of all the NRCFs of TFMs internally stabilized by $C(z)$.

Theorem 1. Assume that $a = \pm 1$. Then, all the NRCFs of TFMs internally stabilizable by $C(z)$ can be parameterized as those of Equation (1) with $L(z) = [\Theta_{22}^T(z) + T(z)\Theta_{21}^T(z)]^{-1}[\Theta_{12}^T(z) + T(z)\Theta_{11}^T(z)]$, in which $T(z)$ is an arbitrary stable TFM satisfying $T(z)T^{\sim}(z) = I_m$.

Compared with the parameter $L(z)$ of Lemma 1, except co-inner, there is no other conditions on $T(z)$. However, unlike Lemma 1 in which NRCFs of a plant internally stabilizable by $C(z)$ are expressed as a linear transformation of inner TFMs, Theorem 1 gives an expression through an LFT of co-inner TFMs. When the parameterization of Theorem 1 is adopted in NRCF identification, the plant output is generally a nonlinear function of the co-inner TFM $T(z)$ that is to be estimated. This is apparently not an appreciable property in system identification.

Owing to the boundary Nevanlinna-Pick interpolation theory and the J-loseless property of the TFM $\begin{bmatrix} \Theta_{11}(z) & \Theta_{12}(z) \\ \Theta_{21}(z) & \Theta_{22}(z) \end{bmatrix}$, the advantages of the parameterizations in Lemma 1 and Theorem 1 can be combined together to get a solution to the nonparametric estimation problem. More precisely, according to the boundary Nevanlinna-Pick interpolation theory, we have the following conclusions.

Lemma 3. There exists a square TFM $T(z)$ in $\mathcal{H}_\infty$ satisfying $T(z)T^{\sim}(z) = I$ and $T(e^{j\omega_k}) = B_k$, $k = 1, 2, \cdots, n$, if and only if $B_k^H B_k = I$.

On the other hand, from the properties of $\Theta_{il}(z)$, $i, l = 1, 2$, the following results can be derived.

Lemma 4. Let $T(z)$ and $L(z)$ be those of Theorem 1. Then, $T^H(e^{j\omega})T(e^{j\omega}) = I$ is equivalent to $L^H(e^{j\omega})L(e^{j\omega}) = I$.

When a nonparametric estimate is required for the plant NRCFs, we are only interested in estimating the value of the plant NRCFs at $z = e^{j\omega}$. From Theorem 1 and Lemmas 1 and 3, this identification problem can be converted to the estimation of a unitary matrix $T(e^{j\omega})$. As the unitariness of $T(e^{j\omega})$ and $L(e^{j\omega})$ are equivalent to each other according to Lemma 4, it is not very difficult to see that estimation of the frequency response of the plant NRCFs can be transformed to that of the inner TFM $L(z)$ in Equation (1), and in the latter estimation, it is not necessary to take into account of the interpolation conditions in Equation (2). This seems to be a significant advantage to divide NCFPMS identification into nonparametric identification and frequency response estimate based parametric identifications, noting that if a parametric NCFPMS is asked to be directly identified from closed-loop experimental data, unitariness of $L(e^{j\omega})$ alone can not guarantee that $N_r(z)$ and $D_r(z)$ of Equation (1) are NRCFs of a plant internally stabilizable by the controller $C(z)$.

3 Frequency Domain Nonparametric Estimation

Concerning the closed-loop system of Figure 1, assume that s identification experiments have been performed. In this paper, superscript $[i]$ is used to clarify that a variable or a function, etc., is related to the ith identification experiment, $1 \le i \le s$. Denote the discrete Fourier transforms (DFT) of the probing signals by $r_1^{[i]}(e^{j\omega_k})$ and $r_2^{[i]}(e^{j\omega_k})$, in which $0 < \omega_k < \pi$, $\omega_k \neq \frac{\pi}{2}$, and $1 \le k \le n$. Let $\hat{e}^{[i]}(e^{j\omega_k})$ represent the DFT of the closed-loop system's response to the external disturbance $e^{[i]}(t)$, while $d_1^{[i]}(e^{j\omega_k})$ and $d_2^{[i]}(e^{j\omega_k})$ the DFTs of the measurement errors. Assume that $d_1^{[i]}(e^{j\omega_k})$, $d_2^{[i]}(e^{j\omega_k})$, $\hat{e}^{[i]}(e^{j\omega_k})$, $d_1^{[l]}(e^{j\omega_f})$, $d_2^{[l]}(e^{j\omega_f})$ and $\hat{e}^{[l]}(e^{j\omega_f})$ are independent of each other whenever $i \neq l$ or $\omega_k \neq \omega_f$, and $d_1^{[i]}(e^{j\omega_k}) \sim \mathcal{N}_c(0, \Sigma_{1k})$, $d_2^{[i]}(e^{j\omega_k}) \sim \mathcal{N}_c(0, \Sigma_{2k})$, $\hat{e}^{[i]}(e^{j\omega_k}) \sim \mathcal{N}_c(0, \Sigma_{0k})$. Here, $\mathcal{N}_c(*, \sharp)$ represents a complex normal distribution with mean $*$ and covariance matrix $\sharp$. Given the measured DFTs of the plant input and output $(u_c^{[i]}(e^{j\omega_k}),\ y_c^{[i]}(e^{j\omega_k}))|_{k=1,i=1}^{k=n,i=s}$, as well as the covariance matrices $\Sigma_{lk}|_{k=1,l=0}^{k=n,l=2}$, our objective is to estimate the frequency response of the NRCFs of $P(z)$ at $\omega_k|_{k=1}^{n}$.

Under a stochastic framework, a popular method in identification is maximum likelihood estimation. On the basis of Theorem 1, an analytic expression has been derived under the condition that $\Xi(z)$, $N_{rx}(z)$ and $D_{rx}(z)$ are provided, for the frequency response of plant NRCFs.

From Theorem 1, we know that successful closed-loop identification experiments imply that NRCFs of $P(z)$ can be represented by a co-inner TFM. This further means that the nonparametric estimation problem can be converted to identifying $T(e^{j\omega_k})$ or $L(e^{j\omega_k})$. On the other hand, denote $D_{lc}(z)D_{rx}(z) + N_{lc}(z)N_{rx}(z)$ and $r_2^{[i]}(z) + C(z)r_1^{[i]}(z)$ respectively by $Q_x(z)$ and $r^{[i]}(z)$. Then, it can be directly proved from Figure 1 that for every $i = 1, 2, \cdots, s$,

$$\begin{aligned}&\begin{bmatrix} u_c^{[i]}(z) \\ y_c^{[i]}(z) \end{bmatrix} + \begin{bmatrix} C(z) \\ -I \end{bmatrix} \hat{e}^{[i]}(z) - \begin{bmatrix} d_1^{[i]}(z) \\ d_2^{[i]}(z) \end{bmatrix} - \begin{bmatrix} D_{rx}(z) \\ N_{rx}(z) \end{bmatrix} Q_x^{-1}(z)D_{lc}(z)r^{[i]}(z) \\ &= \begin{bmatrix} -N_{rc}(z) \\ D_{rc}(z) \end{bmatrix} [D_{lh}^{-1}(z)L(z)M(z) + H(z)]Q_x^{-1}(z)D_{lc}(z)r^{[i]}(z) \qquad (5)\end{aligned}$$

Define $d_{10}^{[i]}(z)$, $w^{[i]}(z)$ and $v^{[i]}(z)$ respectively as

$$d_{10}^{[i]}(z) = u_c^{[i]}(z) + C(z)y_c^{[i]}(z) - r^{[i]}(z) \qquad (6)$$
$$v^{[i]}(z) = M(z)Q_x^{-1}(z)D_{lc}(z)r^{[i]}(z) \qquad (7)$$
$$w^{[i]}(z) = y_c^{[i]}(z) - [N_{rx}(z) + D_{rc}(z)H(z)]Q_x^{-1}(z)D_{lc}(z)r^{[i]}(z) \qquad (8)$$

Then, a necessary and sufficient condition for the existence of a $L(z)$ satisfying Equation (5) whenever $1 \leq i \leq s$, is that $d_1^{[i]}(z) = d_{10}^{[i]}(z) - C(z)d_2^{[i]}(z)$. Moreover, when this condition is satisfied, every solution of Equation (5) also satisfies $L(z)v^{[i]}(z) = D_{lh}(z)D_{rc}^{-1}(z)[w^{[i]}(z) - \hat{e}^{[i]}(z) - d_2^{[i]}(z)]$, and vice versa.

For notational simplicity, denote the value of a function of z, say $x(z)$, at $z = e^{j\omega_k}$ by x_k, afterwards. Moreover, the following six matrices are defined for a concise presentation.

$$D_{1k} = [d_{1k}^{[1]} \ d_{1k}^{[2]} \ \cdots \ d_{1k}^{[s]}], \qquad D_{1k0} = [d_{10k}^{[1]} \ d_{10k}^{[2]} \ \cdots \ d_{10k}^{[s]}] \qquad (9)$$
$$D_{2k} = [d_{2k}^{[1]} \ d_{2k}^{[2]} \ \cdots \ d_{2k}^{[s]}], \qquad W_k = [w_k^{[1]} \ w_k^{[2]} \ \cdots \ w_k^{[s]}] \qquad (10)$$
$$E_k = [\hat{e}_k^{[1]} \ \hat{e}_k^{[2]} \ \cdots \ \hat{e}_k^{[s]}], \qquad V_k = [v_k^{[1]} \ v_k^{[2]} \ \cdots \ v_k^{[s]}] \qquad (11)$$

Then, the above arguments show that the relations among L_k, the experimental data, the external disturbances, and the measurement errors, can be expressed as follows.

$$D_{1k} = D_{1k0} - C_k D_{2k}, \qquad L_k V_k = D_{lhk} D_{rck}^{-1}(W_k - E_k - D_{2k}) \qquad (12)$$

Note that $D_{lh}(z)$ is inner. A direct consequence of Lemmas 2-4 is that $D_{1k} = D_{1k0} - C_k D_{2k}$ and $(W_k - E_k - D_{2k})^H [D_{rck} D_{rck}^H]^{-1} (W_k - E_k - D_{2k}) =$

$V_k^H V_k$ are necessary and sufficient conditions for experimental data regeneration. In other words, these conditions characterize all the external disturbances and measurement errors that are consistent with experimental data, and therefore should be taken into account in minimizing the cost function for nonparametric estimation. Moreover, as the estimate for the NRCFs' frequency response can be directly calculated from Equation (1) when L_k is available, we can declare that in NRCF nonparametric estimation, the actual construction of $\Theta_{il}(z)$, $i,l=1,2$, is not necessary. This fact implies that the selection of α, as well as the decomposition of $D_{lh}(z)$, are also not necessary. On the other hand, from the assumptions on the external disturbances and measurement errors, it is not difficult to understand that the logarithm of the likelihood function can be defined as

$$J((D_{1k},D_{2k},E_k)|_{k=1}^{n}) = \sum_{k=1}^{n}[||\Sigma_{1k}^{-1/2}D_{1k}||_F^2 + ||\Sigma_{2k}^{-1/2}D_{2k}||_F^2 + ||\Sigma_{0k}^{-1/2}E_k||_F^2] \tag{13}$$

From these relations, the following results are derived, which establishes a relation between the external disturbances E_k and the measurement errors D_{2k} that are most likely to happen.

Lemma 5. Denote the optimal D_{2k} and E_k respectively by D_{2k}^* and E_k^*. Then $E_k^* = \Sigma_{0k}[C_k^H\Sigma_{1k}^{-1}C_k + \Sigma_{2k}^{-1}]D_{2k}^* - \Sigma_{0k}C_k^H\Sigma_{1k}^{-1}D_{1k0}$.

From this result, it can be shown that the MLE of the NRCFs' frequency response has the following expression.

Theorem 2. Define matrices F_k and X_k respectively as $F_k = C_k^H\Sigma_{1k}^{-1}C_k + \Sigma_{2k}^{-1}$ and $X_k = D_{rck}^H(I+F_k\Sigma_{0k})^{-1}(F_kW_k - C_k\Sigma_{1k}^{-1}D_{1k0})V_k^H$. Assume that X_k is of full rank and $V_kV_k^H = a_kI_m$ with a_k a nonzero scalar. Then, the MLE for the frequency response of the plant NRCFs at ω_k, denote them by $\hat{N}_{rk}$ and $\hat{D}_{rk}$, can be expressed as

$$\begin{bmatrix}\hat{N}_{rk}\\ \hat{D}_{rk}\end{bmatrix} = \left(\begin{bmatrix}N_{rxk}\\ D_{rxk}\end{bmatrix} + \begin{bmatrix}D_{rck}\\ -N_{rck}\end{bmatrix}[H_k + X_k(X_k^HX_k)^{-\frac{1}{2}}M_k]\right)\Xi_k^{-1} \tag{14}$$

When X_k is rank deficient, the optimal estimate can be obtained through singular value decompositions. On the other hand, denote $[r_{ik}^{[1]}\ r_{ik}^{[2]}\ \cdots\ r_{ik}^{[s]}]$ by R_{ik}, $i=1,2$. Then, from the definitions of $v^{[i]}(z)$ and V_k in Equations (7) and (11), it is obvious that $V_k = M_kQ_{xk}^{-1}D_{lck}(R_{2k}+C_kR_{1k})$. Therefore, $V_kV_k^H = a_kI_m$ is equivalent to $(R_{2k}+C_kR_{1k})(R_{2k}+C_kR_{1k})^H = \alpha_k[D_{lck}^HQ_{xk}^{-H}M_k^HM_kQ_{xk}^{-1}D_{lck}]^{-1}$. This implies that the condition $V_kV_k^H = a_kI_m$ can be simply satisfied through identification experiment design.

Let $N_{rp}(z)$ and $D_{rp}(z)$ denote NRCFs of plant $P(z)$. Define $Q_p(z)$ and X_{k0} respectively as

$$Q_p(z) = D_{lc}(z)D_{rp}(z) + N_{lc}(z)N_{rp}(z) \tag{15}$$
$$X_{k0} = D_{rck}^H(I+F_k\Sigma_{0k})^{-1}F_k[N_{rpk}Q_{pk}^{-1}Q_{xk} - (N_{rxk}+D_{rck}H_k)]M_k^{-1} \tag{16}$$

Then, the nonparametric estimate of Equation (14) can also be expressed as follows.

$$\begin{bmatrix} \hat{N}_{rk} \\ \hat{D}_{rk} \end{bmatrix} = \left(\begin{bmatrix} N_{rxk} \\ D_{rxk} \end{bmatrix} + \begin{bmatrix} D_{rck} \\ -N_{rck} \end{bmatrix} \{H_k + Z_k(Z_k^H Z_k)^{-1/2} M_k\} \right) \Xi_k^{-1} \quad (17)$$

Here, $Z_k = X_{k0} + \Delta_k$ and Δ_k is a random matrix independent of the frequency. Moreover, all of its columns are independent and have a complex normal distribution $\mathcal{N}_c(0,\ \Sigma_k)$ with $\Sigma_k = D_{rck}^H(I+F_k\Sigma_{0k})^{-1}(F_k\Sigma_{0k}F_k+C_k^H\Sigma_{1k}^{-1}C_k+\Sigma_{2k}^{-1})(I+F_k\Sigma_{0k})^{-H}D_{rck}$.

This result makes it clear that the statistical properties of the estimator are completely determined by the random matrix Z_k after normalization. In principle, all the statistical properties of $\hat{N}_{rk}$ and $\hat{D}_{rk}$ can be derived from Equation (17). However, more efforts are still required to derive a simple form for them. On the other hand, Equation (17) also makes it clear that the covariance matrix of the estimation error is always with magnitude bounded elements. This differs significantly from plant frequency response estimation, and is an appreciable property in the subsequent parametric identification [13, 15]. Moreover, from this result, the following conclusion can also be achieved.

Theorem 3. The estimator of Theorem 2 is correct if and only if $Q_{xk} = Q_{pk}\Xi_k$.

4 Time Domain Nonparametric Estimation

Let $r_{1k}^{[i]}$, $r_{2k}^{[i]}$, $u_{ck}^{[i]}$ and $y_{ck}^{[i]}$, $i = 1, 2, \cdots, s$; $k = 0, 1, \cdots, N-1$, denote respectively the sampled values of the probing signals and plant input/output measurements in s identification experiments. In this section, an estimate is given for the frequency response of the plant NRCFs, on the time domain closed-loop experimental data. Moreover, some of its statistical properties are also briefly discussed.

Compared with the frequency domain experimental data based estimation problem, there are no assumptions on the statistical properties of the external disturbances and measurement errors. And it is also clear that in addition to the usually adopted assumptions in closed-loop system identification, no other requirements are put on the plant, the controller, as well as the disturbances and measurement errors. On the other hand, it is worthwhile to mention that while frequency domain experimental data is easier to be prefiltered and combined from different identification experiments, time domain experimental data is generally much easier to be obtained [7, 9].

Note that $N_{rc}(z)$ and $D_{rc}(z)$ are NRCFs of the controller. There exist stable TFMs $X_{rc}(z)$ and $Y_{rc}(z)$ such that $X_{rc}(z)N_{rc}(z) + Y_{rc}(z)D_{rc}(z) = I$. Multiplying both sides of Equation (5) from the left by $[-X_{rc}(z)\ \ Y_{rc}(z)]$, we then have that

$$\begin{aligned} L(z)M(z)Q_x^{-1}(z)D_{lc}(z)r^{[i]}(z) &= D_{lh}(z)\{-X_{rc}(z)u_c^{[i]}(z) + Y_{rc}(z)y_c^{[i]}(z) + \\ &\quad [X_{rc}(z)D_{rx}(z) - Y_{rc}(z)N_{rx}(z) - H(z)]Q_x^{-1}(z)D_{lc}(z)r^{[i]}(z)\} + \\ &\quad D_{lh}(z)\{X_{rc}(z)d_1^{[i]}(z) - Y_{rc}(z)d_2^{[i]}(z) - [X_{rc}(z)C(z) + Y_{rc}(z)]\hat{e}^{[i]}(z)\} \quad (18) \end{aligned}$$

Equation (18) suggests that identification of $L(e^{j\omega})$ is in fact an open identification problem. Recalling that $L(e^{j\omega})$ is required to be unitary, this identification problem is a little different from the conventional open-loop frequency response estimation problem that has been extensively discussed in, for example, [7, 9], etc. On these observations, the following procedure is proposed for estimating the frequency response of plant NRCFs.

Nonparametric Estimation Algorithm:

Step 1: Let q denote the right shift operator. Define $u_k^{[i]}$, $y_k^{[i]}$, U_k, Y_k, $i = 1, 2, \cdots, s$, $k = 0, 1, \cdots, N-1$, as follows.

$$u_k^{[i]} = M(q)Q_x^{-1}(q)D_{lc}(q)r_k^{[i]} \tag{19}$$

$$y_k^{[i]} = D_{lh}(q)\{-X_{rc}(q)u_{ck}^{[i]} + Y_{rc}(z)y_{ck}^{[i]} + [X_{rc}(q)D_{rx}(q) - Y_{rc}(q)N_{rx}(q) - H(q)]Q_x^{-1}(q)D_{lc}(q)r_k^{[i]}\} \tag{20}$$

$$U_k = [u_k^{[1]} \ u_k^{[2]} \ \cdots \ u_k^{[s]}], \qquad Y_k = [y_k^{[1]} \ y_k^{[2]} \ \cdots \ y_k^{[s]}] \tag{21}$$

Step 2: Define $U(j\omega)$ and $Y(j\omega)$ the DFT of $U_k|_{k=0}^{N-1}$ and $Y_k|_{k=0}^{N-1}$, respectively. That is,

$$U(j\omega) = \frac{1}{\sqrt{N}}\sum_{k=0}^{N-1} U_k e^{-j\omega k}, \qquad Y(j\omega) = \frac{1}{\sqrt{N}}\sum_{k=0}^{N-1} Y_k e^{-j\omega k} \tag{22}$$

Step 3: Let the estimate of $L(e^{j\omega})$, denoted by $\hat{L}(e^{j\omega})$, be the minimizer of a cost function $J(L(e^{j\omega}))$ defined as $J(L(e^{j\omega})) = ||Y(j\omega) - L(e^{j\omega})U(j\omega)||_F^2$ under the condition that $L^H(e^{j\omega})L(e^{j\omega}) = I$. On the basis of standard algebraic operations, it can be easily proved that when $Y(j\omega)U^H(j\omega)$ is of full column rank, the estimate of $L(e^{j\omega})$, that is, $\hat{L}(e^{j\omega})$, can be expressed as follows.

$$\hat{L}(e^{j\omega}) = Y(j\omega)U^H(j\omega)\left[U(j\omega)Y^H(j\omega)Y(j\omega)U^H(j\omega)\right]^{-1/2} \tag{23}$$

Step 4: Denote the estimate of the frequency response of the plant NRCFs at the angular frequency ω by $\begin{bmatrix}\hat{N}_{rp}(e^{j\omega})\\ \hat{D}_{rp}(e^{j\omega})\end{bmatrix}$. And let

$$\begin{bmatrix}\hat{N}_{rp}(e^{j\omega})\\ \hat{D}_{rp}(e^{j\omega})\end{bmatrix} = \left(\begin{bmatrix}N_{rx}(e^{j\omega})\\ D_{rx}(e^{j\omega})\end{bmatrix} + \begin{bmatrix}D_{rc}(e^{j\omega})\\ -N_{rc}(e^{j\omega})\end{bmatrix} \times [D_{lh}^{-1}(e^{j\omega})\hat{L}(e^{j\omega})M(e^{j\omega}) + H(e^{j\omega})]\right)\Phi^{-1}(e^{j\omega}) \tag{24}$$

In the above estimation procedure, the TFM $H(z)$ is generally not stable. However, $D_{lh}(z)H(z)$ always belongs to $\mathcal{H}_\infty$ from their constructions. This implies that when the excitation signals are of finite magnitude, both $u_k^{[i]}$ and $y_k^{[i]}$; $i = 1, 2, \cdots, s$, $k = 0, 1, \cdots, N-1$, are also of finite magnitude, and $U(j\omega)$ and $Y(j\omega)$ are well defined.

When $Y(j\omega)U^H(j\omega)$ is not of full column rank, an analytic expression can also be derived for the optimal $L(e^{j\omega})$ through singular value decomposition. In this case, however, matrix $Y(j\omega)U^H(j\omega)$ is not right invertible and the optimal $L(e^{j\omega})$ is not unique, which means that the estimation problem is not well posed [7]. Therefore, in practical engineering, this case should be avoided as much as possible.

On the other hand, it has been proved that the nonparametric estimate of Equation (24) for plant NRCFs is asymptotically unbiased only if $Q_x(e^{j\omega}) = Q_p(e^{j\omega})\Phi(e^{j\omega})$. Moreover, if $Q_x(z) = Q_p(z)\Phi(z)$ and matrix $U(j\omega)$ is right invertible while its smallest nonzero singular value increases at a speed N^γ with $\gamma > 0$, then, the estimate is certainly asymptotically unbiased.

5 Auxiliary Plant Selection

From the stochastic properties of the estimators, it is obvious that in order to get an asymptotically unbiased estimate for the frequency response of the plant NRCFs from closed-loop experimental data, some knowledge on the NRCFs is required. There are, however, some slight differences in the conditions between the frequency domain data based estimation and the time domain data based estimation. In detail, the satisfaction of $Q_x(e^{j\omega}) = Q_p(e^{j\omega})\Phi(e^{j\omega})$ at all the frequencies to be estimated is a necessary and sufficient condition for the estimate to be correct when frequency domain experimental data is available, while $Q_x(z) = Q_p(z)\Phi(z)$ is required to guarantee the consistency of the estimate when time domain experimental data is available, and the latter condition is apparently stronger than its former counterpart. In fact, it is not out of imagination that as long as the Youla-Kucera parameterization is used in NRCF identification, $Q_x(e^{j\omega}) = Q_p(e^{j\omega})\Phi(e^{j\omega})$ is always required for an identification algorithm to be asymptotically unbiased. Moreover, when there is a great amount of frequencies at which the frequency response of plant NRCFs is asked to be estimated, it can be proved on the basis of interpolation theories that the differences between the conditions $Q_x(e^{j\omega}) = Q_p(e^{j\omega})\Phi(e^{j\omega})$ and $Q_x(z) = Q_p(z)\Phi(z)$ are not very significant.

While the aforementioned condition seems to be a great impediment to NRCF identification, it is relatively easy to have an estimate for $Q_p(z)$ from closed-loop time domain experimental data. Note that from Figure 1, we can prove that the TFM from the probing signals $r_1(t)$ and $r_2(t)$ to the plant input-output measurement $[u_c^T(t) \;\; y_c^T(t)]^T$, denote it by $\Psi(z)$, can be expressed as follows.

$$\Psi(z) = \begin{bmatrix} D_{rp}(z) \\ N_{rp}(z) \end{bmatrix} Q_p^{-1}(z)[N_{lc}(z) \;\; D_{lc}(z)] \tag{25}$$

As $N_{rp}(z)$ and $D_{rp}(z)$ are NRCFs of the plant, and $N_{lc}(z)$ and $D_{lc}(z)$ are NLCFs of the controller, we therefore have that

$$[Q_p(z)Q_p^{\sim}(z)]^{-1} = \{\Psi(z)[N_{lc}(z) \;\; D_{lc}(z)]^{\sim}\}^{\sim}\{\Psi(z)[N_{lc}(z) \;\; D_{lc}(z)]^{\sim}\} \tag{26}$$

This further implies that

$$Q_p(z)Q_p^{\sim}(z) = \left\{[N_{lc}(z) \ \ D_{lc}(z)]\Psi^{\sim}(z)\Psi(z)\begin{bmatrix} N_{lc}^{\sim}(z) \\ D_{lc}^{\sim}(z) \end{bmatrix}\right\}^{-1} \tag{27}$$

On the other hand, the internal stability of the closed-loop system implies that $Q_p(z)$ is square and outer. As $\Psi(z)$ can be easily estimated from closed-loop experimental data using open-loop identification methods, these relations mean that when the controller $C(z)$ is known, $Q_p(z)$ can also be identified. With an estimate for $Q_p(z)$, denote it by $\hat{Q}_p(z)$, it is possible to construct the desirable $N_{rx}(z)$ and $D_{rx}(z)$ such that $Q_x(z) = \hat{Q}_p(z)\Phi(z)$. The detailed constructions are given in following Lemma 6 and Theorem 4.

When the condition $Q_x(z) = Q_p(z)\Phi(z)$ is not satisfied by the NRCFs of the auxiliary plant $P_x(z)$, it can be shown that with increasing the data length N, the estimation bias converges to a constant, while the estimation variance decreases monotonically to zero, provided that $U(j\omega)$ is of full row rank and its smallest nonzero singular value increases at a speed N^{γ} with $\gamma > 0$.

The above conclusions can be further enhanced to that the stochastic properties of the time domain nonparametric estimation are completely determined by $Q_x(z)$ and $X_{rc}(z)$, as well as $Y_{rc}(z)$. Therefore, the selection of the auxiliary plant itself is not very important.

Now, we discuss how to obtain stable TFMs $N_{rx}(z)$ and $D_{rx}(z)$ such that they are NRCFs of a plant internally stabilizable by the controller $C(z)$ and satisfies $Q_x(z) = \hat{Q}_p(z)\Phi(z)$. To achieve this objective, the following results are at first established.

Lemma 6. $N_{rx}(z)$ and $D_{rx}(z)$ are NRCFs of a plant internally stabilizable by the controller $C(z)$ and satisfy the condition $D_{lc}(z)D_{rx}(z) + N_{lc}(z)N_{rx}(z) = \hat{Q}_p(z)\Phi(z)$, if and only if there exists a TFM $S(z)$ such that

$$\begin{bmatrix} D_{rx}(z) \\ N_{rx}(z) \end{bmatrix} = \begin{bmatrix} D_{lc}^{\sim}(z) \\ N_{lc}^{\sim}(z) \end{bmatrix}\hat{Q}_p(z)\Phi(z) + \begin{bmatrix} -N_{rc}(z) \\ D_{rc}(z) \end{bmatrix}S(z)W(z) \tag{28}$$

$$S^{\sim}(z)S(z) = I, \qquad D_{rx}(z) \in \mathcal{H}_{\infty}, \qquad N_{rx}(z) \in \mathcal{H}_{\infty} \tag{29}$$

Here, $W(z)$ is a stable TFM satisfying $W^{\sim}(z)W(z) = I - \Phi^{\sim}(z)\hat{Q}_p^{\sim}(z)\hat{Q}_p(z)\Phi(z)$.

Assume that $W(z) = W_{in}(z)W_{out}(z)$, $[D_{lc}(z) \ \ N_{lc}(z)]^{\sim}\hat{Q}_p(z)\Phi(z)W_{out}^{-1}(z) = N_{ra}(z)D_{ra}^{-1}(z)$. Here, $W_{in}(z)$ and $W_{out}(z)$ are respectively inner and outer TFMs of $\mathcal{H}_{\infty}$, and $N_{ra}(z)$ and $D_{ra}(z)$ are right coprime over $\mathcal{H}_{\infty}$ with $D_{ra}^{\sim}(z)D_{ra}(z) = I$. Denote $\begin{bmatrix} -N_{rc}(z) \\ D_{rc}(z) \end{bmatrix} S(z)W_{in}(z)D_{ra}(z)$ by $\hat{S}(z)$. Then,

$$\hat{S}^{\sim}(z)\hat{S}(z) = I, \qquad \begin{bmatrix} D_{rx}(z) \\ N_{rx}(z) \end{bmatrix} = [N_{ra}(z) + \hat{S}(z)]D_{ra}^{-1}(z)W_{out}(z) \tag{30}$$

Therefore, both $N_{rx}(z)$ and $D_{rx}(z)$ of Equation (28) belong to $\mathcal{H}_{\infty}$, only if $\hat{S}(z)$ belongs to $\mathcal{H}_{\infty}$. This further implies that $||\hat{S}(z)||_{\infty} \leq 1$.

Denote the Smith-McMillan form of $D_{ra}(z)$ by $D_{ra}(z) = U_a(z)\Sigma_a(z)V_a(z)$ with $\Sigma_a(z) = diag\left\{ \frac{\alpha_{ia}(z)}{\alpha_{ia}(z^{-1})}\Big|_{i=1}^{m} \right\}$, $\alpha_{ia}(z) = \prod_{q=1}^{p_{ia}}(z - z_{iqa})^{k_{iqa}}$, $i = 1, 2, \cdots, m$. As $D_{ra}(z)$ is inner, this decomposition is always possible. Moreover, it is also apparent from the inner property of $D_{ra}(z)$ that $|z_{iqa}| < 1$, $q = 1, 2, \cdots, p_{ia}$; $i = 1, 2, \cdots, m$. Define $\xi_{ia}(z)$ the ith column of $V_a^{-1}(z)$, $i = 1, 2, \cdots, m_a$. Then, the following conclusion can be proved.

Theorem 4. $N_{rx}(z)$ and $D_{rx}(z)$ of Equation (28) belong to $\mathcal{H}_\infty$, if and only if $\hat{S}(z) \in \mathcal{H}_\infty$ and for all $l = 0, 1, \cdots, k_{iqa} - 1$ and $i = 1, 2, \cdots, m$,

$$\left.\frac{d^l(\hat{S}(z)\xi_{ia}(z))}{dz^l}\right|_{z=z_{iqa}} = -\left.\frac{d^l(N_{ra}(z)\xi_{ia}(z))}{dz^l}\right|_{z=z_{iqa}} \tag{31}$$

Based on Lemma 6 and Theorem 4, all the desirable $N_{rx}(z)$ and $D_{rx}(z)$, that is, those satisfy the conditions that $N_{lc}(z)N_{rx}(z) + D_{lc}(z)D_{rx}(z) = \hat{Q}_p(z)\Phi(z)$ and the closed-loop system $\{N_{rx}(z)D_{rx}^{-1}(z),\ C(z)\}$ is internally stable, can be parameterized by an LFT of an inner TFM, through an application of the two-sided Nudelman interpolation theory.

In addition to $N_{rx}(z)$ and $D_{rx}(z)$, $X_{rc}(z)$ and $Y_{rc}(z)$ are also required in the suggested time domain procedure for estimating the frequency response of plant NRCFs. To have an appropriate choice for these stable TFMs, let $d_1^{[i]}(j\omega)$, $d_2^{[i]}(j\omega)$ and $\hat{e}^{[i]}(j\omega)$ represent respectively the DFTs of the measurement errors and external disturbances in the ith identification experiments, $i = 1, 2, \cdots, s$. Then, from the internal stability of the closed-loop system and the stability of $D_{lh}(z)$, $X_{rc}(z)$ and $Y_{rc}(z)$, as well as the requirement that $L(z)$ is an inner TFM, it can be easily proved that when the external disturbances and the measurement errors are of finite magnitude,

$$\lim_{N\to\infty} \hat{V}(j\omega) = L^H(j\omega)D_{lh}(j\omega)\{X_{rc}(j\omega)D_1(j\omega) - Y_{rc}(j\omega)D_2(j\omega) - [X_{rc}(j\omega)C(j\omega) + Y_{rc}(j\omega)]\hat{E}(j\omega)\} \tag{32}$$

and the convergence rate is at least $\frac{1}{\sqrt{N}}$. Here, $D_1(j\omega) = [d_1^{[1]}(j\omega)\ \ d_1^{[2]}(j\omega)\ \ \cdots\ \ d_1^{[s]}(j\omega)]$, $D_2(j\omega) = [d_2^{[1]}(j\omega)\ \ d_2^{[2]}(j\omega)\ \ \cdots\ \ d_2^{[s]}(j\omega)]$ and $\hat{E}(j\omega) = [\hat{e}^{[1]}(j\omega)\ \ \hat{e}^{[2]}(j\omega)\ \ \cdots\ \ \hat{e}^{[s]}(j\omega)]$.

To simplify discussions, it is assumed afterwards, without too much loss of generality, that for every $i = 1, 2, \cdots, s$, $d_1^{[i]}(j\omega) \in As\mathcal{N}_c(0,\ \Sigma_1(\omega))$, $d_2^{[i]}(j\omega) \in As\mathcal{N}_c(0,\ \Sigma_2(\omega))$ and $\hat{e}^{[i]}(j\omega) \in As\mathcal{N}_c(0,\ \Sigma_0(\omega))$ and the external disturbances and measurement errors of different identification experiments are independent of each other. Here, $As\mathcal{N}_c(\star,\ \sharp)$ stands for the set of sequences of complex random vectors which converges in distribution to the complex normal distribution with mean $\star$ and covariance matrix $\sharp$. These assumptions are quite natural when the environments have not changed significantly during identification experiments. In this case, all the columns of $\hat{V}(j\omega)$ are independent of each other. Moreover, if we define matrix $\Sigma(j\omega)$ as

$$\Sigma(j\omega) = X_{rc}(j\omega)\Sigma_1(j\omega)X_{rc}^H(j\omega) + Y_{rc}(j\omega)\Sigma_2(j\omega)Y_{rc}^H(j\omega) + [X_{rc}(j\omega)C(j\omega) + Y_{rc}(j\omega)]\Sigma_0(j\omega)[X_{rc}(j\omega)C(j\omega) + Y_{rc}(j\omega)]^H \quad (33)$$

Then, every column of $\hat{V}(j\omega)$ is asymptotically with the complex normal distribution $\mathcal{N}_c(0,\ L^H(e^{j\omega})D_{lh}(e^{j\omega})\Sigma(j\omega)D_{lh}^H(e^{j\omega})L(e^{j\omega})$.

Note that when $Q_x(z) = Q_p(z)\Phi(z)$ is satisfied by the NRCFs of the auxiliary plant, the asymptotic statistical properties of the estimate is completely determined by the random matrix $\hat{V}(j\omega)$, and the smaller the covariance matrix of each column of $\hat{V}(j\omega)$, the better the estimation quality. Therefore, a physically significant choice for $X_{rc}(z)$ and $Y_{rc}(z)$ is to select those that minimize the maximal eigenvalue of $L^H(e^{j\omega})D_{lh}(e^{j\omega})\Sigma(j\omega)D_{lh}^H(e^{j\omega})L(e^{j\omega}$. As both $L^H(e^{j\omega})$ and $D_{lh}(e^{j\omega})$ are unitary matrices, it is obvious that the sets consisting of eigenvalues of $L^H(e^{j\omega})D_{lh}(e^{j\omega})\Sigma(j\omega)D_{lh}^H(e^{j\omega})L(e^{j\omega}$ and those of $\Sigma(j\omega)$ are equal to each other. Hence, it is appropriate to select $X_{rc}(z)$ and $Y_{rc}(z)$ such that the maximal eigenvalue of $\Sigma(j\omega)$ is minimized. More precisely, the desirable $X_{rc}(z)$ and $Y_{rc}(z)$ are chosen as follows. At first, all the $X_{rc}(z)$ and $Y_{rc}(z)$ satisfying $X_{rc}(z)N_{rc}(z)+Y_{rc}(z)D_{rc}(z) = I$ are parameterized by a stable TFM. Afterwards, the desirable $X_{rc}(z)$ and $Y_{rc}(z)$ are determined through selecting the optimal parameter that minimizes $\bar{\sigma}(\Sigma(j\omega))$ by means of square completion.

On the other hand, in practical nonparametric estimation, only an estimate for $Q_p(z)$ is available, and the actual $Q_p(z)$ is generally not known. Assume that $\hat{Q}_p(z) = [I + W_q(z)\Delta_q(z)]Q_p(z)$ with $||\Delta_q(z)||_\infty \leq 1$, in which $W_q(z)$ is an outer TFM that represents the frequency characteristics of the error bound in estimating $Q_p(z)$. Then, when $N_{rx}(z)$ and $D_{rx}(z)$, the NRCFs of the auxiliary plant $P_x(z)$, are chosen to satisfy $Q_x(z) = \hat{Q}_p(z)\Phi(z)$, on the basis of Lemma 6, the following relations can be derived.

$$H(z) = -S(z)W(z) \quad (34)$$

$$D_{lh}(z)[X_{rc}(z)D_{rx}(z) - Y_{rc}(z)N_{rx}(z) - H(z)][Q_x^{-1}(z) - \Phi^{-1}(z)Q_p^{-1}(z)]D_{lc}(z) = -D_{lh}(z)[X_{rc}(z)D_{lc}^{\sim}(z) - Y_{rc}(z)N_{lc}^{\sim}(z)]W_q(z)\Delta_q(z)D_{lc}(z) \quad (35)$$

From these equalities, it is apparent that another physically significant selection for the auxiliary TFMs $X_{rc}(z)$ and $Y_{rc}(z)$ is to choose the stable TFMs that satisfy $X_{rc}(z)N_{rc}(z) + Y_{rc}(z)D_{rc}(z) = I$ and minimize the $\mathcal{H}_\infty$ norm of $D_{lh}(z)[X_{rc}(z)D_{lc}^{\sim}(z)-Y_{rc}(z)N_{lc}^{\sim}(z)]W_q(z)\Delta_q(z)D_{lc}(z)$ for all the stable $\Delta_q(z)$ with $\mathcal{H}_\infty$ norm not greater than 1. Note that

$$\sup_{||\Delta_q(z)||_\infty \leq 1} ||D_{lh}(z)[X_{rc}(z)D_{lc}^{\sim}(z) - Y_{rc}(z)N_{lc}^{\sim}(z)]W_q(z)\Delta_q(z)D_{lc}(z)||_\infty = ||D_{lh}(z)[X_{rc}(z)D_{lc}^{\sim}(z) - Y_{rc}(z)N_{lc}^{\sim}(z)]W_q(z)||_\infty ||D_{lc}(z)||_\infty \quad (36)$$

Moreover, $D_{lh}(z)$ is inner from its construction. It is not very difficult to see that the latter requirement is equal to the minimization of the $\mathcal{L}_\infty$ norm of $[X_{rc}(z)D_{lc}^{\sim}(z)-Y_{rc}(z)N_{lc}^{\sim}(z)]W_q(z)$. Obviously, this selection can be achieved through the same procedure as that for the minimization of $\bar{\sigma}(\Sigma(j\omega))$.

The last TFM required to be determined before the nonparametric estimation is the outer TFM $\Phi(z)$. From Lemma 6, it can be proved through direct algebraic manipulations that when $Q_x(z) = \hat{Q}_p(z)\Phi(z)$ is satisfied,

$$M^{\sim}(z)M(z) = \Phi^{\sim}(z)[I - \hat{Q}_p^{\sim}(z)\hat{Q}_p(z)]\Phi(z) \quad (37)$$

Assume that $M_0(z)$ is an outer TFM satisfying $M_0^{\sim}(z)M_0(z) = I - \hat{Q}_p^{\sim}(z)\hat{Q}_p(z)$. Then, $M(z)$ can be expressed as $M(z) = M_0(z)\Phi(z)$. Therefore, $M(z)Q_x^{-1}(z) = M_0(z)\hat{Q}_p^{-1}(z)$. Moreover,

$$\begin{bmatrix}\hat{N}_{rp}(e^{j\omega})\\ \hat{D}_{rp}(e^{j\omega})\end{bmatrix} = \begin{bmatrix}N_{lc}^H(e^{j\omega})\\ D_{lc}^H(e^{j\omega})\end{bmatrix}\hat{Q}_p(e^{j\omega}) + \begin{bmatrix}D_{rc}(e^{j\omega})\\ -N_{rc}(e^{j\omega})\end{bmatrix}D_{lh}^{-1}(e^{j\omega})\hat{L}(e^{j\omega})M_0(e^{j\omega}) \quad (38)$$

$$[X_{rc}(z)D_{rx}(z) - Y_{rc}(z)N_{rx}(z) - H(z)]Q_x^{-1}(z) = X_{rc}(z)D_{lc}^{\sim}(z) - Y_{rc}(z)N_{lc}^{\sim}(z) \quad (39)$$

From these relations and the definitions of $u_k^{[i]}$ and $y_k^{[i]}$ in Equations (19) and (20), it can be claimed that the frequency response estimate for plant NRCFs, as well as its statistical properties, are independent of $\Phi(z)$. Therefore, except to be outer and to guarantee the existence of the desirable TFM $W(z)$ in Lemma 6, $\Phi(z)$ can be selected arbitrarily.

6 Concluding Remarks

In this paper, estimation procedures are surveyed for the frequency response of plant NRCFs, on the basis of closed-loop time domain and frequency domain experimental data. The plant is of multi-inputs and multi-outputs, and may possibly be unstable.When the NRCFs of the auxiliary plant are appropriately selected, the estimate is asymptotically unbiased and complex normally distributed, provided that the probing signals are periodic. The asymptotic statistical properties of the estimate have also been discussed for the general case. And it has been made clear that under weak conditions, the estimate can asymptotically be expressed by a linear function of a normalized random matrix with all its columns being a complex normal random vector. Influences have also been investigated for the auxiliary TFMs on the estimation quality. Moreover, a computationally tractable procedure is suggested for desirable auxiliary TFM selections.

While the estimators are with a finite covariance matrix, they are generally biased. Especially, the estimation bias may not be acceptable when external disturbances are significant or the time domain experimental data is limited. Efforts are still required to see whether it is possible to achieve this objective through modifying the estimators according to the signal to noise ratio.

Acknowledgments. This research was supported in part by the Trans-Century Training Programme Foundation for the Talents and the Excellent Young Teachers Program of MOE, P.R.C., Tsinghua University under Grant JC2002025, and the National Natural Science Foundation of China under Grant 60174022.

References

1. J.A.Ball, I.Gohberg and L.Rodman, *Interpolation of Rational Matrix Functions. Operator Theory: Advances and Applications, Vol.45*, Basel: Birkhauser, 1990.
2. P.Date and G.Vinnicombe, *An algorithm for identification in the v-gap metric*, in Proc. the 38th Conference on Decision and Control, Phoenix, Arizona, U.S.A., 1999, pp.3230~3235.
3. U.Forssell and L.Ljung, *Closed-loop identification revisted*, Automatica, Vol.35(1999), pp.1215~1241.
4. T.Georgiou and M.C.Smith, *Optimal robustness in the gap metric*, IEEE Transactions on Automatic Control, Vol.35(1990), No.5, pp.673~686.
5. G.X.Gu, *Modeling of normalized coprime factors with v-metric uncertainty*, IEEE Transactions on Automatic Control, Vol.40(1999), No.8, pp.1498~1511.
6. W.P.Heath, *Probability density function of indirect non-parametric transfer function estimates for plants in closed-loop*, in Proc. 2000 SYSID, Santa Barbara, U.S.A., 2000, pp.321~326.
7. L.Ljung, *System Identification: Theory for the User*(the 2nd Edition), Prentice-Hall, Englewood Cliffs, NJ, 1999.
8. D.McFarlane and K.Glover, *Robust Controller Design Using Normalized Coprime Factor Plant Descriptions*, Vol.38, Lecture Notes in Control and Information Sciences, Springer-Verlag, 1990.
9. P.Pintelon and G.Schoukens, *System Identification: a Frequency Domain Approach*, IEEE Press, New York, 2000.
10. P.M.J.Van den Hof, R.J.P.Schrama, R.A.Callafon and O.K.Bosgra, *Identification of normalised coprime factors from closed-loop experimental data*, European Journal of Control, Vol.1(1995), No.1, pp.62~74.
11. M.Vidyasagar, *Control System Synthesis: a Factorization Approach*, MIT Press, Cambridge, MA, 1985.
12. G.Vinnicombe, *Frequency domain uncertainty and the graph topology*, IEEE Transactions on Automatic Control, Vol.38(1993),No.9,pp.1371~1383.
13. J.S.Welsh and G.C.Goodwin, *Finite sample properties of indirect nonparametric closed-loop identification*, IEEE Transactions on Automatic Control, Vol.47(2002),No.8,pp.1277~1292.
14. T.Zhou, *On the computation and statistical properties of the MLE for the frequency response of NCFs*, International Journal of Control, Vol.76(2003),No.15,pp.1547~1559.
15. T.Zhou and H.W.Xing, *Identification of normalized coprime factors through constrained curve fitting*, Automatica, Vol.40(2004), No.9, pp.1591~1601.

Part 4

Applications of Nonlinear and Adaptive Control

Influence of Transfer Conductance on Power System Stability in Control Scheme

Yuanzhang Sun[1], Shan He[1], Romeo Ortega[2] and Martha Galaz[2]

[1] State Key Laboratory of Power System ,Tsinghua University, Beijing, 100080, P.R.China yzsun@mail.tsinghua.edu.cn, heshan00@mails.tsinghua.edu.cn
[2] LSS Supelec, Gif-sur-Yvette, 91192, France, ortega@lss.supelec.fr

1 Introduction

The non-zero transfer conductance appearing in multi-machine power systems during classical reduction of network has significant impact on the control control schemes that enhance power systems stability. This paper gives a detailed discussion on the impact of transfer conductances in a specific energy–based controller developed by the authors. In particular, it investigates the effect of taking into account the transfer conductance in the construction of the closed–loop energy function for the general control scheme. Detailed numerical simulations on a 3-machine power system are used to illustrate these issues.

The adoption of new technologies, (like FACTS technology), and the ever increasing utilization of power electronic converters together with the new "deregulated" market have imposed more stringent requirements on the dynamic behavior of power systems [16], motivating the exploration of new control schemes. As the power systems consist of a large number of generator sets and possess strong nonlinearities, it is expected that controller design approaches based on the nonlinear model will have better transient stability performance than schemes based on linear approximations, see e.g., [1], [2].

Among the novel control ideas, one that has caught people's attention and is becoming increasingly popular in recent days, is to view the power systems controlled dynamics from the "energy" viewpoint, where the energy function has to be shaped and dissipation added. This approach concentrates on the interconnection between the various subsystems that compose the overall power system and its interaction with the surrounding environment. Since the objective is to modify the energy function it requires less precision of system model yielding more robust designs. On the other hand, the energy function of the system has close relationship with the real physical quantities, hence has more practical meaning. A particularly successful approach results from the combination of the Hamiltonian formalism—specifically, the use of Port-Controlled Hamiltonian (PCH) systems—with passivity theory that has led to the development of the controller design methodology of Interconnection and Damping Assignment Passivity–Based Control (IDA–PBC) [17]. PCH systems are more general and its constitutive elements have a precise physical meaning that helps the designer to carry out the energy shaping and damping injection tasks. Consequently, it achieved fruitful outcome when applied to the control of power systems, such as steam valving control and excitation control design [16], [18]. The IDA–PBC method provides a more flexible approach to reflect the transfer conductance compared with the previously proposed approach [15].

Energy functions have been used for a long time to estimate regions of attraction of stable equilibria and is well documented in textbooks like [3] and [4]. However, to the best of the authors' knowledge the existence of transfer conductance has stymied the development of a general energy function for multi-machine systems [6], [7]. A discussion of general energy functions after considering the affect of transfer conductance for a power system with more than 3 machines may be found in [15]. In the textbook [5], for the special case where the system has uniform R/X ratios for all transmission lines, a well-defined energy function reflecting the transfer conductances is derived.

In this paper, we consider the IDA–PBC proposed in [16] and illustrate, with some realistic simulations, that taking into account some practically reasonable approximations to this control law,which is realized by taking part of the transfer conductance, we can obtain enhanced performance and improve the transient stability.The proposed method provides a more flexible approach to reflect the transfer conductance compared with the previously proposed method [15]

The rest of the paper is organized as follows. The mathematical model is formulated in Section 2. The IDA-PBC methodology is briefly discussed in Section 3. Simulation results are given in Section 4. The paper is wrapped–up in Section 5 with some concluding remarks.

2 Dynamic Model of Multi-Machine Power System

The dynamic model for excitation control design of a power system with n generators can be presented as follows:

$$\begin{cases} \dot{\delta}_i = \omega_0 \omega_{ri} \\ T_{Ji}\dot{\omega}_{ri} = P_{mi} - D_i\omega_{ri} - P_{ei} \\ T'_{d0i}\dot{E}'_{qi} = E_{fsi} + u_{fi} - E'_{qi} - (x_{di} - x'_{di})i_d \end{cases} \quad (1)$$

where the subscript i denotes the i–th generator (here,i=1,2,...n),and

$$P_{ei} = E'_{qi}\sum_{j=1}^{n} E'_{qj}(G_{ij}cos\delta_{ij} + B_{ij}sin\delta_{ij})$$

$$i_d = \sum_{j=1}^{n} E'_{qj}(G_{ij} sin\delta_{ij} - B_{ij} cos\delta_{ij})$$

See [9], [16] for a detailed description of this model and Appendix A for a glossary of the symbols.

3 Proposed Controller

The nonlinear controller that we will used here to illustrate the influence of transfer conductance is the IDA-PBC derived in [16]. For the sake of completeness we briefly describe here the basic idea of IDA-PBC and refer the interested reader to [16], [10] for further information.

3.1 The IDA-PBC Procedure

IDA-PBC is a procedure that allows us to design a static state feedback that stabilizes the equilibria of nonlinear systems of the form

$$\dot{x} = f(x) + g(x)u \tag{2}$$

where $x \in \mathbb{R}^n$,is the state vector and $x \in \mathbb{R}^m$, $m < n$, is the control action. The basic idea is to endow the closed-loop with a port-controlled Hamiltonian (PCH) structure

$$\dot{x} = [J_d(x) - R_d(x)]\nabla_x H_d \tag{3}$$

where the matrices $J_d(x) = -J_d^\top(x)$ and $R_d(x) = R_d^\top(x) \geq 0$, which represent the desired interconnection structure and dissipation, respectively, are selected by the designer—hence the name IDA—and $H_d : \mathbb{R}^n \to \mathbb{R}$ is the desired total stored energy. If the latter has an isolated minimum at the desired equilibrium $x_\star \in \mathbb{R}^n$, that is if

$$x_\star = \arg\min H_d(x) \tag{4}$$

then $x_\star$ is stable with Lyapunov function $H_d(x)$. The admissible energy functions are characterized by the following parameterized partial differential equation

$$g^\perp(x)[f(x) - [J_d(x) - R_d(x)]\nabla_x H_d] = 0$$

that should be solved for H_d, where $g^\perp(x)$ is a left annihilator of $g(x)$, i.e. , $g^\perp(x)g(x) = 0$.

The control law is obtained as

$$u = [g^\perp(x)g(x)]^{-1}g^\perp(x)\{[J_d(x) - R_d(x)]\nabla_x H_d - f(x)\} \tag{5}$$

3.2 Application to Power Systems

When applying IDA–PBC to the multi-machine power system (??) the expression of the control becomes extremely complex basically due to the existence of transfer conductances G_{ij} and the corresponding angles $\alpha_{ij} = \arctan \frac{G_{ij}}{B_{ij}}$. This situation is true even for the case of 3-machine power systems with transfer conductances, which is the simplest and most typical model for multi-machine power systems.

To see the impact exerted by the transfer conductances on the control we will make 2 different simplifications to the IDA–PBC that are motivated by the power systems operation principles and dynamic properties.

- First, we will see the effect of completely neglecting the transfer conductance, that is to set $G_{ij} = 0$ together with $\alpha_{ij} = 0$. We denote this as the "Neglect Both G_{ij} and α_{ij}" method, or NB for short.
- Second, we will retain the original value of G_{ij} but set $\alpha_{ij} = 0$. We denote this controller as the "Neglect Only α_{ij}" controller, or NOα for short.

The second simplification is adopted because, after careful consideration of the equations for the 3-machine power systems and review of previous literature, we find that the main difficulties for the construction of the control stem not from the presence of the terms G_{ij}, but from the α_{ij} coefficients, which are usually small.

After tedious calculation, we get the expression of excitation control as follows:

$$
\begin{aligned}
u_1 = & -\tfrac{1}{\gamma_1}[E_2Y_{12}sin(\delta_1-\delta_2)\omega_2+E_3Y_{12}sin(\delta_1-\delta_3)\omega_3] \\
& -\tfrac{\omega_1}{\gamma_1}[E_{2\star}Y_{12}sin(\delta_1-\delta_2)+E_{3\star}Y_{13}sin(\delta_1-\delta_3)+(E_1+E_{1\star})G_{11}] \\
& -r_1\gamma_1(E_1-E_{1\star})+a_1E_1-b_1E_2sin(\delta_1-\delta_2) \\
& -b_1E_3sin(\delta_1-\delta_3)-E_{f1} \\
u_2 = & -\tfrac{1}{\gamma_2}[E_1Y_{12}sin(\delta_1-\delta_2)\omega_1+E_3Y_{23}sin(\delta_2-\delta_3)\omega_3] \\
& +\tfrac{\omega_2}{\gamma_2}[-E_{1\star}Y_{12}sin(\delta_1-\delta_2)+E_{3\star}Y_{23}sin(\delta_2-\delta_3)+(E_2+E_{2\star})G_{22}] \\
& +b_2E_1sin(\delta_1-\delta_2)-r_2\gamma_2(E_2-E_{2\star})+a_2E_2 \\
& -b_2E_3sin(\delta_2-\delta_3)-E_{f2} \\
u_3 = & \tfrac{1}{\gamma_3}[E_1Y_{13}sin(\delta_1-\delta_3)\omega_1+E_2Y_{23}sin(\delta_2-\delta_3)\omega_2] \\
& +\tfrac{\omega_3}{\gamma_3}[-E_{1\star}Y_{13}sin(\delta_1-\delta_3)-E_{2\star}Y_{23}sin(\delta_2-\delta_3)+(E_3+E_{3\star})G_{33}]] \\
& +b_3E_2sin(\delta_2-\delta_3)+b_3E_1sin(\delta_3-\delta_1) \\
& +a_3E_3-r_3\gamma_3(E_3-E_{3\star})-E_{f3}
\end{aligned}
\tag{6}
$$

Here

$$
E_{fi}=\frac{E_{fsi}}{T'_{d0i}},\ u_i=\frac{u_{si}}{T'_{d0i}},\ b_i=\frac{x_{di}-xdi'}{T'_{d0i}}
$$

$$
a_i=\frac{1}{T'_{d0i}}[1-B_{ii}(x_{di}-x'_{di})]
$$

and r_i and γ_i are some free positive parameters.

It is interesting that the IDA–PBC excitation controls under the two aforementioned assumptions have the same expression, but the value of Y_{ij} differ one from each other. Indeed, for the NB controller the coefficients $Y_{ij}=B_{ij}$. On the other hand, for the NOα controller we get $Y_{ij}=\sqrt{B_{ij}^2+G_{ij}^2}$.

3.3 Remarks

From simple observation of the control law in (6), one can observe that:

1. The control law of every generator depends on the parameters and signals of the other generators, so the control law relies on the fully knowledge of the network parameters and structure. Although, in principle, to apply this control law requires precise information of all the generators in

the network, some simulation results presented in the next section reveal strong robustness properties of the IDA–PBC.

2. As indicated above, the parameter r_i and γ_i (i=1,2,3) are chosen freely. Because of the intertwined relationship between these parameters, their selection is usually non–obvious and has to be done based on the specific network.

3. Since the NOα controller contains more information about the system than the NB one, it is expected that it will have a better dynamic control performance, a fact that is corroborated by the simulations below.

4 Simulation Results

4.1 computer simulation

The impact of the controller simplifications described above was tested on an IEEE 9-bus 3-machine system with MATLAB 6.5. The configuration of the studied system is shown in Fig.6, and the related data about generators and transmission line can be found in Appendix B and Tables 2 and 1.

The fault scenario selected for simulation is a three-phase short circuit between Bus 3 and Bus A at 1s with clearing out time of 0.1s and 0.2s respectively. After the fault, the power system recovers its original topology. The simulation time is 5s. The simulation environment is Windows XP.

·The open loop (OL) system, i.e., without field regulation, beginning with the oscillation at 1s, preserves stability with critical clearing out time (T_{cr}) smaller than $0.1s$, and will lose synchronism with $T_{cr} = 0.2$. See Fig.1. So we will choose the $T_{cr} = 0.2s$ scenario to test the two simplified IDA–PBCs.

Simulation results for the NB and NOα IDA—PBCs are depicted in Figs.2. They show that neither of the two control strategies loses synchronism, but NOα has better dynamic response quality than NB, specifically it shows: smoother regulation behavior, smaller rotor angle oscillation magnitude and faster convergence to the stable equilibrium point.

We now proceed to compare the different generators. Generator 3 is the nearest generator to the fault, thus suffers most from the fault and is a good indicator for control performance. From Fig.5–4, we observe that both of the strategies work better for Generator 3 than for the other two generators.

As we stated before, the two proposed control strategies strongly depend on all the parameters in the network. This may lead to control failure when only part of the generators has the controller, which is the usual case for economic operation. To verify this, we also carried out a tested when control is added only to Generator 3 (u_3 in Eqn. (6), while $u_1 = u_2 = 0$). Interestingly, similar results to the ones obtained for the full controller were observed, which testifies to the strong robustness properties of IDA–PBC.

4.2 numeric results

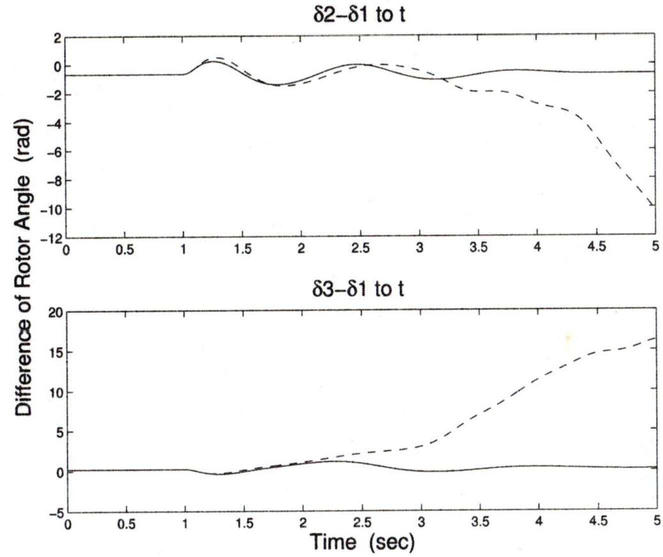

Fig. 1. System performance without controllers (OL), — $T_{cr} = 0.1$, – – – $T_{cr} = 0.2$.

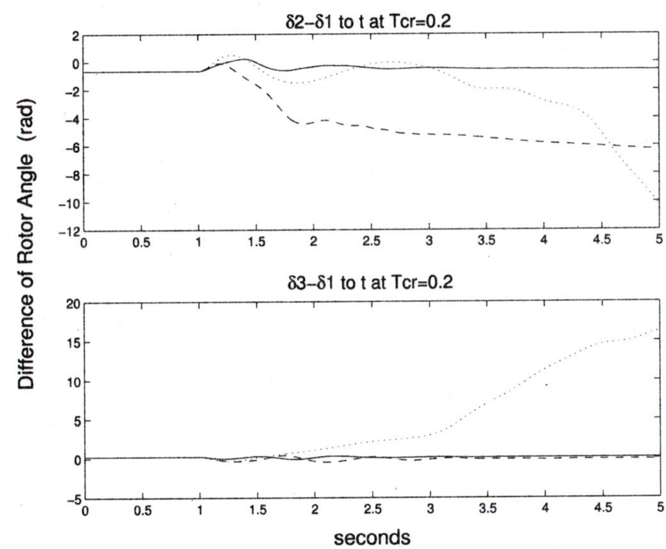

Fig. 2. Rotor angle response of Generator 1, — NOα, – – – NB, - - - - OL

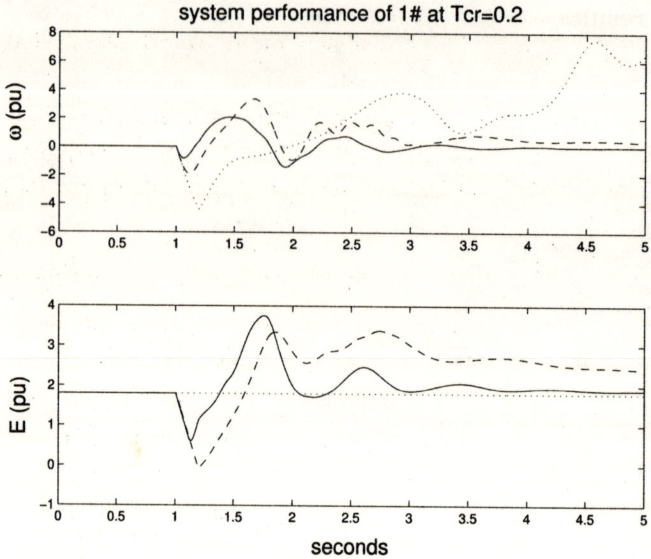

Fig. 3. Performance of Generator 1, — NOα, --- NB, - - - - OL

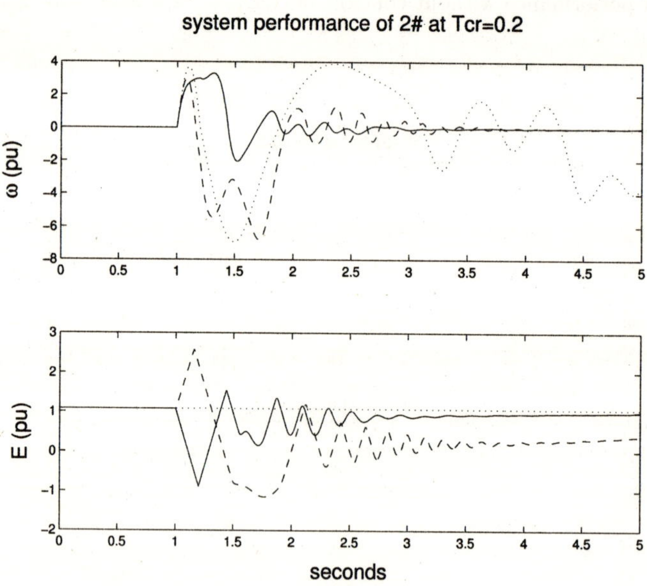

Fig. 4. Performance of Generator 2, — NOα, --- NB, - - - - OL

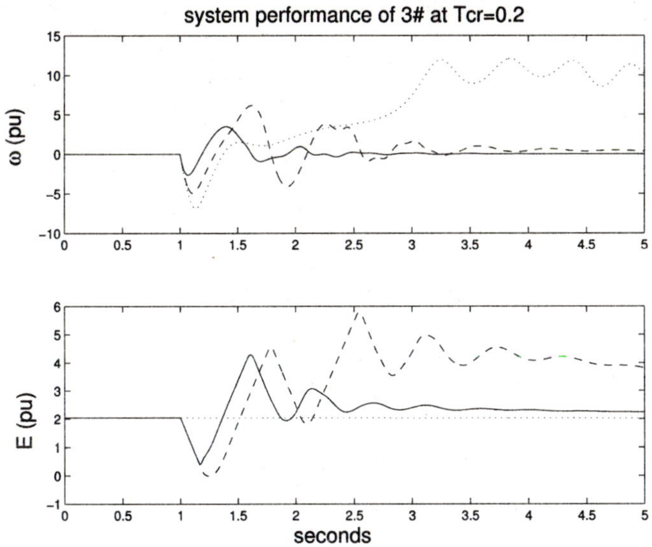

Fig. 5. Performance of Generator 3, — NOα, --- NB, - - - - OL

5 Conclusion

We investigated in this paper the impact of totally, or partially, neglecting the transfer conductances in the design of the nonlinear static state feedback IDA-PBC for multi-machine power system reported in [16]. We compared, via realistic 3-machine simulations, the different dynamics responses of the resulting simplified controllers. The simulation results reveal the efficiency of the proposed excitation control in enhancing transient stability of power system. It also has proven its robustness in a decentralized, partially actuated, scenario a feature that is not contemplated by the theoretical analysis carried out in [16].

Though people have worked for years on the development of energy functions for power systems with transfer conductances, there are still many issues that remain to be solved. We have explored here the idea of considering only the value of G_{ij}, but not the angle α_{ij}, in the energy–shaping IDA–PBC design. The reported results suggest that this idea may be useful for the application of energy–based methods to the real power system with non–negligible transfer conductances.

Appendix

A Glossary of Symbols

δ_i rotor angle, in radian
ω_i rotor speed, in rad./sec.$\omega_0 = 2f_0\pi = 314.159$
f frequency of terminal voltage of a generator, in per unit
P_e electrical power, in per unit
P_m mechanical power, in per unit
D damping constant, in per unit
T_J inertial time constant, in seconds
T'_{d0} direct axis transient short circuit time constant, in seconds
E'_{qi} q-axis internal transient electric potential of the ith generator, in per unit
E_{fs} equivalent EMF in the excitation coil, in per unit
u_f input of the SCR amplifier of the generator, in per unit
i_d d-axis current, in per unit
x_d d-axis synchronous reactance of a generator, in per unit
x'_d d-axis transient reactance, in per unit
δ_{ij} $\delta_{ij} = \delta_i - \delta_j$
α_{ij} $\alpha_{ij} = \arctan \frac{G_{ij}}{B_{ij}}$
B_{ij} ith row and jth column element of the nodal susceptance matrix, which is symmetric, at the internal nodes after eliminating all physical buses, in per unit

B IEEE 9-bus System

Bus I	Bus J	R1	X1	B1/2
Gen2	STATION A	0.01	0.085	0.044
Gen2	STATION B	-0.017	0.092	0.0395
STATION A	Gen3	0.032	0.161	0.0765
STATION B	Gen1	0.039	0.17	0.0895
Gen3	STATION C	0.085	0.072	0.03725
STATION C	Gen1	0.0119	0.1008	0.05225

Table 1. Line data (in per unit.)

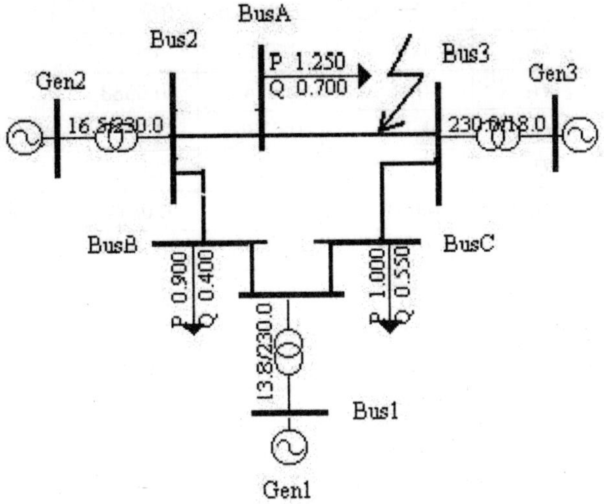

Fig. 6. IEEE benchmark 9-bus 3-machine system, (base power=100MVA)

Generator	Voltage reference	T_j	X_d'	X_q'	X_d	X_q	X_{d0}'
Gen1	13.8	20	0.0969	0.0969	0.146	0.0969	8.96
Gen2	16.5	20	1.258	0.25	1.313	1.258	5.89
Gen3	18	20	0.8645	0.1969	0.8958	0.8645	6
Generator	Voltage reference	T_{q0}'	X_l	X_d''	X_q''	T_{d0}''	T_{d0}'
Gen1	13.8	0.6	0.0742	0.107	0.107	0.033	0.07
Gen2	16.5	20	1.258	0.25	1.313	1.258	5.89
Gen3	18	20	0.8645	0.1969	0.8958	0.8645	6

Table 2. Generator data, (in per unit except voltage reference in kV, T_j, T_{q0}',T_{d0}'' and T_{d0}' in second.)

References

1. Q. Lu, Y. Sun, Z. Xu, T. Mochizuki, decentralized nonlinear optimal excitation control, IEEE Transactions on power systems, Vol.11, No4, November 1996.
2. C. Sun, Z. zhao, Y. Sun, Q. Lu, Design of nonliear robust excitation control for multimachine power systems, IEE Proc. -Gener. Transm. Distrib, Vol.143, No.3, May, 1996.
3. M. A. Pai, Power System Stability, North-Holland Publishing Co., Amsterdam, 1981.
4. M. A. Pai, Energy Function Analysis for Power System Stability, Kluwer Academic, Boston, 1989.

	Before fault	During fault	After fault
G_{11}	0.04	0.3877	0.04
B_{11}	-0.658	-2.4743	-0.6583
G_{22}	1.189	0.6336	1.189
B_{22}	-2.217	-3.7225	-2.2171
G_{33}	0.0814	0.5172	0.0814
B_{33}	-0.881	-3.3693	-0.8806
G_{12}	0.1498	0.1676	0.1498
B_{12}	0.3477	1.1787	0.3477
G_{13}	0.0351	0.1081	0.0351
B_{13}	0.09	0.8659	0.09
G_{23}	0.2227	0.3917	0.2227
B_{23}	0.4872	1.884	0.4872

Table 3. Parameter data, before and after the fault, (in per unit.)

5. Young-Hyun Moon, Heon-Su Ryu, Byoung-Hoon Cho, Baik Kim, Second-kind Energy Function of Power System and Its Applications, 2001 IEEE.
6. MA. Pai, P. G. Murthy, On Lyapunov Functions for Power Systems with Transfer Conductance, IEEE Trans. On Automatic Control, VOLAC-18, April 1973.
7. U. Gudaru, A general Lyapunov function for multimachine power systems with transfer conductance, M, J cont , VO121, pp.333-343, 1975.
8. Prabha Kundur, John Paserba, Venkat Ajjarapu, Goran Andersson , Anjan Bose , Claudio Canizares, Nikos Hatziargyriou, David Hill, Alex Stankovic, Carson Taylor, Thierry Van Cutsem, and Vijay Vittal, IEEE/CIGRE Joint Task Force on Stability Terms and Definitions, Definition and Classification of Power System Stability, IEEE Transactions on Power Systems, 0885-8950/04, 2004, page 4-5.
9. Lu. Q and Sun. Y. Z: The nonlinear control of power system, (Academic Press, Beijing, 1993).
10. R. Ortega, A. van der Schaft, B. Maschke and G. Escobar, Interconnection and damping assignment passivity-based control of port-controlled hamiltonian systems, Automatica, Vol. 38, No. 4, 2002.
11. O. Dahl, Electric Power Circuits: Theory and Applications, McGraw-Hill, NY, 1938.
12. P. C. Magnusson, The transient-energy method of calculating stability, AIEE Transactions, Vol. 66, pp. 747-755, 1947.
13. Q. Lu, Y. Z. Sun and S. Mei, Nonlinear Excitation Control of Large Synchronous Generators, Kluwer Academic Publishers, Boston, MA., 2001.
14. P. M. Anderson and A. A. Fouad, Power Systems Control and Stability, The Iowa State University Press, 1977.
15. H.D.Chiang, Study of the Existence of Energy Function for Power Systems with loss, IEEE Trans. On Circuits and Systems, Vol. CAS-36, p1423-1429, Nov. 1989.

16. R. Ortega, M. Galaz, A. Astolfi,Y.Z. Sun, T. Shen, Transient Stabilization of Multimachine Power Systems with Nontrivial Transfer Conductances, *IEEE Trans. Automat. Contr.*, Vol 50, number 1, pages 60–75, January 2005.
17. H-D. Chiang, C-C. Chu and G. Cauley, Direct stability analysis of electric power systems using energy functions: theory, applications, and perspective, Proceedings of the IEEE, Vol. 83, No. 11, pp. 1497-1529, 1995.
18. T. Shen, Y. Sun, R. Ortega and Shengwei Mei, Energy-shaping Control of Synchronous Generators with Exciter-Governor Dual Control Loops, *Int J of Control*, 2004.
19. F. H. J. R. Silva, L. F. C. Alberto and N. G. Bretas, Extended Lyapunove function for power systems with transmission losses, 2003 IEEE Bologna PowerTech Conference, June 23-26, Bologna, Italy.

Nonlinear Robust Velocity Control for Water Hydraulic Servomotor System with Uncertainty

Kazuhisa Ito

Fluidics Laboratory, Mechanical Engineering Department, Faculty of Science and Technology, SOPHIA University, Kioicho, Chiyoda-ku, Tokyo 1028554 Japan
kazu-ito@me.sophia.ac.jp

1 Introduction

Water hydraulic system is an attractive technology for safety environment. Water as a pressure medium gives many benefits compared with mineral oil, including easy maintenance, good availability and so forth. In the early stage of development, many literatures reported the comparison between water and mineral oil on physical and dynamical properties in use[1]-[5] while these days' reports give us the results of control techniques of water hydraulic devices[6]-[12]. Linjima *et.al.* studied the position control of water hydraulic cylinder taking advantage of the low compressibility of water and low-cost low-pressure valves[6]-[9]. In [10], the position and velocity control of water hydraulic cylinder is realized introducing the profile generator which generates smooth control signal to attenuate a large overshoot. The rotational angle control and the speed control of water hydraulic servomotor are also tested with conventional PID controller in [11],[12]. These results are mainly based on the conventional technique for mineral oil hydraulic system and there is a need to be discussed to consider the properties of water hydraulic devices. These days some modern nonlinear robust control methods are applied to hydraulic devices, however, almost all such papers are on a control of the mineral oil hydraulic systems. An adaptive controller, which compensates uncertainties of inertia load and some physical parameters, is proposed in [13]. A feedback linearization method and backstepping design of mineral oil hydraulic motor are proposed in [14] while the solution for parameter uncertainty is an open question in the paper.

In the last decade, water hydraulic servo valve bacame arailable. In water hydraulic cylinder case, the controller for water hydraulic pressing machine is constructed by applying the H_∞ mixed-sensitivity problem to compensate parameter uncertainties for the spool clearance and the coefficient of flow rate of servo valve and the bulk modulus of water[15]. However, the results on water hydraulic servomotor have been reported in few papers.

In this paper, we present a development of a design of robust rotational velocity controller for water hydraulic servomotor for the uncertain parameters and the load fluctuation. And we develop a robust controller without linearizing the flow equation which includes highly nonlinear term. To be more precise, we apply the Lyapunov-based recursive design method to construct the virtual input and Lyapunov-like function to ensure stability of each subsystem compensating the effect of parameter uncertainty. Generally speaking, we can achieve the exponential stability for such error system by using the signature function to design a controller. The resulting controller, however, generates non-smooth signal, therefore it requires large discontinuous control input which may occur undesirable behaviour for system. The proposed controller ensures to drive the state error to the ultimate set specified by the designer *a priori* without using signature function. The ultimate set to which error converges can be chosen arbitrary small. This is called the uniformly ultimately bounded stability (UUBS)[16] of the system. This scheme can be extended to the rotational angle control system.

In Section 2, we derive mathematical model of water hydraulic servomotor. Then we define the regulation problem to be solved. In Section 3, we construct a robust controller for uncertain parameters and load fluctuation with Lyapunov-based recursive design. We show some simulation results of the proposed controller in Section 4. Finally we summarize our results.

2 Problem formulation

The schematic diagram of the water hydraulic servo motor system is shown in Fig.1.

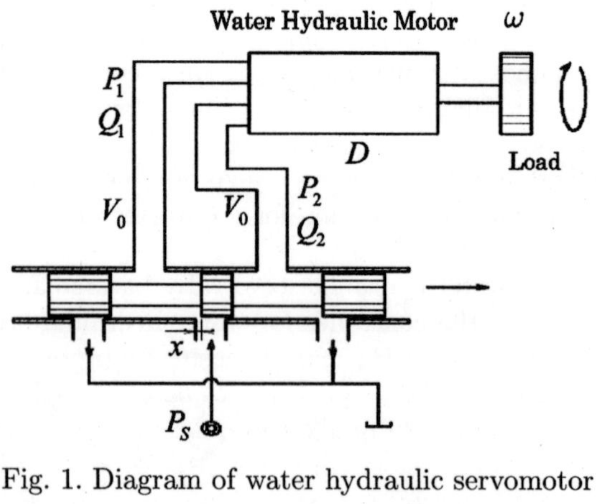

Fig. 1. Diagram of water hydraulic servomotor

In general, the stark differences between mineral oil hydraulic system and water hydraulic one are the strong mechanical friction (Coulomb friction) and impulsive surge pressure. The former depends on lower viscosity while the latter on higher bulk modulus of water. Low viscosity requires strong liquid seal to prevent the leakage and this leads to the larger starting torque. It is known from experiments that the Coulomb friction of water hydraulic motor is very large comparing to the mineral oil hydraulic motor.

From Fig.1, the equation of motion of the water hydraulic servomotor is expressed as follows.

$$\dot{\omega} = -c_e \frac{D\mu}{2\pi I}\omega + (1 - c_f)\frac{D}{2\pi I}P_L \tag{1}$$

where $P_L (= P_1 - P_2)$ is load pressure. Considering the effect of leakage flow, the actuator dynamics can be written as

$$\dot{P}_L = -\frac{DK}{\pi V_0}\omega - c_s P_L + \frac{2K}{V_0}Q_L \tag{2}$$

while the equations for load discharge Q_L is given by

$$Q_L = c_d w x \sqrt{\frac{P_S - \mathrm{sgn}(x)P_L}{\rho}} \tag{3}$$

The spool valve displacement x is related to the control input voltage u with first order system given by

$$\dot{x} = -\frac{1}{\tau}x + \frac{K_v}{\tau}u \tag{4}$$

where u is the input voltage to the servo valve. We introduce the scaling factor S_p, S_x because the rotational speed ω, the load pressure P_L and valve spool displacement x are of completely different orders

$$\bar{P}_L = \frac{P_L}{S_p},\ \bar{P}_S = \frac{P_S}{S_p},\ \bar{x} = \frac{x}{S_x} \tag{5}$$

With scaling factors and equations (1)-(4), we have system equation

$$\begin{aligned}\dot{\omega} &= -c_e \tfrac{D\mu}{2\pi I}\omega + (1 - c_f)\tfrac{DS_p}{2\pi I}\bar{P}_L\\ \dot{\bar{P}}_L &= -\tfrac{DK}{S_p \pi V_0}\omega - c_s \bar{P}_L + c_d \tfrac{2K}{S_p V_0} w S_x \bar{x}\sqrt{\tfrac{\bar{P}_S - \mathrm{sgn}(\bar{x})\bar{P}_L}{\rho}}\\ \dot{\bar{x}} &= -\tfrac{\bar{x}}{\tau} + \tfrac{K_v}{S_x \tau}u\end{aligned} \tag{6}$$

The control objective is to maintain the rotational velocity ω to operation point ω_e specified by the designer. In this paper, we assume that the motor shaft does not change the rotational direction and ω_e is given as a positive constant. This is a reasonable assumption and implies that spool displacement in servo valve is not necessary to move both directions around zero, therefore we have $\mathrm{sgn}(\bar{x}) = 1$ in (6). It is shown by straightforward that the equilibrium point $(\omega_e, \bar{P}_{Le}, \bar{x}_e)$ of system is

$$\begin{aligned} \omega_e &: \text{arbitrary specified by designer} \\ \bar{P}_{Le} &= \frac{\bar{c}_e \mu}{S_p(1-\bar{c}_f)}\omega_e \\ \bar{x}_e &= \frac{\frac{DK}{S_p \pi V_0}\omega_e + \bar{c}_s \bar{P}_{Le}}{\bar{c}_d \frac{2K}{S_p V_0} w S_x \sqrt{\frac{S_p}{\rho}(\bar{P}_S - \bar{P}_{Le})}} \end{aligned} \tag{7}$$

while the control input u_e to keep ω to the equilibrium point ω_e is

$$u_e = \frac{S_x}{K_v}\bar{x}_e \tag{8}$$

where the $\bar{c}_e, \bar{c}_f, \bar{c}_s, \bar{k}_q, \bar{k}_c$ stand for the nominal values of these parameters:

$$\begin{aligned} c_e &= \bar{c}_e + \Delta c_e, \ 0 < c_e^- \le \bar{c}_e + \Delta c_e \le c_e^+ \\ c_f &= \bar{c}_f + \Delta c_f, \ 0 < c_f^- \le \bar{c}_f + \Delta c_f \le c_f^+ \\ c_s &= \bar{c}_s + \Delta c_s, \ 0 < c_s^- \le \bar{c}_s + \Delta c_s \le c_s^+ \\ c_d &= \bar{c}_d + \Delta c_d, \ 0 < c_d^- \le \bar{c}_d + \Delta c_d \le c_d^+ \\ I &= I_0 + \Delta I, \ 0 < l_1 \le I_0 + \Delta I \le l_2 \end{aligned} \tag{9}$$

Note that it is not necessary to have an information of load inertia to get the equilibrium point x_e.

With (8), we introduce following state variables

$$e_1 = \omega - \omega_e, \quad e_2 = \bar{P}_L - \bar{P}_{Le}, \quad e_3 = \bar{x} - \bar{x}_e \tag{10}$$

and nominal control input

$$u = u_e + \frac{S_x \tau}{K_v} v \tag{11}$$

where v is the robust control input to be decided, then we have the error system

$$\dot{e} = A(\Delta)e + \Delta A + bv \tag{12}$$

where

$$e = [e_1, e_2, e_3]^T, \ b = [0, 0, 1]^T,$$

$$A(\Delta) = \begin{bmatrix} -c_e \frac{D\mu}{2\pi I} & (1-c_f)\frac{DS_p}{2\pi I} & 0 \\ -\frac{DK}{S_p \pi V_0} & -(c_s + M(e_2)) & c_d \frac{2K}{S_p V_0} w S_x \sqrt{\frac{S_p}{\rho}(\bar{P}_S - e_2 - \bar{P}_{Le})} \\ 0 & 0 & -\frac{1}{\tau} \end{bmatrix},$$

$$\Delta A = \begin{bmatrix} -\frac{D}{2\pi I}(\Delta c_e \mu \omega_e + \Delta c_f S_p \bar{P}_{Le}) \\ \Delta c_d \frac{2K}{S_p V_0} w S_x \bar{x}_e \sqrt{\frac{S_p}{\rho}(\bar{P}_S - e_2 - \bar{P}_{Le})} - \Delta c_s \bar{P}_{Le} \\ 0 \end{bmatrix}$$

and

$$M(e_2) = \frac{\bar{c}_d \frac{2K}{S_p V_0} w S_x \bar{x}_e}{\sqrt{\frac{S_p}{\rho}(\bar{P}_S - e_2 - \bar{P}_{Le})} + \sqrt{\frac{S_p}{\rho}(\bar{P}_S - \bar{P}_{Le})}} > 0 \ \forall \ e_2$$

To clarify the problem to be solved, we introduce following definition [16].

Definition 2.1 (Uniformly Ultimately Bounded Stable) *Consider the system* $\dot{x} = f(x)$, $x(t_0) = x_0$. *A solution* $x{:}\mathbf{R}^+ \rightarrow \mathbf{R}^n$, *is said to be uniformly ultimately bounded stable (UUBS) with respect to a set* $W \in \mathbf{R}^\mathbf{n}$ *containing the origin if there is a nonnegative constant* $T = T(x_0, W) < \infty$, *possibly dependent on* x_0 *and* W *but not on* t_0, *such that for all* $t \geq t_0 + T(x_0, W)$

$$\|x(t_0)\| < \eta \quad \Longrightarrow \quad x(t) \in W \tag{13}$$

□

Here we put a natural assumption on uncertain parameters.

Assumption 2.1 *For uncertain parameters in (9), each upper and lower bound, that is,* $(.)^-$ *and* $(.)^+$ *are known.* □

Remark 2.1 *We can obtain upper and lower bounds of uncertainties in (9) through experiments.* □

3 Controller Design

In this section, we construct a robust controller for uncertainties and load fluctuations with recursive Lyapunov-based design. To be more specified, we design the virtual input and Lyapunov-like function to ensure UUB stability of each sub-system recursively compensating the effect of uncertain parameters under the Assumption 2.1. Before designing controller, we set some controller parameters evaluating some bounds of elements in (12).

$$\varepsilon_0^- := (1 - c_f^+)\frac{DS_p}{2\pi l_2}, \quad \varepsilon_0^+ := (1 - c_f^-)\frac{DS_p}{2\pi l_1},$$

$$\varepsilon_1 := \frac{D}{2\pi l_1}\left[(c_e^+ - c_e^-)\mu\omega_e + (c_f^+ - c_f^-)S_p\bar{P}_{Le}\right], \tag{14}$$

$$\varepsilon_2 := \frac{DK}{S_p\pi V_0}, \quad \varepsilon_3 := c_d^- \frac{2K}{S_pV_0} wS_x\sqrt{\frac{S_p}{\rho}}, \tag{15}$$

$$\varepsilon_4 := c_d^+ \frac{2K}{S_pV_0} wS_x\sqrt{\frac{S_p}{\rho}}$$

$$\varepsilon_5 := (c_d^+ - c_d^-)\frac{2K}{S_pV_0} wS_x\bar{x}_e\sqrt{\frac{S_p}{\rho}}, \quad \varepsilon_6 := c_s^+\bar{P}_{Le} \tag{16}$$

[Step 1] Defining new state variable $\tilde{e}_1, \tilde{e}_2$ and positive definite function V_1

$$\tilde{e}_1 = e_1, \quad \tilde{e}_2 = e_2 - \beta_1(\tilde{e}_1), \quad V_1 = \frac{1}{2}\tilde{e}_1^2 \tag{17}$$

the time derivative of V_1 along the trajectories of the system (12) satisfies

$$\dot{V}_1 \le -c_e\frac{D\mu}{2\pi I}V_1 + \tilde{e}_1\left[(1-c_f)\frac{DS_p}{2\pi I}(\tilde{e}_2 + \beta_1) - \frac{D}{2\pi I}(\Delta c_e\mu\omega_e + \Delta c_f S_p\bar{P}_{Le})\right] \tag{18}$$

For positive constant ξ_1, we choose virtual input β_1 for e_1-subsystem as

$$\beta_1(\tilde{e}_1) = -\frac{\varepsilon_1}{\varepsilon_0^-}\frac{\tilde{e}_1}{|\tilde{e}_1| + \xi_1} \tag{19}$$

This implies

$$\dot{V}_1 \le -m_1V_1 + f_1(\tilde{e}_1) + (1-c_f)\frac{DS_p}{2\pi I}\tilde{e}_1\tilde{e}_2 \tag{20}$$

where

$$m_1 = c_e^-\frac{D\mu}{2\pi l_2}, \quad f_1(\tilde{e}_1) = \frac{\xi_1\varepsilon_1|\tilde{e}_1|}{|\tilde{e}_1| + \xi_1}$$

The last term in the right hand side of (20) will be canceled in Step 2.
[Step 2] We introduce a new variable

$$\tilde{e}_3 = e_3 - \beta_2(\tilde{e}_1, \tilde{e}_2) \tag{21}$$

then we obtain

$$\begin{aligned}\dot{\tilde{e}}_2 &= \dot{e}_2 - \dot{\beta}_1(\tilde{e}_1)\\ &= -\frac{DK}{S_p\pi V_0}\tilde{e}_1 - (c_s + M(e_2))(\tilde{e}_2 + \beta_1) - \Delta c_s\bar{P}_{Le}\\ &\quad + c_d\frac{2K}{S_pV_0}wS_x\sqrt{\frac{S_p}{\rho}(\bar{P}_S - e_2 - \bar{P}_{Le})}(\tilde{e}_3 + \beta_2)\\ &\quad + \Delta c_d\frac{2K}{S_pV_0}wS_x\bar{x}_e\sqrt{\frac{S_p}{\rho}(\bar{P}_S - e_2 - \bar{P}_{Le})} - \dot{\beta}_1\end{aligned} \tag{22}$$

We take a positive definite function for whole error system

$$V_2 = V_1 + \frac{1}{2}\tilde{e}_2^2 \tag{23}$$

The time derivative of V_2 along the trajectories of the system (12) satisfies

$$\begin{aligned}
\dot{V}_2 &= \dot{V}_1 + \tilde{e}_2\dot{\tilde{e}}_2 \\
&\leq -m_2V_2 + f_1(\tilde{e}_1) \\
&\quad +\tilde{e}_2\Bigg[(1-c_f)\frac{DS_p}{2\pi I}\tilde{e}_1 - \frac{DK}{S_p\pi V_0}\tilde{e}_1 - (c_s + M(e_2))\beta_1 \\
&\quad -c_s\bar{P}_{Le} + c_d\frac{2K}{S_pV_0}wS_x\beta_2\sqrt{\frac{S_p}{\rho}(\bar{P}_S - e_2 - \bar{P}_{Le})} \\
&\quad +\Delta c_d\frac{2K}{S_pV_0}wS_x\bar{x}_e\sqrt{\frac{S_p}{\rho}(\bar{P}_S - e_2 - \bar{P}_{Le})} - \dot{\beta}_1\Bigg] \\
&\quad +c_d\frac{2K}{S_pV_0}wS_x\sqrt{\frac{S_p}{\rho}(\bar{P}_S - e_2 - \bar{P}_{Le})}\tilde{e}_2\tilde{e}_3
\end{aligned} \tag{24}$$

where $m_2 = m_1 + c_s^-$. Therefore we choose the virtual control input β_2 as

$$\begin{aligned}
\beta_2(\tilde{e}_1, \tilde{e}_2) = &-\frac{1}{\varepsilon_3\sqrt{\bar{P}_S - e_2 - \bar{P}_{Le}}}\frac{\tilde{e}_2}{|\tilde{e}_2| + \xi_2}\Big[(\varepsilon_0^+ + \varepsilon_2)|\tilde{e}_1| \\
&+(c_s^+ + M(e_2))|\beta_1| + |\dot{\beta}_1| + \varepsilon_5\sqrt{\bar{P}_S - e_2 - \bar{P}_{Le}} + \varepsilon_6\Big]
\end{aligned} \tag{25}$$

where ξ_2 is positive constant design parameter. Then

$$\dot{V}_2 \leq -m_2V_2 + f_2(\tilde{e}_1, \tilde{e}_2) + c_d\frac{2K}{S_pV_0}wS_x\sqrt{\frac{S_p}{\rho}(\bar{P}_S - e_2 - \bar{P}_{Le})}\tilde{e}_2\tilde{e}_3 \tag{26}$$

where

$$\begin{aligned}
f_2(\tilde{e}_1, \tilde{e}_2) = f_1(\tilde{e}_1) + &\frac{\xi_2|\tilde{e}_2|}{|\tilde{e}_2| + \xi_2}\Big[(\varepsilon_0^+ + \varepsilon_2)|\tilde{e}_1| \\
&+(c_s^+ + M(e_2))|\beta_1| + |\dot{\beta}_1| + \varepsilon_5\sqrt{\bar{P}_S - e_2 - \bar{P}_{Le}} + \varepsilon_6\Big]
\end{aligned}$$

[Final Step] We take a positive definite function for whole error system

$$V = V_2 + \frac{1}{2}\tilde{e}_3^2 \tag{27}$$

We obtain the time derivative of V along the trajectories of the system (12) as

$$\begin{aligned}\dot{V} &= -\dot{V}_2 + \tilde{e}_3\dot{\tilde{e}}_3 \\ &\leq -mV + f_2 + \tilde{e}_3\left[v - \frac{\beta_2}{\tau} - \dot{\beta}_2\right. \\ &\left.+c_d\frac{2K}{S_pV_0}wS_x\sqrt{\frac{S_p}{\rho}(\bar{P}_S - e_2 - \bar{P}_{Le})}\tilde{e}_2\right]\end{aligned} \tag{28}$$

where $m = m_2 + \frac{1}{\tau} > 0$. Therefore we choose a robust control input v as

$$v = \frac{\beta_2}{\tau} + \dot{\beta}_2 - \frac{\varepsilon_4\tilde{e}_3}{|\tilde{e}_3| + \xi_3}|\tilde{e}_2|\sqrt{\bar{P}_S - e_2 - \bar{P}_{Le}} \tag{29}$$

where ξ_3 is positive constant design parameter. Finally we have

$$\dot{V} \leq -mV + f_3(\tilde{e}_1, \tilde{e}_2, \tilde{e}_3) \tag{30}$$

where

$$f_3(\tilde{e}_1, \tilde{e}_2, \tilde{e}_3) = f_2(\tilde{e}_1, \tilde{e}_2) + \frac{\xi_3|\tilde{e}_2||\tilde{e}_3|}{|\tilde{e}_3| + \xi_3}\varepsilon_4\sqrt{\bar{P}_S - e_2 - \bar{P}_{Le}}$$

Applying LaSalle's invariant theorem [17] to this result, we have desired stability property (UUBS). It is easy to show that the residual set

$$D_s = \left\{(\tilde{e}_1, \tilde{e}_2, \tilde{e}_3) \;\middle|\; \frac{\tilde{e}_1^2 + \tilde{e}_2^2 + \tilde{e}_3^2}{2} \leq f_3(\tilde{e}_1, \tilde{e}_2, \tilde{e}_3)\right\} \tag{31}$$

is a compact set. Moreover by taking the design parameters ξ_1, ξ_2 small, the set D_s will be arbitrary small. Then we have main result:

Theorem 1. *Consider the system (12) under the assumption 2.1. The robust controller (11), (29) ensures the UUBS of the origin of (12), therefore it achieves the robust velocity regulation of water hydraulic servomotor. Moreover the state error approaches compact set D_s defined in (31).* □

Remark 3.1 *As we see from the equation (12), the coefficient of e_2 which is input signal to the e_1-subsystem has the uncertainty I, c_f. In this case, we construct theoretically the stabilizing controller by means of* sgn(·) *function in general, however, it requires large control input practically. To avoid the case, we define the virtual input β_1 as smooth function by introducing design parameters ξ_1. For remaining two design steps, we apply same ides. In fact, we have* sgn *function as boundary function by taking ξ_1, ξ_2 and ξ_3 as* 0. □

4 Simulation Results

To illustrate the proposed robust controller above, simulation results are obtained for a system (Table 1).

It is set that each friction coefficient, leak coefficient and load inertia exist in the following interval:

$$\begin{aligned}
1 \times 10^4 = c_e^- \leq c_e \leq c_e^+ = 1 \times 10^7 \\
0.2 = c_f^- \leq c_f \leq c_f^+ = 0.7 \\
0.1 = c_s^- \leq c_s \leq c_s^+ = 0.4 \\
0.45 = c_d^- \leq c_d \leq c_d^+ = 0.55 \\
0.124 = l_1 \leq I \leq l_2 = 0.284
\end{aligned}$$

which are obtained from pilot study.

Table 1. Physical parameters for numerical example

parameter	value
D	1.6×10^{-5}(m^3/rev)
μ	0.797×10^{-3}(Pa·s) at 30°C
ρ	995.7(kg/m^3)
K	2250×10^6(Pa)
P_S	5×10^6(Pa)
V_0	1.272×10^{-4}(m^3)
w	12.5×10^{-3}(m)
k_v	1×10^{-4}(m/V)
τ	0.01(1/s)

We give the nominal values of these parameters as its average. The desired rotational velocity is set to $\omega_e = 20$(rad/s), then this implies $P_{Le} =$ 0.145(MPa), $x_e = 0.825$(mm), $u_e = 0.825(V)$. The design parameters are chosen as $S_p = 1 \times 10^6$, $S_x = 1 \times 10^{-5}$, $\xi_1 = \xi_2 = \xi_3 = 0.1$ and the initial condition as $\omega(0) = 0\,(\text{rad/s}), P_L(0) = 0\,(\text{MPa}), x = 0\,(\text{mm})$. Moreover we consider the input saturation of servo valve, $\pm 10(V)$.

In this simulation, we consider a PI controller as well as robust controller to compare the results of these controllers. By try and error, we choose PI controller parameters as

$$u_{PI} = u_e - 0.05e_1 - 0.3 \int_0^t e_1 dt, \quad e_1 = \omega - \omega_e$$

so that no undesirable vibration appears.

The first simulation is for the case that all uncertain parameters are periodically time varying, for example

$$c_e(t) = \frac{1}{2}(c_e^+ - c_e^-) + (c_e^- - c_e^-) \sin(t)$$

Fig.2 is a simulation result and shows a time response of the rotational velocity ω. The solid line is for the robust control results and the dotted for PI control. From this figure, it is observed that the rotational velocity with the proposed robust controller is kept to neighborhood of reference velocity and attenuates the effects of parameter uncertainties while the PI controller shows large vibration. Fig.3 shows the corresponding input signals. It is noted that the proposed controller does not need large input signal.

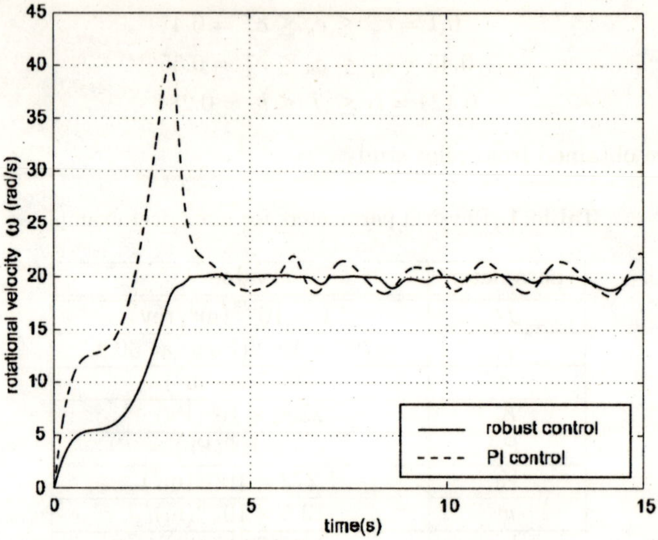

Fig. 2. rotational velocity ω ($\omega_e = 20$(rad/s))

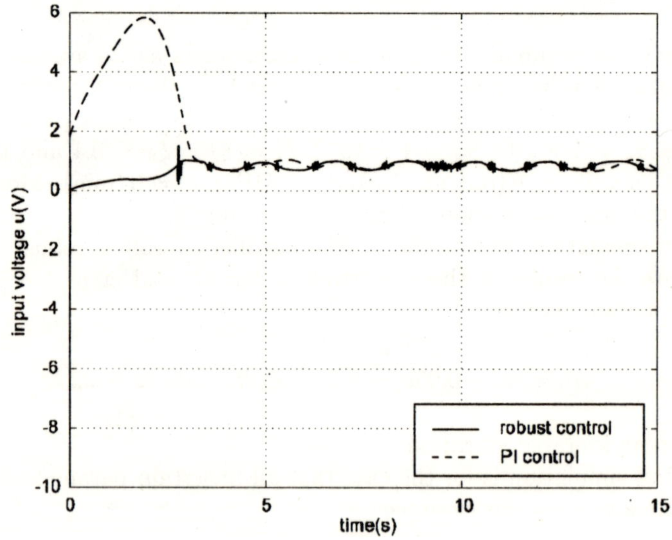

Fig. 3. input voltage u

In the second simulation, the coefficients for leakage flow and Coulomb friction and viscous friction values are changed step-wise from nominal value to its lowest value at 5(sec), 8(sec) and 12(sec), respectively. The result is shown in Fig.4. In this case, we can observe that the proposed controller keeps ω to the given reference velocity for uncertain parameter jump. With fixed PI controller also can achieve asymptotic convergence of regulation error, however, it needs longer time for good regulation. Fig.5 shows the corresponding input signals.

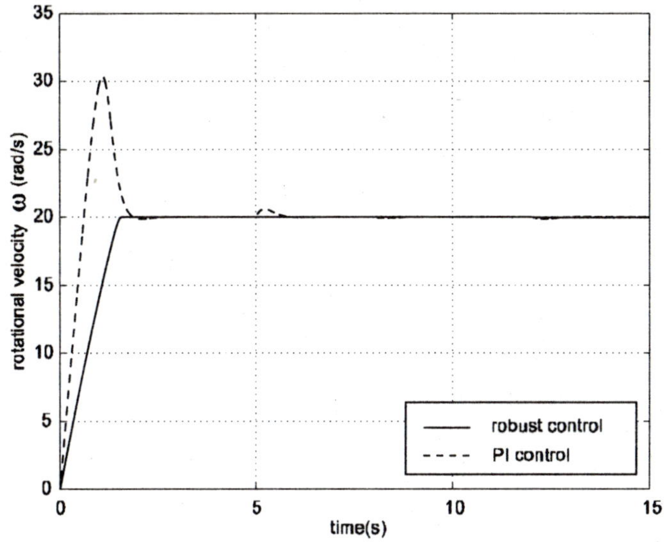

Fig. 4. rotational velocity ω ($\omega_e = 20$(rad/s))

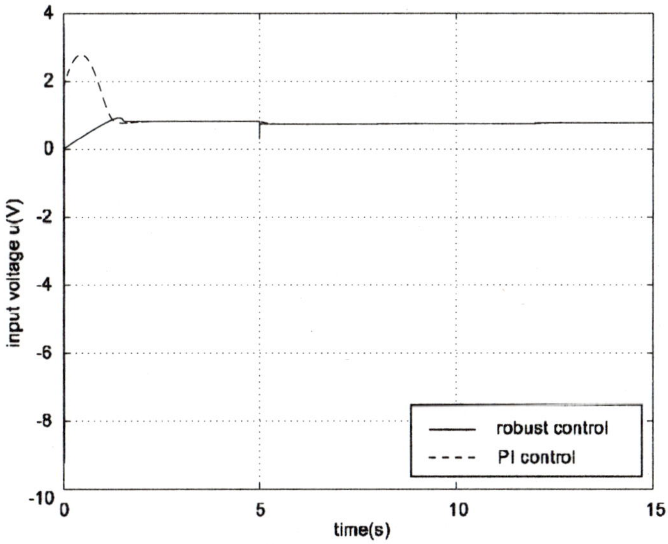

Fig. 5. input voltage u

5 Conclusion

In this paper, we proposed a method to construct a robust rotational velocity controller for water hydraulic servomotor. The parameter uncertainties describing viscous and Coulomb friction, leakage of the water hydraulic motor, linearizing error of flow equation and load fluctuation, are considered in controller design. By using Lyapunov-based recursive method, the uniformly ultimately stabilizing problem with input uncertainty is solved. The resulting controller ensures the convergence of the rational velocity error to the compact set that can be arbitrary small by design parameter. The numerical examples show the effectiveness of proposed robust controller. We examine the controller experimentally in future work. Moreover we exploit some robust control strategy for the influence of dead zone of water hydraulic valve.

Nomenclature

c_e : coefficient of viscous friction
c_f : coefficient of Coulomb friction
c_s : coefficient of motor leacage
c_d : coefficient of discharge
I : moment of inertia
D : motor displacement
ω : angular velocity
x : valve spool displacement
μ : viscosity of water
ρ : density of water
w : valve port width
V_0 : average volume of piping between valve and motor
K : bulk modulus of water
P_S : supply pressure
K_v : valve input voltage gain
τ : valve time constant

Acknowledgements

The author would like to acknowledge the financial support supplied by the Fluid Power Technology Promotion Foundation.

References

1. Nessie TM, Wasserhydraulik arbeitet mit Klarwasser, 38-12, pp.45-46, 1994
2. E.Trostmann, Water Hydraulic Control Technology, Marcel Dekker, Inc., 1995

3. E.Urata, S.Miyakawa and C.Yamashina, Hydrostatic Support of Spool for Water Hydraulic Servovalves, Proc. 4th Scandinavian International Conference on Fluid Power, 2-2, pp.910-929, 1995
4. F.Conrad, B.Hilbrecht, H.Jepsen and Gröndal: Design of Environmental-Friendly Water Hydraulic Motion Control Systems for Industrial Applications, The proceedings of 7th Scandinavian Internatinal Conference on Fluid Power, SICFP'01, Vol.3, pp.205-225, May 30-June 1, Linköping, Sweden, 2001
5. N.Ookubo, S.Miura, Y.Sakurai, S.Ikeo and K.Takahashi: Characteristics of Water Hydraulic Servo System, Fluid Power. Forth JHPS international Symposium, pp.715-720, 1999
6. M.Linjama, S.Oshima, K.T.Koskinen and M.Vilenius: High-Pressure On/Off Position Control of Low-Pressure Water Hydraulic Cylinder, The proceedings of 7th Scandinavian Internatinal Conference on Fluid Power, SICFP'01, Vol.2, pp.41-55, May 30-June 1, Linköping, Sweden, 2001
7. M.Linjama, K.T.Koskinen and M.Vilenius: Adaptive on/off Position Control of Low-Pressure Water Hydraulic Cylinder, The Conference Proceedings of 3rd International Fluid Power Conference, pp.315-326, March 5-6, Aachen, 2002
8. M.Linjama, J.Tammisto, K.T.Koskinen and M.Vilenius: On/Off Position control of Low-Pressure Water Hydraulic Cylinder Using Low-Cost Valves, FLUCOME 2000 The proceedings of 6th Technical Symposium on Fluid Control, Measurement and Visualization, pp.41-55, May 30-June 1, Linköping, Sweden, 2001
9. M.Linjama, J.Tammisto, K.T.Koskinen and M.Vilenius: Pseudo-Prpportional Position Control of Water Hydraulic Cylinder Using On/Off Valves, Proceedings of the 5th JFPS International Symposium on Fluid Power, Vol.1, pp.155-160, November 12-15, Nara, Japan, 2002
10. E.Mäkinen and T.Virvalo: On the Motion of a Water Hydraulic Servo Cylinder Drive, The proceedings of 7th Scandinavian Internatinal Conference on Fluid Power, SICFP'01, Vol.1, pp.109-123, May 30-June 1, Linköping, Sweden, 2001
11. Y.Mochizuki and C.Yamashina: Servo Control of a Water Hydraulic Axial Piston Motor, Ebara engineering review, No.184, pp.17-23, 1997
12. Dai Terasaka, Kazuhisa Ito and Shigeru Ikeo: PID-Control of Water Hydraulic Servomotor System, Proceedings of the 5th JFPS International Symposium on Fluid Power, Vol.1, pp.143-148, November 12-15, Nara, Japan, 2002
13. B.Yao, F.Bu and G.T.C.Chiu: Non-linear adaptive robust control of electrohydrailic systems driven by double-rod actuators, International Journal of Control, Vol.74, No.8, pp.761-775, 2001
14. M.Jovanović, Nonlinear Control of an Electrohydrailic Velicity Servosystem, Proceedings of the 2002 American Control Conference, pp.588-593, 2002
15. K.Sanada: A method of designing a robust force controller of a water-hydraulic servo system, Proc. Instn. Mech. Eng. Vol.216, Part I:J Systems and Control Engineering, pp.135-141, 2002
16. Z.Qu: Robust Control of Nonlinar Uncertain Systems, John Wiley & Sons, 1998
17. H.Khalil: Nonlinear Systems, *Prentice Hall*, 1996

An Approximate Design Method of Pneumatic Servo Systems Based on MRAC and Neural Network Techniques

Kanya Tanaka, Yuji Wakasa, Yoshiki Mizukami and Jinhua LI

Department of Electrical and Electronic Engineering, Faculty of Engineering, Yamaguchi University, 2-16-1, Tokiwadai, Ube, Yamaguchi 755-8611, Japan
ktanaka@eee.yamaguchi-u.ac.jp

1 Introduction

Pneumatic systems are widely used in various industries for the reason that they are safe and economically efficient. Especially, the pneumatic systems utilize air as their transfer medium, and therefore, have no risks of ignition, explosion or pollution unlike hydraulic ones. On the other hand, the pneumatic systems are intrinsically nonlinear due to large compressibility of air. Moreover, their dynamic characteristics usually change with load mass and friction. Therefore, in order to realize accurate control of the pneumatic systems, it is important to take account of these nonlinear elements and time varying properties.

To this end, model reference adaptive control (MRAC) combined with the so-called delta operator technique has been applied to the pneumatic systems [1, 2, 5]. The MRAC method is one of well-known adaptive control methods, and is only applied to a minimum phase system. In digital implementation, however, the discrete-time model of the pneumatic system tends to be non-minimum phase due to the sampling period, the load mass and so on even when the continuous-time plant is minimum phase. To cope with this difficulty, the delta-operator technique has been applied to the pneumatic system, and it has been shown that the stable discrete-time model can be regarded as a minimum phase system if the sampling period is sufficiently small [1, 2, 5]. In this way, although the MRAC method can compensate for the time varying properties of the pneumatic systems, there is still room for improvement of accurate control because the nonlinear elements are not explicitly taken into account in the MRAC method mentioned above.

In this paper, we propose an MRAC scheme combined with a Neural Network (NN) technique [4] to realize a more accurate control. Certainly, the primitive version of such a control scheme has been proposed by Tsuji *et al* [6]. It is however inapplicable to non-minimum phase systems. Our design method has an advantage that the MRAC with an NN is performed for reasonably approximated minimum phase model, and the NN compensates for the nonlinearity of the pneumatic servo systems. To demonstrate the effectiveness of the proposed method, we give some experimental results.

2 The Pneumatic Servo System

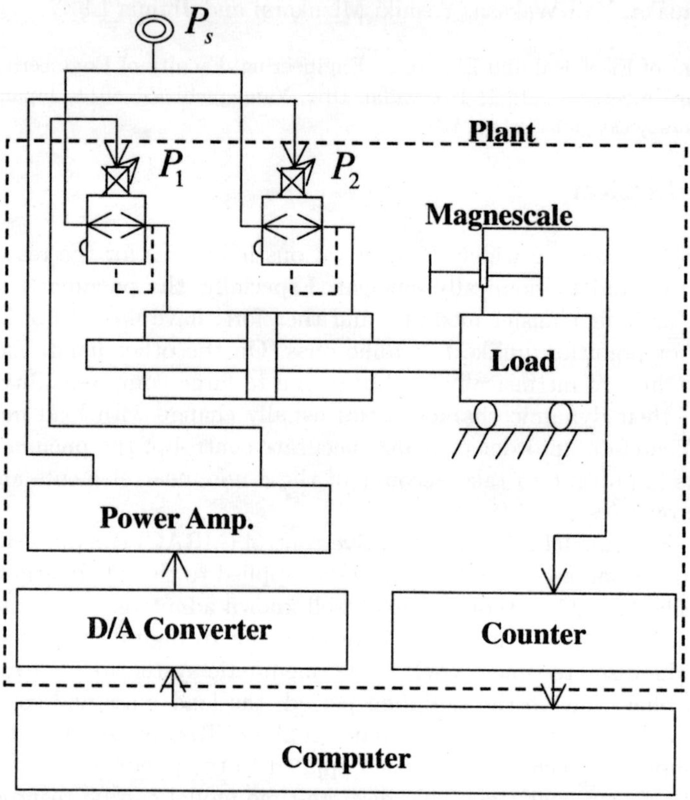

Fig. 1. Scheme diagram of the pneumatic servo system

Figure 1 shows a schematic diagram of a pneumatic servo system considered in this paper. The air cylinder is a both-rods type with the inside diameter of 55mm, the stroke of 170mm, and the rod-diameter of 22mm. The piston and rod are sealed with the labyrinth packing and the air bearings, respectively. The cylinder is placed horizontally and is constructed so that the load can be moved in the horizontal direction. The control valves are three-directional electro-pneumatic proportional pressures ones. The magnescale sensor is used for the detection of the load position. The signal from the magnescale sensor is inputted into the computer through the counter. The control signal is calculated in the computer and transmitted to the control valves through the D/A converter and the power amplifiers. The air pressure adjusted by the control valves move the cylinder.

3 Modeling of the Plant

The portion surrounded with the dashed line in Fig. 1 is considered as a plant in this paper. From our preliminary experiment, the time-delay of this plant can be regarded as 0.04sec. Then, the linearized discrete-time model is described as

$$y\left(k\right)=\frac{B(z^{-1})}{A(z^{-1})}u\left(k-d\right), \tag{1}$$

$$\left.\begin{aligned} A\left(z^{-1}\right)&=1+a_1z^{-1}+a_2z^{-2}+a_3z^{-3}, \\ B\left(z^{-1}\right)&=b_0z^{-1}+b_1z^{-2}+b_2z^{-3},\ d=L/T, \end{aligned}\right\} \tag{2}$$

where L and Tdenote the time delay and the sampling period, respectively; and $u(k)$ and $y(k)$ express the control signal and the output of the plant, respectively. We should note that this discrete-time model may become a non-minimum phase system. For such a plant, the MRAC cannot be constituted. However this system can be regarded as a minimum phase system for a sufficiently small sampling period. To show this reasonable approximation, we adopt the delta-operator [1, 2, 5] described as

$$\delta=\frac{z-1}{T}. \tag{3}$$

By using the delta-operator, Eq.(1) can be rewritten as

$$y\left(k\right)=\frac{D\left(\delta\right)}{C\left(\delta\right)}u\left(k-d\right), \tag{4}$$

where

$$\left.\begin{aligned} C(\delta)&=\delta^3+\alpha_1\delta^2+\alpha_2\delta, \\ D(\delta)&=\beta_0\delta^2+\beta_1\delta+\beta_2, \\ d&=0.04/T, \end{aligned}\right\} \tag{5}$$

$$\left.\begin{aligned} &\alpha_1=\frac{3+a_1}{T}, \quad \alpha_2=\frac{3+2a_1+a_2}{T^2}, \\ &\beta_0=\frac{b_0}{T}, \quad \beta_1=\frac{2b_0+b_1}{T^2}, \quad \beta_2=\frac{b_0+b_1+b_2}{T^3}. \end{aligned}\right\} \tag{6}$$

In Eq.(6), when $T\rightarrow 0$, the plant parameters β_0 and $\beta_1$converge to 0 owing to the characteristics of b_0 and b_1. Then, if we choose a sampling period of 0.02sec., β_0 and β_1 are regarded as 0. Therefore, the approximate model is described as follows:

$$y(k) = \frac{\beta_2}{C(\delta)} u(k-2). \tag{7}$$

This model is a minimum phase system when it is stable, and therefore, the MRAC is applicable to it. In the next section, our design scheme of the MRAC for the approximate model is described.

4 Design Scheme of The MRAC

The following equation is considered as the reference model for the approximate model shown by Eq.(7).

$$y_M(k) = \frac{\beta_M}{A_M(\delta)} r(k-2), \tag{8}$$

where

$$A_M(\delta) = \delta^3 + \alpha_{M1}\delta^2 + \alpha_{M2}\delta + \alpha_{M3}. \tag{9}$$

In Eq.(8), $r(k)$ is a bounded reference input and $A_M(\delta)$ is a Hurwits polynomial. The framework of the MRAC gives the control signal adaptively so that the output of the plant $y(k)$ follows that of the reference model $y_M(k)$. Then, the construction of the MRAC which attains this control purpose can be described as follows. The following Hurwits polynomial $F(\delta)$, $\bar{F}(\delta)$ and $H(\delta)$ are introduced:

$$\left.\begin{aligned} F(\delta) &= \bar{F}(\delta)(\delta + f)^2 \\ &= \delta^4 + \sum_{i=1}^{4} f_i \delta^{4-i}, \\ \bar{F}(\delta) &= \delta^2 + \bar{f}_1\delta + \bar{f}_2, \quad f = \frac{1}{T}, \\ H(\delta) &= \delta^3 + h_1\delta^2 + h_2\delta + h_3. \end{aligned}\right\} \tag{10}$$

Note that the polynomials $R(\delta)$, $S(\delta)$ which satisfy the following relation for $C(\delta)$ of Eq.(7) exist.

$$F(\delta)H(\delta) = C(\delta)R(\delta) + S(\delta), \tag{11}$$

where

$$\left.\begin{aligned} R(\delta) &= \delta^4 + r_1\delta^3 + r_2\delta^2 + r_3\delta + r_4, \\ S(\delta) &= s_0\delta^2 + s_1\delta + s_2. \end{aligned}\right\} \tag{12}$$

Multiplying both sides of Eq.(11) by $y(k+2)$ and using Eq.(7) to it, we have

$$\begin{aligned} y(k+2) &= \frac{1}{H(\delta)} \left\{ \frac{\beta_2 R(\delta)}{F(\delta)} u(k) + \frac{S(\delta)(1+T\delta)^2}{F(\delta)} y(k) \right\} \\ &= \frac{1}{H(\delta)} \phi^T \xi(k), \end{aligned} \tag{13}$$

where

$$\phi = [\phi_1, \cdots, \phi_8]^T, \tag{14}$$

$$\left.\begin{aligned} \phi_1 &= \beta_2, \quad \phi_{i+1} = \beta_2 (r_i - f_i) ; i = 1, \cdots, 4, \\ \phi_6 &= \frac{s_0}{f^2}, \quad \phi_{6+i} = \frac{s_i - s_0 f_i}{f^2} ; i = 1, 2, \end{aligned}\right\} \tag{15}$$

$$\left.\begin{aligned} &\xi (k) = [\xi_1 (k), \cdots, \xi_8 (k)]^T, \\ &\xi_1 (k) = u (k), \quad \xi_{i+1} (k) = \frac{\delta^{4-i}}{\bar{F} (\delta)} u (k); \ i = 1, \cdots, 4, \\ &\xi_6 (k) = y (k), \quad \xi_{6+i} (k) = \frac{\delta^{2-i}}{\bar{F} (\delta)} y (k) ; i = 1, 2. \end{aligned}\right\} \tag{16}$$

To synthesize the control input, replace $y(k)$ and ϕ with $y_M(k)$ and the variable parameter $\hat{\phi}(k) = \left[\hat{\phi}_1(k), \cdots, \hat{\phi}_8(k)\right]^T$, respectively. Then, we obtain the following relation.

$$\hat{\phi}^T(k)\xi(k) = y_{MH}(k+2),$$

$$u (k) = \frac{y_{MH} (k + 2) - \sum_{i=2}^{8} \hat{\phi}_i (k) \xi_i (k)}{\hat{\phi}_1 (k)}, \tag{17}$$

where

$$y_{MH}(k) = H(\delta) y_M(k). \tag{18}$$

In practice, since parameters of the plant are unknown, it is required to identify these unknown parameters and to decide the plant parameters by using the estimated parameters. Then, the identification method of the plant parameters is shown below. Equation (1) is rewritten as

$$y (k) = \theta^T \zeta (k), \tag{19}$$

where

$$\theta = [a_1, a_2, a_3, b_0, b_1, b_2]^T, \tag{20}$$

$$\zeta(k) = [\zeta_1(k), \cdots, \zeta_6(k)]^T, \tag{21}$$

$$\zeta_i(k) = -y(k-i), \quad \zeta_{3+i}(k) = u(k-2-i); \quad i = 1, \cdots, 3. \tag{22}$$

According to Eq.(19), the identification model is considered as

$$\hat{y} (k) = \hat{\theta}^T (k) \zeta (k), \tag{23}$$

where

$$\hat{\theta}(k) = \left[\hat{a}_1(k), \hat{a}_2(k), \hat{a}_3(k), \hat{b}_0(k), \hat{b}_1(k), \hat{b}_2(k)\right]^T. \tag{24}$$

Define the identification error as

$$\varepsilon(k) = \hat{y}(k) - y(k) = \zeta^T(k)\hat{\theta}(k) - y(k). \tag{25}$$

Under the above settings, the estimated parameter $\hat{\theta}(k)$ is identified by means of the constant trace algorithm as follows [3]:

$$\hat{\theta}(k) = \hat{\theta}(k-1) - \frac{\Gamma(k-1)\zeta(k)\varepsilon(k)}{1+\zeta^T(k)\Gamma(k-1)\zeta(k)}, \tag{26.a}$$

$$\hat{\Gamma}(k-1) = \Gamma(k-1) - \frac{\Gamma(k-1)\zeta(k)\zeta^T(k)\Gamma(k-1)}{1+\zeta^T(k)\Gamma(k-1)\zeta(k)}, \tag{26.b}$$

$$\lambda(k) = \frac{tr\hat{\Gamma}(k-1)}{tr\Gamma(0)}, \tag{26.c}$$

$$\Gamma(k) = \frac{\hat{\Gamma}(k-1)}{\lambda(k)}. \tag{26.d}$$

Based on the estimated parameter $\hat{\theta}(k)$ obtained from Eq.(26) as the true values, control parameters are calculated as follows:

$$\left.\begin{aligned}
&\hat{\phi}_1(k) = \hat{\beta}_2(k),\\
&\hat{\phi}_{i+1}(k) = \hat{\beta}_2(k)\left(\hat{r}_i(k) - f_i\right); \quad i = 1,\cdots,4,\\
&\hat{\phi}_6(k) = \frac{\hat{s}_0(k)}{f^2},\\
&\hat{\phi}_{6+i}(k) = \frac{\hat{s}_i(k) - \hat{s}_0(k)\bar{f}_i}{f^2}; \quad i = 1,2,
\end{aligned}\right\} \tag{27}$$

$$\left.\begin{aligned}
&\hat{r}_1(k) = f_1 + h_1 - \hat{\alpha}_1(k),\\
&\hat{r}_2(k) = f_2 + f_1h_1 + h_2 - \hat{\alpha}_2(k) - \hat{\alpha}_1(k)\hat{r}_1(k),\\
&\hat{r}_3(k) = f_3 + f_2h_1 + f_1h_2 + h_3 - \hat{\alpha}_2(k)\hat{r}_1(k) - \hat{\alpha}_1(k)\hat{r}_2(k),\\
&\hat{r}_4(k) = f_4 + f_3h_1 + f_2h_2 + f_1h_3 - \hat{\alpha}_2(k)\hat{r}_2(k) - \hat{\alpha}_1(k)\hat{r}_3(k),\\
&\hat{s}_0(k) = f_4h_1 + f_3h_2 + f_2h_3 - \hat{\alpha}_2(k)\hat{r}_3(k) - \hat{\alpha}_1(k)\hat{r}_4(k),\\
&\hat{s}_1(k) = f_4h_2 + f_3h_3 - \hat{\alpha}_2(k)\hat{r}_4(k),\\
&\hat{s}_2(k) = f_4h_3.
\end{aligned}\right\} \tag{28}$$

$$\left.\begin{aligned}
&\hat{\alpha}_1(k) = \frac{3+\hat{a}_1(k)}{T}, \quad \hat{\alpha}_2(k) = \frac{3+2\hat{a}_1(k)+\hat{a}_2(k)}{T^2},\\
&\hat{\beta}_2(k) = \frac{\hat{b}_0(k)+\hat{b}_1(k)+\hat{b}_2(k)}{T^3}.
\end{aligned}\right\} \tag{29}$$

5 Design Scheme of the MRAC with the NN

In practice, since the pneumatic servo systems contain nonlinear elements caused by compressibility of air, various frictions and so on, it is difficult to accomplish satisfactory control performance. Thus, we propose an approximate design method that constitutes the controller of the MRAC and an NN in parallel illustrated in Fig. 2. In this design, the NN trains so as to compensate for the modeling errors and the nonlinearity of the pneumatic servo systems. The control input $u(k)$ is synthesized as the sum of the output of the adaptive controller $u_M(k)$ and the output of the NN $u_N(k)$ as follows:

$$u(k) = u_M(k) + u_N(k). \tag{30}$$

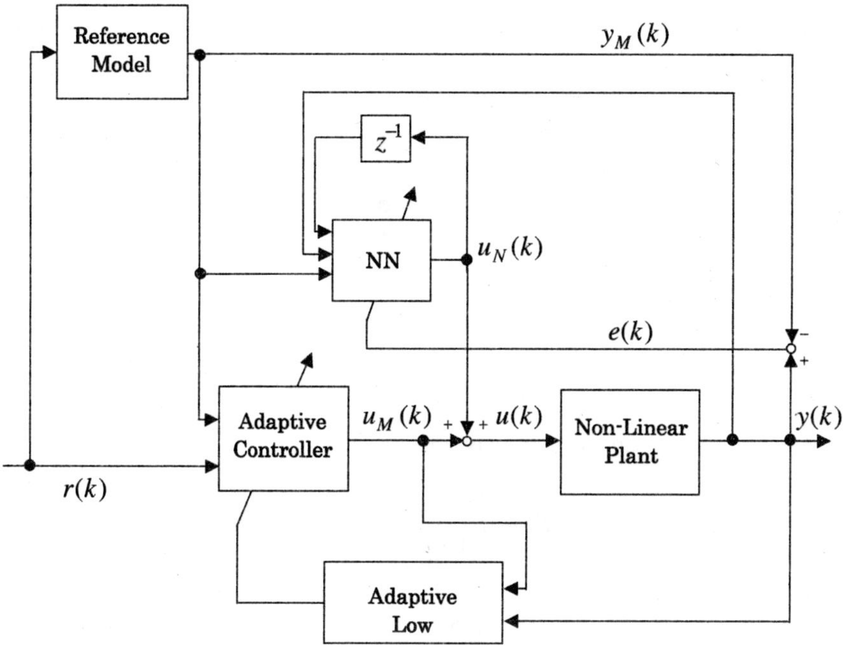

Fig. 2. Schematic diagram of the MRAC with the NN

We use the architecture of the 3-layer type NN in Fig. 3. The input-layer consists of 8 units and their output signals are described as

$$\{x_i(k)\} = \{y_{MH}(k+2), \varsigma_2(k), \varsigma_3(k), \varsigma_4(k), \varsigma_5(k), \varsigma_6(k), \varsigma_7(k), \varsigma_8(k)\}, \tag{31}$$

where

$$\left.\begin{aligned} &\varsigma_{i+1}(k) = \frac{\delta^{4-i}}{F(\delta)} u_N(k);\ i = 1, \cdots, 4, \\ &\varsigma_6(k) = y(k), \quad \varsigma_{6+i}(k) = \frac{\delta^{2-i}}{\bar{F}(\delta)} y(k);\ i = 1, 2. \end{aligned}\right\} \tag{32}$$

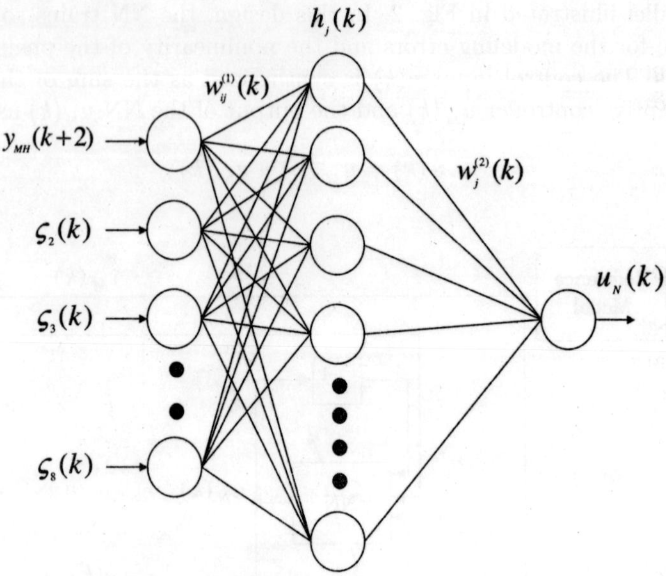

Fig. 3. Architecture of the NN

The hidden-layer consists of 20 units. The output signals in the hidden and output-layer are, respectively, described as

$$h_j\left(k\right) = f\left(\sum_{i=1}^{8} x_i\left(k\right) w_{ij}^{(1)}\left(k\right)\right), \tag{33}$$

$$u_N\left(k\right) = \sum_{j=1}^{20} w_j^{(2)}\left(k\right) h_j\left(k\right), \tag{34}$$

where

$$f(x) = \frac{1}{1+\exp(-x)}. \tag{35}$$

Moreover, define the output-error as

$$e(k) = y(k) - y_M(k). \tag{36}$$

Using the back-propagation (BP) algorithm of the NN in the same way as in [6], the weights $w_{ij}^{(1)}(k), w_j^{(2)}(k) (i = 1, \cdots, 8, j = 1, \cdots, 20)$ are updated so as to decrease the square of the output-error. That is

$$\begin{aligned} &w_{ij}^{(1)}(k+1) \\ &= w_{ij}^{(1)}(k) - \eta e(k) \frac{\partial y(k)}{\partial u(k-3)} w_j^{(2)}(k-3) h_j(k-3)(1-h_j(k-3)) x_i(k-3), \end{aligned} \tag{37}$$

$$w_j^{(2)}(k+1) = w_j^{(2)}(k) - \eta e(k) \frac{\partial y(k)}{\partial u(k-3)} h_j(k-3), \tag{38}$$

where $\eta(>0)$ is the learning rate of the algorithm. The value of Jacobian, $\partial y(k)/\partial u(k-3)$, is given by the following equation:

$$\frac{\partial y(k)}{\partial u(k-3)} = \hat{b}_0(k). \tag{39}$$

6 Experimental Results

This section shows the experimental results obtained by using the existent pneumatic servo system to illustrate the effectiveness of the proposed design scheme. The parameters used in the control algorithm are chosen as

$$\begin{aligned}
&\alpha_{M1} = 30.30, \alpha_{M2} = 270.09, \alpha_{M3} = 710.87, \beta_M = 710.87, \\
&\bar{f}_1 = 100, \bar{f}_2 = 2500, f = 50, \Gamma(0) = 10I, \\
&h_1 = 30.30, h_2 = 270.09, h_3 = 710.87.
\end{aligned}$$

The initial values of the controller parameters are decided as

$$\begin{aligned}
&\hat{a}_1(0) = -2.3, \hat{a}_2(0) = 1.9, \hat{a}_3(0) = -0.5, \\
&\hat{b}_0(0) = 0.1, \hat{b}_1(0) = 0.1, \hat{b}_2(0) = -0.1.
\end{aligned}$$

The learning rate η is chosen as 0.01, and the initial values of the weights $w_{ij}^{(1)}(0)$ and $w_{ij}^{(2)}(0)$ are given by the random numbers at the range of $[-0.001, 0.0001]$. We adopted a rectangle wave as the reference input $r(k)$ which has the amplitude of ±25mm and the period of 30sec.

Figures 4 and 5 show the transient and stationary responses of the MRAC with the NN. In these figures, (a) shows the reference model output $y_M(k)$ and the plant output $y(k)$, and (b) and (c) show the estimated parameters $\hat{a}_i$ and $\hat{b}_i$, respectively. From these results, we see that the estimated parameters in Fig. 4 are not constant and it is difficult for the output of the plant to accurately track to that of the reference model in the transient response. However, in the stationary response, the estimated parameters converge to the constant values and the output of the plant tracks to that of the reference model as shown in Fig. 5.

Figures 6 and 7 show the histograms of 30 times iterative positioning experimental results measured at 12sec. after the change of the sign of the reference input $r(k)$. Figure 6 shows the experimental results of the conventional design scheme, i.e., the MRAC without the NN. It can be seen from Fig. 6 that the positioning errors are within ±0.12mm. On the other hand, Fig. 7 shows those of the proposed design scheme, i.e., the MRAC with the NN. From Fig. 7, we can see that the positioning errors are within ±0.08mm, which shows that the proposed method has the superior positioning accuracy in comparison with the conventional method. These results clarify that the NN compensates for the modeling errors and the nonlinearity of the pneumatic servo systems.

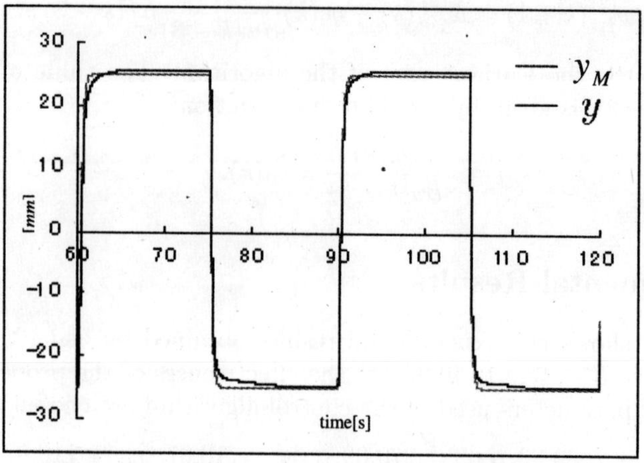

(a) output of the reference model and plant.

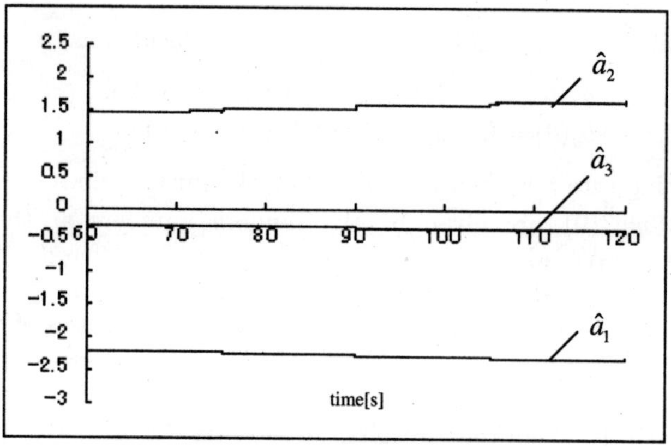

(b) estimated parameters $\hat{a}_i$ $(1 \le i \le 3)$.

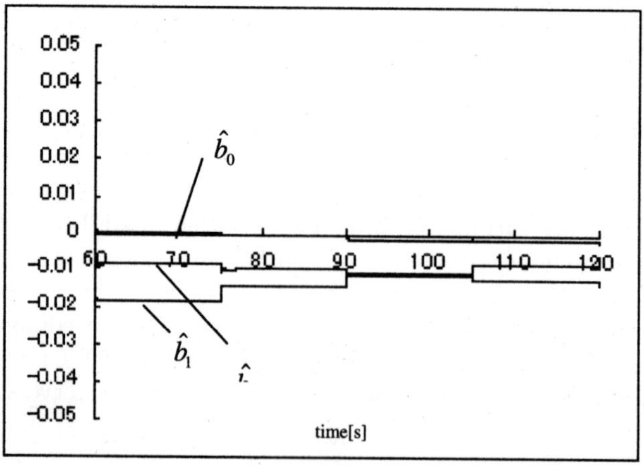

(c) estimated parameters $\hat{b}_i$ $(0 \le i \le 2)$.

Fig. 4. Transient responses of the MRAC using the delta operator with the NN

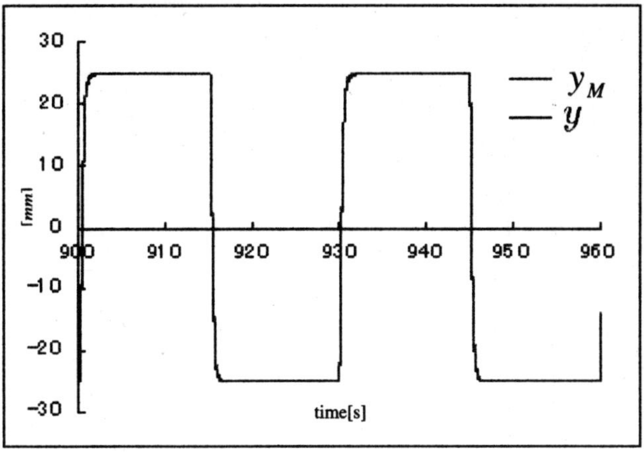

(a) output of the reference model and plant.

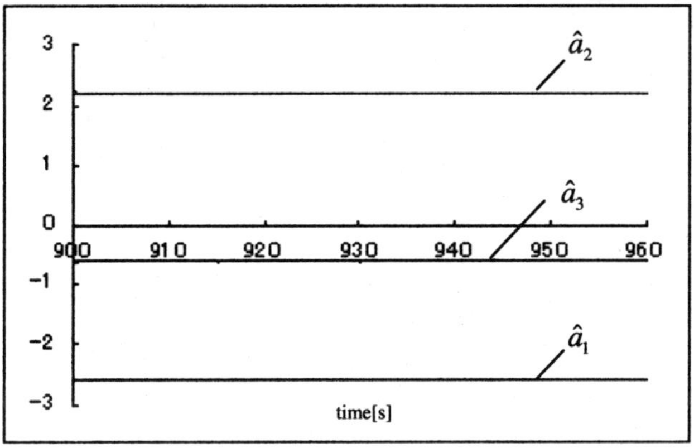

(b) estimated parameters $\hat{a}_i$ $(1 \leq i \leq 3)$.

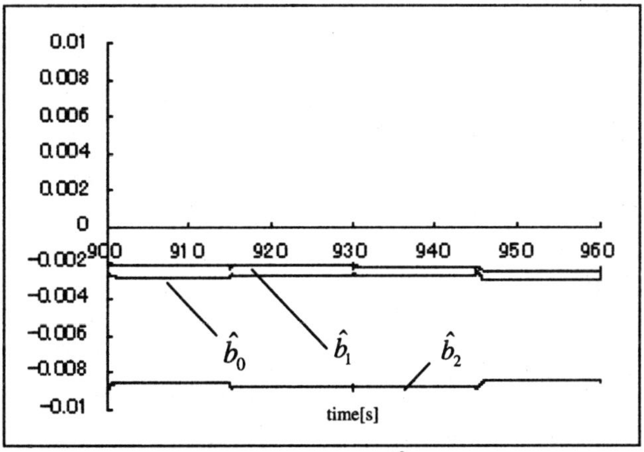

(c) estimated parameters $\hat{b}_i$ $(0 \leq i \leq 2)$.

Fig. 5. Stationary responses of the MRAC using the delta operator with the N

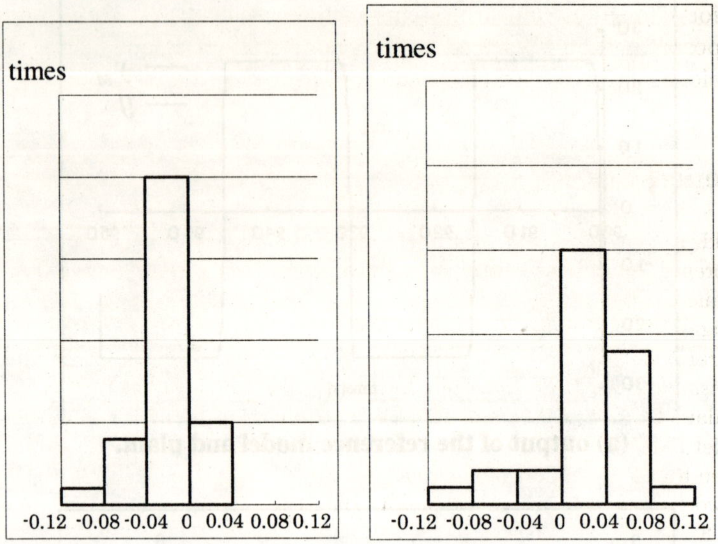

Fig. 6. The position accuracy of the conventional MRAC using the delta operator

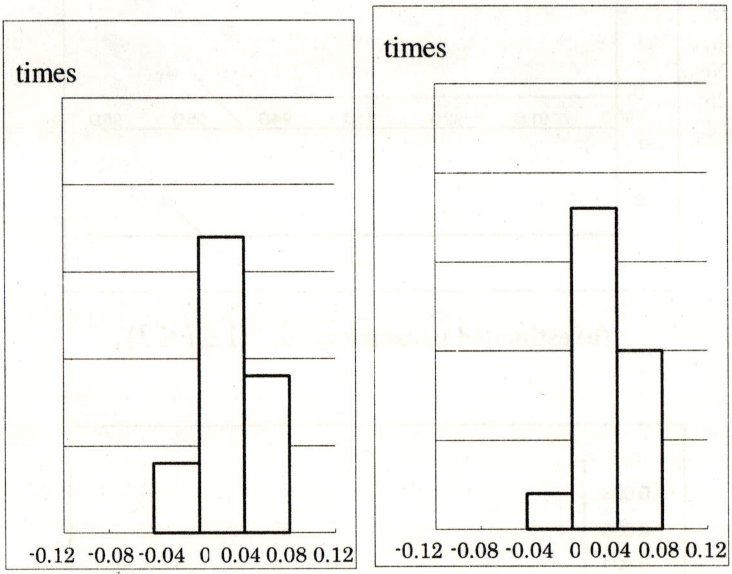

Fig. 7. The position accuracy of the MRAC using the delta operator with the NN

7 Conclusion

In this paper, we have proposed the design method of the MRAC with the NN for the pneumatic servo systems. The proposed method can deal with the nonlinearity of the system, so that the accuracy of control is significantly improved for strongly nonlinear systems such as a pneumatic system. Although the proposed method is applied to the pneumatic systems in this paper, it is applicable and effective for the other nonlinear systems.

References

1. Araki, K., Yamakawa, Y., Waji, H., and Yumuro, A. (1994), δ-Operator Model Reference Adaptive Control of a Pneumatic Servo, *Fluid Power System*, (in Japanese), Vol. 23, 91-98.
2. Kurigami, M., Tanaka, K., and Shimizu, A. (1997), Multirate MRACS Using δ-Operator for Electro-Pneumatic Servo Systems with Additive External Forces, *Transactions on IEE Japan*, (in Japanese), Vol. 177-D, 1069-1076.
3. Landau, I. D., and Lozano, R. (1981), Unification of Discrete Time Explicit Model Reference Adaptive Control Designs, *Automatica*, Vol. 17, 593-611.
4. Narendra, K. S., and Mukhopadhyay, S. (1997), Adaptive Control Using Neural Networks and Approximate Models, *IEEE Transactions on Neural Networks*, Vol. 8, 475-485.
5. Tanaka, K., Shimizu, A., Shibata, S., Senoo, M., and Kurigami, M. (1995), MRAC Using δ-Operator for Electro-Pneumatic Servo Systems, *Transactions on IEE Japan*, (in Japanese), Vol. 115-D, 605-611.
6. Tsuji, J., Ohmori, H., and Sano, A. (1994), Adaptive Control Incorporating Neural Network, *Transactions of the Society of Instrument and Control Engineers*, (in Japanese), Vol. 30, 295-302.

Design Effective FACTS Controller with Time-delay Signals

Quanyuan Jiang and Yijia Cao

College of Electrical Engineering, Zhejiang University, Hangzhou 310027, P. R. China yijiacao@zju.edu.cn

1 Introduction

In the last decade, some major utilities had shown an interest in the synchronous phasor measurement technology. PMU technology is attractive since it can provide synchronized, real-time measurements of voltage, incident current phasors, rotor angle and electrical power at the system buses and lines. Once these real-time signals of the whole system are available in the form of synchronized phasors, the operators can give on-line monitoring of power systems operating conditions [1]. Moreover, phasor measurement unit (PMU) is also introduced for the use of stability assessment and wide-area control of the power system [2],[3]. Due to the large scale and intricate structure of modern power system, the demand for structuring wide-area measurement systems (WAMS) has been increasing. WAMS provides a dynamic coverage of the wide-area power network and is also able to handle cascaded outages through coordinated and optimized stabilizing actions. Emerging techniques such as computer technology, communication technology and PMU technology are being used in WAMS and form the basis for real-time dynamic monitoring, on-line security assessment and wide-area stability control of power system. It can be concluded that WAMS will play a vital role for interconnected power system in the next decades.

Communication links used by WAMS include both wired (telephone lines, fiber-optics, power lines) and wireless (micro-wave, satellites) options. The propagation delay is dependent of the medium characters and the physical route distance in WAMS. For a local controller, the time delay of feedback signals is very small (less than $10ms$), so the small time delay is often ignored in the controller design. However, for a wide-area controller, the time delay in a interconnected power system can vary from tens to several hundred milliseconds or more. An experimental research in [4] has shown that the time delays caused by different communication links are different, but all of the delays are more than $100ms$. In the case of satellite link, the propagation delay could be as high as above $700ms$. There could be a larger delay when a large number of signals are routed and signals from different areas are waiting for synchronizing. Such large time delays can invalidate many controllers that work well in no delayed-input systems and even cause disaster accidents [5]. The impact of time delay on controller performance has been ignored for a long time in power systems, but it will become a significant topic in the wide-area control.

This paper addresses the design of a TCSC controller including feedback signal delay, using a theoretical approach based on time-delay dynamic system combined with LMI technique. More precisely, the interest here is to find a delay-output feedback controller such that the closed-loop power system is asymptotically stable for any delay less than a given value. The paper is organized as follows: firstly, the effects of time delay in WAMS on the dynamic performance of power system are analyzed. Then, a TCSC controller considering time delay of the feedback signals is designed by LMI technique. Finally, simulation results are used to verify the validity of the derived TCSC controller under different feedback delays.

2 The impact of time delay on TCSC controller

From mathematical viewpoint, the interconnected power system including WAMS delays is a typical nonlinear and multi-delay dynamical system. Commonly, the stability analysis and control of dynamical systems with delays are more complex and difficult than ordinary differential systems. In the traditional power system analysis, controller designs are usually based on local signals and their time delay is small enough to be ignored. In the wide area system, the stability analysis and control of power system rely on the acquisition, transmission and action of some key signals from the communication system. The reliability and time lag of the communication system have a vital effect on real-time on-line control. In order to consider the effect of WAMS on the dynamic performance of the power system, non-linear deferential equations with delays should be used. The following example shows that delays in the feedback signals can lead to instability of the system even if the controller works well for the delay-free system.

Figure 1 describes a single-machine infinite-bus (SMIB) system with a TCSC. TCSC is modeled as a controlled capacitor whose reactance is limited to the range of $[-0.2, 0.2]$. The system parameters are given as follows: transmission line and transformer:$x_T = 0.1$, $x_{L1} = 0.4$, $x_{L2} = 0.6$. Generator: $x_d = 1.2$, $x'_d = 0.2$, $x_q = 1.2$, $x'_q = 0.1$, $M = 6.0$, $T'_{d0} = 10.0$. The following non-linear equations are used to denote the system:

$$\begin{cases} \dot{\delta} = \omega_0(\omega - 1) \\ \dot{\omega} = \frac{T_m - T_e}{M} \\ \dot{E}'_q = \frac{1}{T'_{d0}}[E_{fd} - E'_q - (x_d - x'_d)i_d] \\ \dot{E}_{fd} = \frac{1}{T_e}[K_e(V_{ref} - V_t) - E_{fd}] \end{cases} \tag{1}$$

in which $T_e \approx P_e = \frac{E'_q V_0 \sin\delta}{x_e + x'_d} + \frac{(x'_d - x_q)V_0^2 \sin 2\delta}{2(x_e + x'_d)(x_e + x_q)}$, $i_d = \frac{E'_q - V_0 \cos\delta}{x_e + x'_d}$, $i_q = \frac{V_0 \sin\delta}{x_e + x_q}$, $x_e = x_T + x_{L1} + x_{L2} - x_c$, $V_d = x_q i_q$, $V_q = E'_q - x'_d i_d$, $V_t = \sqrt{V_d^2 + V_q^2} = \sqrt{\left(\frac{x_q}{x_e + x_q} V_0 \sin\delta\right)^2 + \left(\frac{x_e}{x_e + x'_d} E'_q + \frac{x'_d}{x_e + x'_d} V_0 \cos\delta\right)^2}$

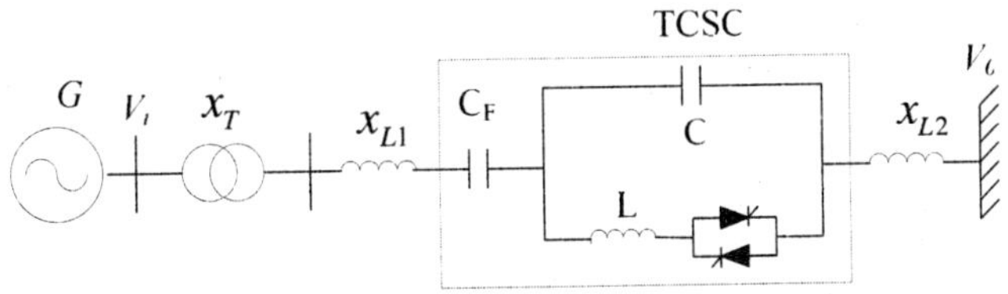

Fig. 1. SMIB system with a TCSC

For this SMIB system, reference [6] designs a nonlinear adaptive TCSC controller which takes two remote signals, generator power angle δ and rotor speed ω, as feedback signals, i.e

$$x_c = \frac{E'_q V_0 \sin\delta}{P_m - D\frac{\Delta\omega}{\omega_0} + \frac{M}{\omega_0}(k_1 \Delta\delta + k_2 \Delta\omega)} - x_{L\Sigma}$$

in which $x_{L\Sigma} = x_T + x_{L1} + x_{L2}$. Obviously, time delay of the feedback signals δ and ω is inevitable in the remote transmission and communication system. Our research results demonstrate that the time delay of feedback signals affects greatly the performance of the derived TCSC nonlinear controller. The dynamic responses of the system to a disturbance of a $0.1s$ $0.15p.u.$ step change in the generator rotor angle under different feedback signal delays are investigated. Results are given in Fig. 2. From the simulation results of Fig. 2, we can see that the dynamic performance of nonlinear TCSC controller deteriorates sharply with the feedback delay's increament. To improve the TCSC controller performance, an LMI approach is used to design a TCSC controller, which is not sensitive to time delay.

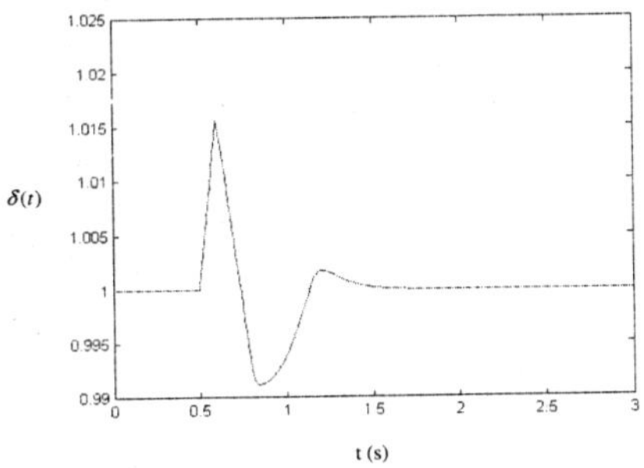

(a) time delay=$0ms$

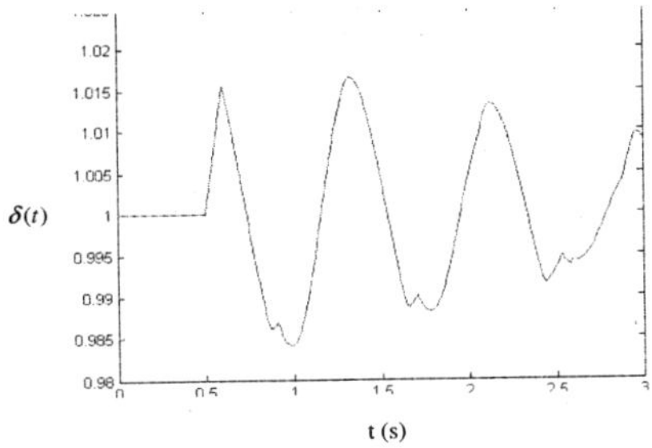

(b) timc delay=$50ms$

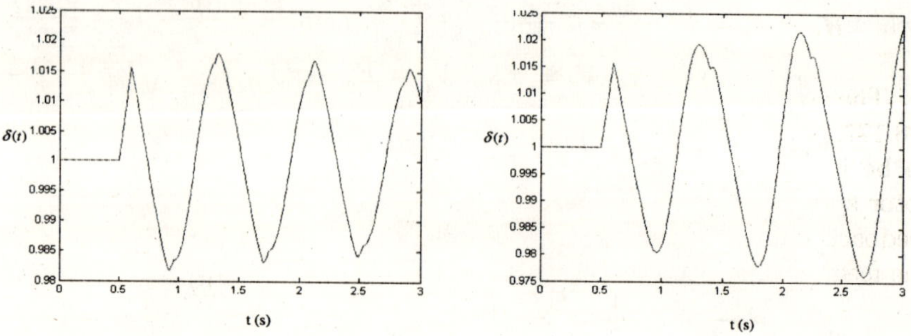

(c) time delay=100ms (d) time delay=150ms

Fig. 2. Dynamic response of the TCSC controller with different delays

3 TCSC controller design based on LMI technique

When time delays in WAMS are considered, the power system should be modeled into differential-algebraic equations with delays instead of ordinary differential-algebraic equations. LMI technique is one of the basic method used in delay differential dynamic system. In this paper, LMI technique is applied to designing the TCSC controller. Firstly, linearize equation (1) and the state space model of the power system is given as follows:

$$\begin{cases} \dot{x}(t) = Ax(t) + Bu(t) \\ y(t) = Cx(t) \end{cases} \tag{2}$$

in which $x = [\Delta\delta\ \Delta\omega\ \Delta E'_q\ \Delta E_f d]^T$, $u = \Delta x_c$, $y = [\Delta\delta\ \Delta\omega\ \Delta E'_q]^T$,

$$A = \begin{bmatrix} 0 & \omega_0 & 0 & 0 \\ -\frac{K_1}{M} & 0 & -\frac{K_2}{M} & 0 \\ -\frac{K_4}{T'_{d0}} & 0 & -\frac{K_5}{T'_{d0}} & \frac{1}{T'_{d0}} \\ -\frac{K_e K_7}{T_e} & 0 & -\frac{K_e K_8}{T_e} & -\frac{1}{T_e} \end{bmatrix}, \quad B = \begin{bmatrix} 0 \\ \frac{K_3}{M} \\ \frac{K_6}{T'_{d0}} \\ \frac{K_e K_9}{T_e} \end{bmatrix}, \quad C = \begin{bmatrix} 1\,0\,0\,0 \\ 0\,1\,0\,0 \\ 0\,0\,1\,0 \end{bmatrix}$$

$$K_1 = \frac{E'_{q0} V_0 \cos\delta_0}{x'_d + x_e} + \frac{(x'_d - x_q) V_0^2 \cos 2\delta_0}{(x'_d + x_e)(x_q + x_e)}, \quad K_2 = \frac{V_0 \sin\delta_0}{(x'_d + x_e)},$$

$$K_3 = -\frac{E'_{q0} V_0 \sin\delta_0}{(x'_d + x_e)^2} + \frac{V_0^2 \sin 2\delta_0}{2} \left[\frac{1}{(x'_d + x_e)^2} - \frac{1}{(x_q + x_e)^2} \right],$$

$$K_4 = \frac{x_d - x'_d}{x'_d + x_e} V_0 \sin\delta_0, \quad K_5 = \frac{x_d + x_e}{x'_d + x_e}, \quad K_6 = -\frac{x_d - x'_d}{(x'_d + x_e)^2} (E'_{q0} - V_0 \cos\delta_0)$$

$$K_7 = \frac{V_{d0} x_q}{V_{t0}(x_q + x_e)} V_0 \cos\delta_0 - \frac{V'_{q0} x'_d}{V_{t0}(x'_d + x_e)} V_0 \sin\delta_0, \quad K_8 = \frac{V_{q0} x_e}{V_{t0}(x'_d + x_e)},$$

$$K_9 = -\frac{V_{d0}x_q}{V_{t0}(x_q + x_e)^2}V_0 \sin\delta_0 + \frac{V_{q0}x'_d}{V_{t0}(x'_d + x_e)^2}(E'_{q0} - V_0 \cos\delta_0).$$

The eigenvalues of the open-loop system (2) are: -42.545, $0.2487 \pm 6.803i$, -8.177. So the open-loop system is unstable, a feedback TCSC controller has to be introduced to ensure the stability of the system. The rotor angle δ, rotor speed ω and the generator q-axis transient voltage E'_q are used as the feedback signals. Assuming these feedback signals measured by PMU, then the feedback law has the following representation:

$$u(t) = -Ky(t-d) \tag{3}$$

in which $K = [k_\delta \; k_\omega \; k_{E'}]$, d is the time delay of the feedback signals, $d > 0$. In order to simplify the analysis, it is assumed that all the feedback signals have the same time delay d . If equation (3) is adopted as TCSC controller's law, then the closed-loop system can be described as follows:

$$\dot{x}(t) = Ax(t) + A_d x(t-d) \tag{4}$$

in which $A_d = -BKC$. The TCSC controller design problem is addressed as follows: find a set of controller parameters K such that the resulting closed-loop system (4) is asymptotically stable for any delay d less than a given positive scalar d_0.

The stability for linear system (4) can be classified into two types: delay-independent stability or delay-dependent stability. The definitions are given as follows:

Definition 1: *Delay-independent stability: when the time-delay dynamical system is stable for any positive time delay d, then we say the system has delay-independently stability.*

Definition 2: *Delay-dependent stability: when the time-delay dynamic system is stable only for some given time delay $0 < d < d_0$, where d_0 is a given positive constant, then we say the system has delay-dependently stability.*

Reference [7] addressed the necessary and sufficient conditions of asymptotical stability for the linear dynamic system (4). However, this condition requires a complete understanding of all the eigenvalues of the system (4), and this condition is commonly difficult to be satisfied. In the last decades, many simple conditions have been developed to judge the stability of such system. Usually, the results based on the time-independent criteria are too conservative and in some cases we cannot get the controller parameters. In most engineering problems, the estimation of the maximum delay time to keep system stability is enough. So the control problem above is addressed now: find a controller of the form (3) so that the resulting closed-loop system (4) is asymptotically stable for any time delay d less than a given positive scalar d_0.

In order to solve the above output feedback controller problem, we cite here a result which will be used as a theorem.

Theorem [8]: *Consider the system $\dot{x}(t) = Ax(t) + A_d x(t-d)$, given a positive scalar d_0, the system is asymptotically stable for any time delay $0 < d < d_0$ if there exist positive matrixes P, Q, V and a matrix W such that the following LMI is satisfied:*

$$\begin{bmatrix} S & -W^T A_d & A^T A_d^T V & d_0(W+P) \\ -A_d^T W & -Q & A_d^T A_d^T V & 0 \\ VA_dA & VA_dA_d & -V & 0 \\ d_0(W^T+P) & 0 & 0 & -V \end{bmatrix} < 0 \tag{5}$$

in which $S = (A+A_d)^T P + P(A+A_d) + W^T A_d + A_d^T W + Q$.

The LMI (5) can be transformed to the following form:

$$\begin{bmatrix} S & -W^T A_d & A^T A_d^T V & 0 \\ -A_d^T W & -Q & A_d^T A_d^T V & 0 \\ VA_dA & VA_dA_d & -V & 0 \\ 0 & 0 & 0 & -V \end{bmatrix} < -d_{\max} \begin{bmatrix} 0 & 0\,0 & (W+P) \\ 0 & 0\,0 & 0 \\ 0 & 0\,0 & 0 \\ (W^T+P) & 0\,0 & 0 \end{bmatrix} \tag{6}$$

The LMI (6) is easy to be solved using LMI toolbox in Matlab so that the optimal TCSC controller is designed to permit the largest time delay. For the power system in Fig. 1, we set $d_0 = 800ms$, then the control parameters of the optimal TCSC are: $K_* = [-0.5\ \ 5\ \ -0.8]$.

4 Simulation Results

The optimal TCSC controller designed in section 3 is used in linear time-delay system (4) and the simulation results are given in Fig. 3.

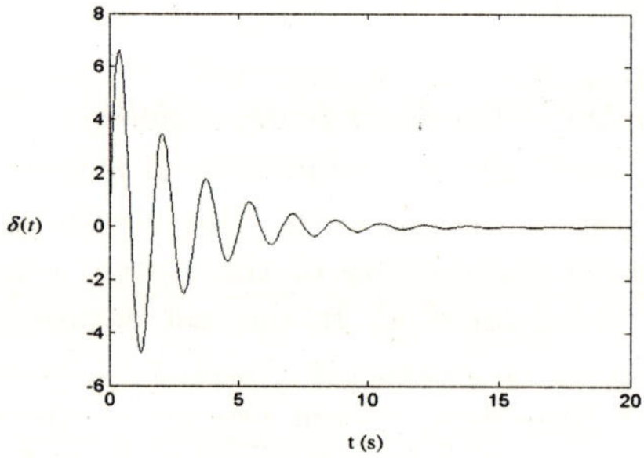

(a) $d = 10ms$

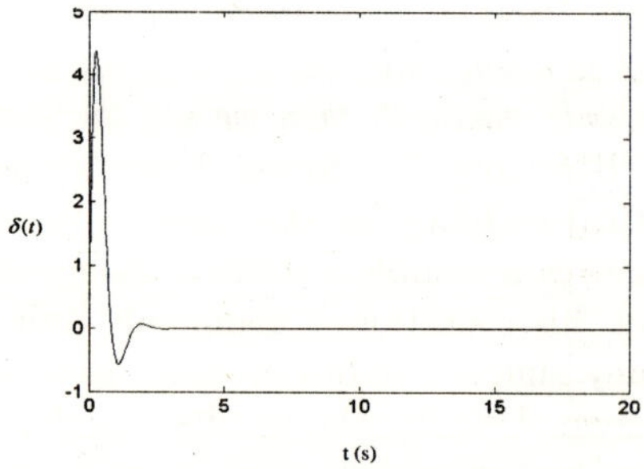

(b) $d = 100ms$

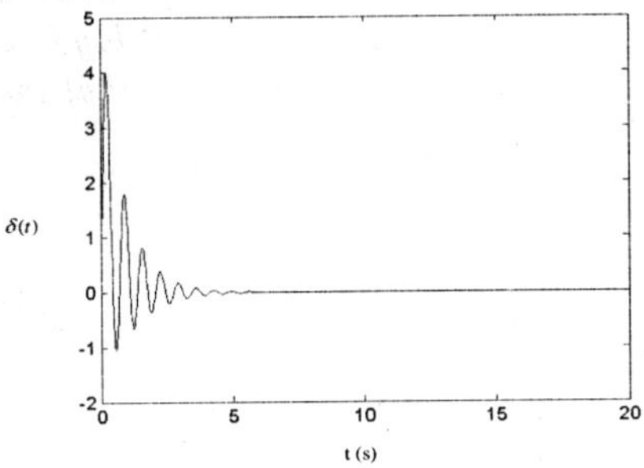

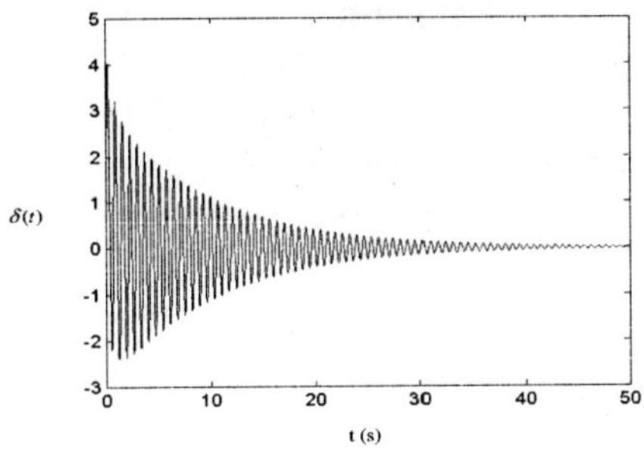

(c) $d = 250ms$ (d) $d = 300ms$

Fig. 3. Power angle response of the linear time-delay system

The optimal TCSC controller designed in section 3 is also used in nonlinear time-delay system (1). Considering the limiting segments of TCSC reactance adjustment and excitation adjustment, simulation results are given in Fig. 4. Fig 3 and Fig. 4 show the TCSC controller effectively ensure the stability of the power system even when the time lag is so large as $300ms$.

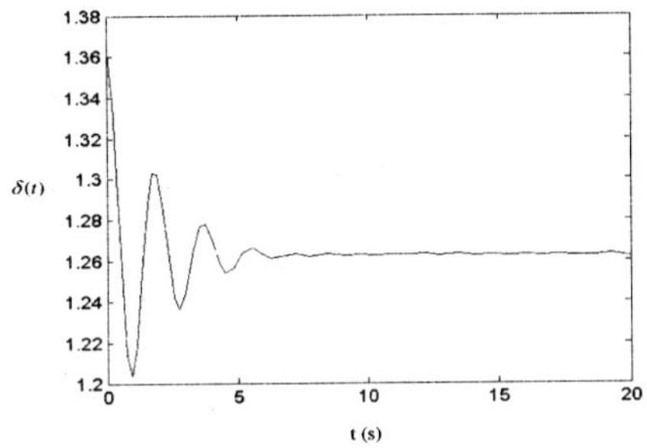

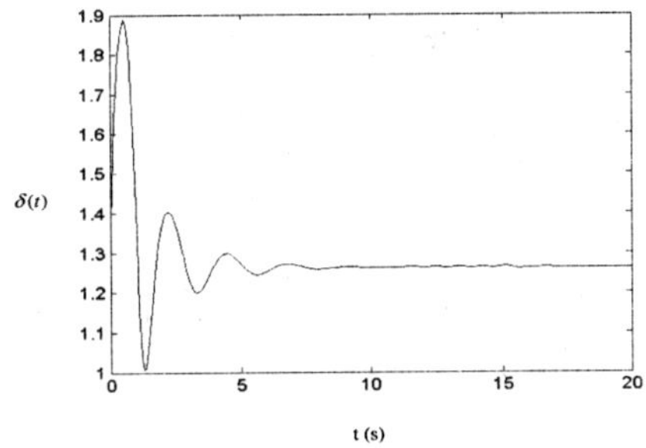

(a) $d = 0ms$ (b) $d = 100ms$

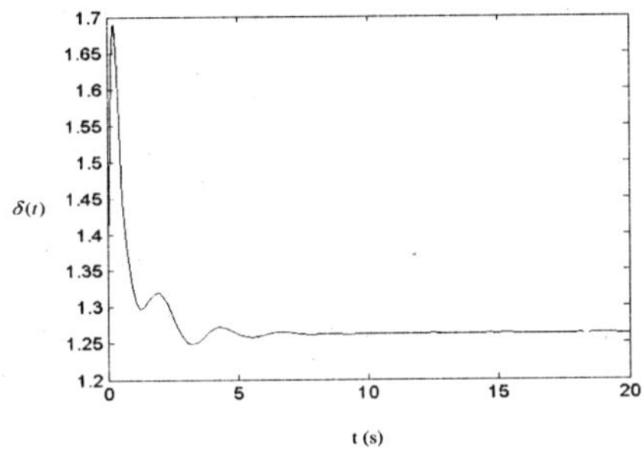

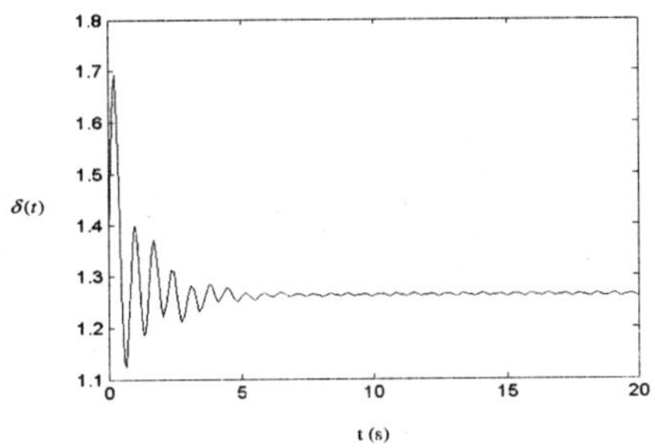

(c) $d = 200ms$ (d) $d = 300ms$

Fig.4. Power angle response of the nonlinear time-delay system

5 Conclusions

A TCSC control scheme that is not sensible to time delays is proposed in this paper to stabilize the power system and achieve good closed-loop performance. The given evaluation criteria, expressed in terms of linear matrix inequalities, has been tested with MATLAB on both linear and nonlinear systems to illustrate the performance of the controller. Simulation results have shown that the designed TCSC controller is efficient even when the time delay is as large as $300ms$.

6 Acknowledgment

The authors would like appreciate the financial support from the Key Research Fund from MOE, China.

References

1. Taylor, C.W. (2000), The future in On-line Security Assessment and Wide-area Stability Control, *IEEE Power Engineering Society Winter Meeting*, Vol. 1, 78-83
2. Hui, N., G, T, Heydt, Lamine Mili. (2002), Power system stability agents using robust wide area control. *IEEE Transactions on Power Systems*, 17(4), 1123-1131
3. I. Kamwa, R.Grondin, Y.Hebert. (2001), Wide-area measurement based stabilizing control of large power systems-a decentralized/hierarchical approach. *IEEE Transactions on Power Systems*, 16(1), 136-153
4. B. Naduvathuparambil, M.C.Valenti, A. Feliachi. (2002), Communication Delays in Wide Area Measurement Systems. *Proceedings of the 34th Southeastern Symposium on System Theory*, Vol. 1, 118-122
5. V.B. Kolmanovskii, S.I. Niculescu, K. Gu. (1999), Delay Effects on Stability, a survey. *Proceedings of the 38th IEEE Conference on Decision and Control*, Vol. 2 , 1993-1998
6. Zhou, X. , Liang, J., (1999), Nonlinear Adaptive Control of TCSC to improve the Performance of Power Systems. *IEE Proc. Part C*, 146(3), 301-305
7. T. Mori, H. Kokame. (1989), Stability of $\dot{x}(t) = Ax(t) + Bx(t-\tau)$. *IEEE Transactions on Automatic control*, 34(4), 460-462
8. Yu, L. (2002), Robust Control-Linear Matrix inequality method, *Tsinghua University Press*, Beijing.

Co-ordinated Design of Multiple Robust Stabilizers in Multi-machine Power Systems

Haifeng Wang

University of Bath, U.K. eeshw@bath.ac.uk

1 Introduction

The installation of a Power System Stabilizer (PSS) in power systems is an effective measure to improve system oscillation stability [1][2]. With recent development in Flexible AC Transmission Systems (FACTS) [3][4], the FACTS-based stabilizers have also shown great potential in damping power system oscillations [5]-[10]. In a multi-machine power system, the co-ordinated design of multiple stabilizers installed in the system is important for their efficient operation. This problem has been studied for multiple PSS's [11]-[16]. Recently, the co-ordination among FACTS-based stabilizers and PSS's is attracting much attention [17].

However, stabilizers are designed on power system model linearized at an operating condition. In the multi-machine power system installed with multiple stabilizers to damp multi-mode oscillations, the changing pattern of the effectiveness of individual stabilizer to the variations of system operating conditions may be different. So in some cases it will be difficult to select a proper common operating condition at which all stabilizers can be designed simultaneously and their robustness will be ensured. In the literature, this problem of co-ordinated design of multiple robust stabilizers has rarely been studied.As the power systems consist of a large number of generator sets and possess nonlinearity, the design approach based on the nonlinear model may have better system performance and improve the transient stability of power systems[1][2]. Meanwhile, the adoption of new technologies, like FACT technology and the ever increasing utilization of power electronic converters together with the new "deregulated" market have also imposed more stringent requirements on the dynamic behavior of power systems[17] and urge the recent enthusiasm on exploring new nonlinear control ideas.

Therefore, in this paper, a method for the co-ordinated design of multiple robust FACTS-based stabilizers and PSS's in multi-machine power systems is proposed. The method is developed by using technique of damping torque analysis, selection of the "least effective" operating condition and phase compensation. The co-ordination is achieved by sequentially setting the phase of transfer function of every stabilizer and simultaneously tuning the gains of all stabilizers. The robustness of the stabilizers is ensured by the selection of the "least effective" operating condition. In the paper, an example three-machine power system with two-mode oscillations is presented, where a robust PSS and TCSC (Thyristor Controlled Series Compensator)-based stabilizer are successfully designed by the method proposed.

Moreover, the marriage of Hamiltonian system(specifically, Port-Controlled Hamiltonian System with Dissipation) with passivity control theory has further motivated the study in this area. Their combination has more general and more precise physical meaning and the special structure can overcome the difficulty of energy forming and damping injection "naturally". Consequently, it achieved fruitful outcome when applied to the power system control, such as steam valving control and excitation control design.

One basic idea in this list of methods is to use the "energy functions" to estimate regions of attraction for stable equilibria. The use of "energy functions" to estimate regions of attraction for stable equilibria in power system stability analysis has a long history, well documented in texts [3] and [4].

Though the "energy function" been studied for years, still, there's a generally encountered problem that is the existence of transfer conductance, namely G_{ij}, hampers the realization of general energy function for multi-machine systems. Or say, it hampers the realization of the Hamiltonian structure for our popular PCH theory. The problem of how to treat transfer conductance should always be noted in all "energy function" based control method. During the derivation of energy conservation law, a complex integral approach has been introduced and this complex-integral yields two kinds of energy integral expressions: a regular and an adjoint energy integrals for its imaginary and real parts respectively. The former provides a structure-preserving energy function, while the latter with path-independent integral terms related to the transfer conductance.

Many efforts have made for the development of an energy function to reflect transfer conductances[6][7]. Chiang[16] presented a paper on the discussion of general energy functions for any lossy system with more than 3 machines. And in text[5], in a special case where the system has uniform R/X ratios for all transmission lines, we can derive a well-defined energy function reflecting the transfer conductance.

Here in this paper, we will see that by taking the weighted part of the transfer conductance into the control law, it is possible to generate a proper energy function and have good control effects. The proposed method provides a more flexible approach to reflect the transfer conductance compared with the previously proposed method [16], and it can give some hints for solving the difficulty after considering transfer conductance.

The rest of the paper is organized as follows. The mathematic model is formulated in section 2. And the specific nonlinear control idea to illustrate our findings is presented in section 3, from preliminaries to Hamiltonian structure and IDA-PBC methodology to the realization of excitation control. Then comes the simulation in section 4. The paper is concluded in section 5.

2 dynamic model

2.1 dynamic model of multi-machine power system

The dynamic model for power system excitation control design can be presented as follows:

$$\begin{cases} \dot{\delta}_i = \omega_0 \omega_{ri} \\ T_{Ji}\dot{\omega}_{ri} = P_{mi} - D_i \omega_{ri} - P_{ei} \\ T'_{d0i}\dot{E}'_{qi} = E_{fsi} + u_{fi} - E'_{qi} - (x_{di} - x'_{di})i_d \end{cases} \tag{1}$$

for a system with n generators, i=1,2,...n, subscription i denoted the ith generator. In Eqn.1

$$P_{ei} = E'_{qi} \sum_{j=1}^{n} E'_{qj}(G_{ij}cos\delta_{ij} + B_{ij}sin\delta_{ij})$$

$$i_d = \sum_{j=1}^{n} E'_{qj}(G_{ij}sin\delta_{ij} - B_{ij}cos\delta_{ij})$$

2.2 dynamic model of 3-machine power system

For the convenience of the latter illustration, we will put forward the mathematic model for the 3-machine power system here, that is to set i=1,2,3 for the above equation.

3 A specific nonlinear control idea

The specific nonlinear control idea in our research to illustrate how to treat the influence of transfer conductance is a Lyapunov-like energy-based method, that is the IDA-PBC (Interconnection and Damping Assignment Passivity Based Control). The IDA-PBC is firstly introduced into literature review by R. Ortega around 1999, see texts for more details.

3.1 preliminaries to IDA-PBC [17]

IDA-PBC is a procedure that allows us to design a static state feedback that stabilizes the equilibria of nonlinear systems of the form.

$$\dot{x} = f(x) + g(x)u \tag{2}$$

where $x \in \mathbb{R}^n$,is the state vector and $x \in \mathbb{R}^m$,$m < n$,is the control action, endowing the closed-loop with a port-controlled Hamiltonian (PCH) structure.

$$\dot{x} = [J_d(x) - R_d(x)]\nabla_x H_d \tag{3}$$

where the matrices $J_d(x) = -J_d^{\top}(x)$ and $R_d(x) = R_d^{\top}(x) \geq 0$, which represent the desired interconnection structure and dissipation, respectively, are selected by the designerhence the name IDA—and $H_d : \mathbb{R}^n \to \mathbb{R}$ is the desired total stored energy. If the latter has an isolated minimum at the desired equilibrium $x_\star \in \mathbb{R}^n$, that is if

$$x_\star = argminH_d(x) \tag{4}$$

then $x_\star$ is stable with Lyapunov function $H_d(x)$. The admissible energy functions are characterized by a parameterized partial differential equation (PDE).

3.2 control law [17]

Then solve the following equation.

$$f(x) + g(x)u = [J_d(x) - R_d(x)]\nabla_x H_d$$

That is to solve a set of PDEs, then the proposed excitation control is, where $g^{\perp}(x)$ is a left annihilator of $g(x)$, i.e. , $g^{\perp}(x)g(x) = 0$.

$$u = [g^{\perp}(x)g(x)]^{-1}g^{\perp}(x)\{[J_d(x) - R_d(x)]\nabla_x H_d - f(x)\} \tag{5}$$

3.3 realization of control law

In the expression of the control strategy for multi-machine power systems, the existence of transfer conductance becomes the main burden to construct the Hamiltonian function as we have mentioned before, even though using IDA-PBC. It is still very hard to obtain the excitation controllers even for the 3-machine power systems with transfer conductance, which is the simplest and most typical model for multi-machine power systems. In the light of power systems' operation principles and dynamic properties, we will make 2 different simplifications and assumptions to see the impact exerted by the transfer conductance on control effects. And here we will select a 3-machine model as a real case to study for ease of convenience. This process is called the realization of the control law.

ignore transfer conductance

First we will see the control effect if we neglect the value of transfer conductance, that is to set $G_{ij} = 0$ together with $\alpha_{ij} = arctan\frac{G_{ij}}{B_{ij}} = 0$. Here we will have the following assumptions:

- set $G_{ij} = 0$
- set $\alpha_{ij} = 0$

We denote this as the "neglect BOTH G_{ij} and α_{ij}" method. After tedious calculation, we get the expression of excitation control as follows:

$$\begin{cases} u_1 = -\frac{E_2Y_{12}sin(\delta_1-\delta_2)\omega_2+E_3Y_{12}sin(\delta_1-\delta_3)\omega_3}{\gamma_1} \\ \quad -\frac{[E_{2\star}Y_{12}sin(\delta_1-\delta_2)+E_{3\star}Y_{13}sin(\delta_1-\delta_3)+(E_1+E_{1\star})G_{11}]\omega_1}{\gamma_1} \\ \quad -r_1\gamma_1(E_1-E_{1\star})+a_1E_1-b_1E_2sin(\delta_1-\delta_2) \\ \quad -b_1E_3sin(\delta_1-\delta_3)-E_{f1} \\ u_2 = -\frac{E_1Y_{12}sin(\delta_1-\delta_2)\omega_1+E_3Y_{23}sin(\delta_2-\delta_3)\omega_3}{\gamma_2} \\ \quad +\frac{[-E_{1\star}Y_{12}sin(\delta_1-\delta_2)+E_{3\star}Y_{23}sin(\delta_2-\delta_3)+(E_2+E_{2\star})G_{22}]\omega_2}{\gamma_2} \\ \quad +b_2E_1sin(\delta_1-\delta_2)-r_2\gamma_2(E_2-E_{2\star})+a_2E_2 \\ \quad -b_2E_3sin(\delta_2-\delta_3)-E_{f2} \\ u_3 = \frac{E_1Y_{13}sin(\delta_1-\delta_3)\omega_1+E_2Y_{23}sin(\delta_2-\delta_3)\omega_2}{\gamma_3} \\ \quad +\frac{[-E_{1\star}Y_{13}sin(\delta_1-\delta_3)-E_{2\star}Y_{23}sin(\delta_2-\delta_3)+(E_3+E_{3\star})G_{33}]\omega_3}{\gamma_3} \\ \quad +b_3E_2sin(\delta_2-\delta_3)+b_3E_1sin(\delta_3-\delta_1) \\ \quad +a_3E_3-r_3\gamma_3(E_3-E_{3\star})-E_{f3} \end{cases} \tag{6}$$

Here

$$E_{fi} = \frac{E_{fsi}}{T'_{d0i}}, u_i = \frac{u_{si}}{T'_{d0i}}, b_i = \frac{x_{di} - xdi'}{T'_{d0i}}$$

$$a_i = \frac{1}{T'_{d0i}}[1 - B_{ii}(x_{di} - x'_{di})]$$

together with r_i and γ_i being free parameters.

taking transfer conductance into account

After fully consideration of the equation for 3-machine power systems and review of previous literature, we find that the main difficulties during the Hamiltonian constructions of the mathematic formulation is not the existence of G_{ij}, but α_{ij} to be more precise. According to power systems' operation principles and dynamic properties, here we will make some reasonable assumptions:

- remain the original value of G_{ij}
- set $\alpha_{ij} = 0$

We denote this approach as the "neglect ONLY α_{ij}". After some lengthy calculation, we will arrive at the same excitation control as Eqn.6.

Though the excitation control in these two means have the same expression, but the value of Y_{ij} differs from each other. In the general approach that is to "neglect BOTH G_{ij} and α_{ij}", $Y_{ij} = B_{ij}$, while in our new attempt namely to "neglect ONLY α_{ij}", $Y_{ij} = \sqrt{B_{ij}^2 + G_{ij}^2}$

3.4 features of the control strategy

From simple observation of the control law in Eqn.6, one can observe that:

1. The control law of each generator in the power system has relationships with the parameters and outputs of the rest generators, so the control law depends on the network parameters and structure. To apply this control law, it calls for precise information of all the generators in the network.
2. From the expression of excitation control (Eqn.6), the parameter r_i and γ_i (i=1,2,3) are free parameters. Because of the intertwined relationship between these parameters, the value for them should be selected based on the real network.
3. Though these two different approaches of excitation control design solve the difficulties during the Hamiltonian construction, but they all neglect system parameters to some extend. Therefore the control strategies we got are based on the approximation of system model. In these two approach, Y_{ij} equals to B_{ij} and $\sqrt{B_{ij}^2 + G_{ij}^2}$ respectively. With a bare glance at their mathematic expression, the "neglect ONLY α_{ij}" method should have better approximation to the original power system, hence has better dynamic performance than the "neglect BOTH G_{ij} and α_{ij}"approach.

4 simulation

4.1 computer simulation

The effectiveness of the controller constructed in this paper is tested on an IEEE 9-bus 3-machine system with MATLAB 6.5. The configuration of the studied system is shown in Fig.6, and the related data about generators and transmission line can be found in Appendix.A.2 in Table.2 and Table.1.

The fault scenario selected for simulation is a three-phase short circuit between Bus 3 and Bus A at 1s with clearing out time 0.1s and 0.2s respectively. After the fault, the power system retains to its original topology. The simulation time is 5s. The simulation environment is Windows XP.

In the base case(without the regulation of controllers), beginning with the oscillation at 1s, the system can keep stable with $T_{cr} = 0.1$s(T_{cr},critical clearing out time), and will lose synchronism with $T_{cr} = 0.2$. So we will choose the 0.2s scenario to be inspected on. (see Fig.1)

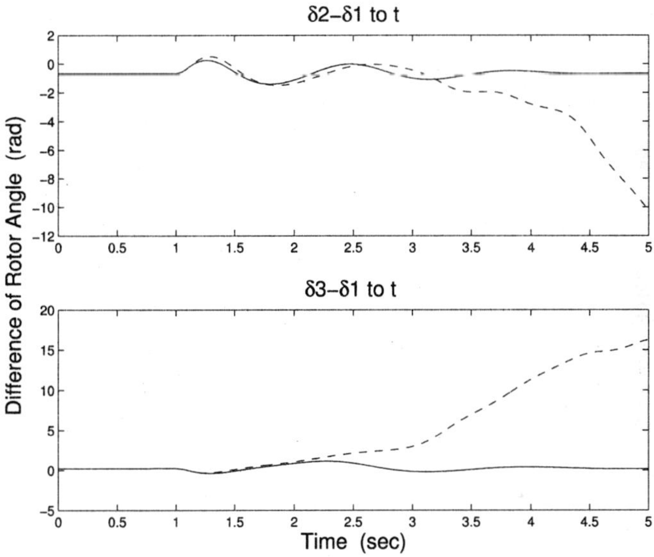

Fig. 1. system performance without controllers

—— CS with $T_{cr} = 0.1$

- - - CS with $T_{cr} = 0.2$

Here in this section, the simulation will be towards three cases:

Case 1

The control system with "neglect ONLY α_{ij}"method (NOCS)

Case 2

The control system with "neglect BOTH G_{ij} and α_{ij}" method (BCS)

Case 3

System performance without the regulation of controllers. (CS)

4.2 numeric results

- The comparison of NOCS and BCS :
  Simulation results recorded in Figs.2 show that, during the transient procession, neither of the two control strategy lose synchronism, but NOCS has better dynamic response quality than BCS: more smooth regulation behavior, smaller rotor angle oscillation magnitude, faster rate of arriving at the stable equilibrium point.

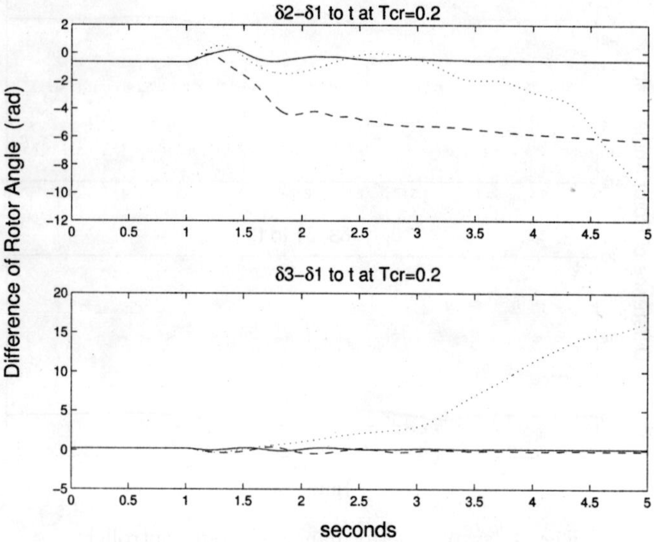

Fig. 2. rotor angle response respectable to 1♯ ($T_{cr} = 0.2$)

—— NOCS
– – – BCS
- - - - CS

- The comparison between different generators :
  Generator 3 is the nearest generator to the fault, thus suffers most from the fault and is a good indicator for control performance. From Fig.3-5, we observe that both of the two strategy work better for Gen 3 than the other two generators.
  As we stated before, these two control strategy have very close interconnection with all the parameters in the network. This may lead to control

failure when only part of the generators has the controller, which is the usual case for economical operation. Consequently, we also tested the efficiency when only adding controller to Gen 3 (u_3 in Eqn.6). Similar results are obtained when the same simulation procedure is performed on the system.

5 conclusion

We investigated in this paper possibility to consider the impact of transfer conductance partially in the design of a nonlinear static state feedback controller for multi-machine power system from "energy" aspects using IDA-PBC, and compared the difference of with and without the transfer conductance being taken into account. Simulation results of the 3-machine case indicate the efficiency of the proposed excitation control in enhancing transient stability of power system.

Though people have worked for years on energy function considering transfer conductance, there is still a long way to go. The idea of only taking the value of G_ij but not the angle α_{ij} into control design combined with the original energy function may give good hints to the application of energy function to real system after consider the influence of transfer conductance.

Appendix

A IEEE 9-bus system

A.1 system model

A.2 system data

See Tab.1 for line data and Tab.2 for generator data.

Bus I	Bus J	R1	X1	B1/2
Gen2	STATION A	0.01	0.085	0.044
Gen2	STATION B	-0.017	0.092	0.0395
STATION A	Gen3	0.032	0.161	0.0765
STATION B	Gen1	0.039	0.17	0.0895
Gen3	STATION C	0.085	0.072	0.03725
STATION C	Gen1	0.0119	0.1008	0.05225

Table 1. line data

(All in per unit.)

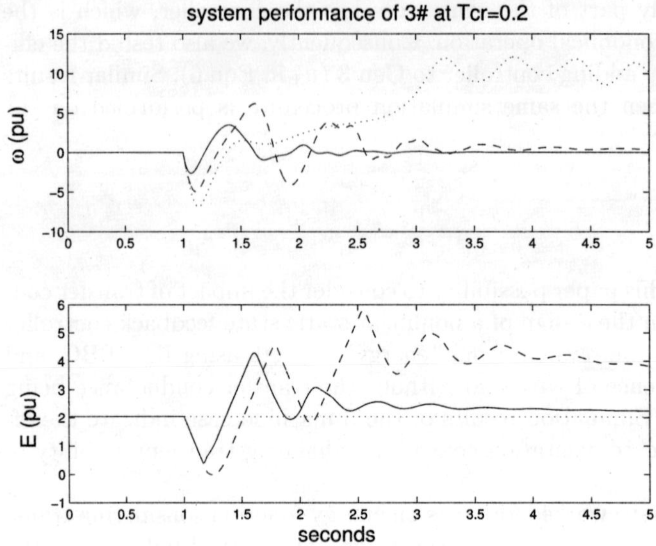

Fig. 3. system performance of 3 ♯ ($T_{cr} = 0.2$)

—— NOCS
– – – BCS
- - - - CS

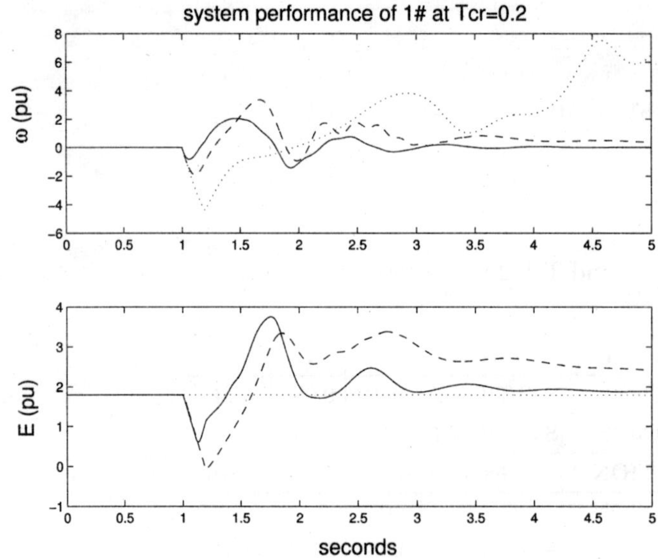

Fig. 4. system performance of 1 ♯ ($T_{cr} = 0.2$)

—— NOCS
– – – BCS
- - - - CS

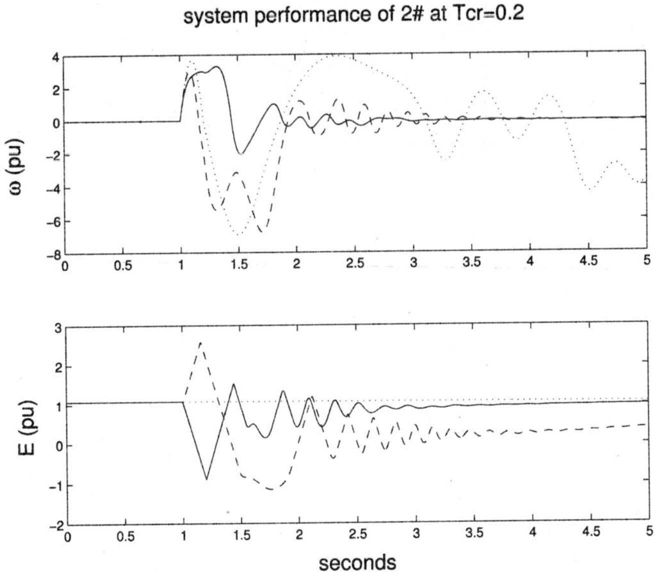

Fig. 5. system performance of 2 ♯ ($T_{cr} = 0.2$)

—— NOCS
– – – BCS
- - - - CS

Generator	Voltage reference	T_j	$X_d^{'}$	$X_q^{'}$	X_d	X_q	$X_{d0}^{'}$
Gen1	13.8	20	0.0969	0.0969	0.146	0.0969	8.96
Gen2	16.5	20	1.258	0.25	1.313	1.258	5.89
Gen3	18	20	0.8645	0.1969	0.8958	0.8645	6
Generator	Voltage reference	$T_{q0}^{'}$	X_l	$X_d^{''}$	$X_q^{''}$	$T_{d0}^{''}$	$T_{d0}^{'}$
Gen1	13.8	0.6	0.0742	0.107	0.107	0.033	0.07
Gen2	16.5	20	1.258	0.25	1.313	1.258	5.89
Gen3	18	20	0.8645	0.1969	0.8958	0.8645	6

Table 2. generator data

(All in per unit except Voltage reference in kV, T_j, $T_{q0}^{'}$,$T_{d0}^{''}$ and $T_{d0}^{'}$ in second.)

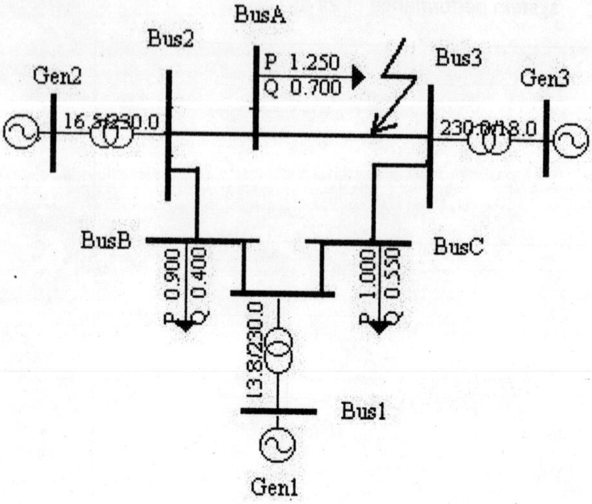

Fig. 6. 9-bus 3-machine system
(from IEEE, base power=100MVA)

B parameter data

See Tab.3 for system parameter data

	Before fault	During fault	After fault
G_{11}	0.04	0.3877	0.04
B_{11}	-0.658	-2.4743	-0.6583
G_{22}	1.189	0.6336	1.189
B_{22}	-2.217	-3.7225	-2.2171
G_{33}	0.0814	0.5172	0.0814
B_{33}	-0.881	-3.3693	-0.8806
G_{12}	0.1498	0.1676	0.1498
B_{12}	0.3477	1.1787	0.3477
G_{13}	0.0351	0.1081	0.0351
B_{13}	0.09	0.8659	0.09
G_{23}	0.2227	0.3917	0.2227
B_{23}	0.4872	1.884	0.4872

Table 3. parameter data before and after the fault

(All in per unit.)

glossary of symbols

δ_i rotor angle,in radian
ω_i rotor speed, in rad./sec.$\omega_0 = 2f_0\pi = 314.159$
f frequency of terminal voltage of a generator, in per unit
P_e electrical power, in per unit
P_m mechanical power, in per unit
D damping constant, in per unit
T_J inertial time constant, in seconds
$T_{d0}^{'}$ direct axis transient short circuit time constant, in seconds
$E_{qi}^{'}$ q-axis internal transient electric potential of the ith generator, in per unit
E_{fs} equivalent EMF in the excitation coil, in per unit
u_f input of the SCR amplifier of the generator, in per unit
i_d d-axis current, in per unit
x_d d-axis synchronous reactance of a generator, in per unit
$x_d^{'}$ d-axis transient reactance, in per unit
δ_{ij} $\delta_{ij} = \delta_i - \delta_j$
α_{ij} $\alpha_{ij} = arctan\frac{G_{ij}}{B_{ij}}$
B_{ij} ith row and jth column element of the nodal susceptance matrix, which is symmetric, at the internal nodes after eliminating all physical buses, in per unit

References

1. Q. Lu, Y. Sun, Z. Xu, T. Mochizuki, decentralized nonlinear optimal excitation control, IEEE Transactions on power systems, Vol.11, No4, November 1996
2. C. Sun, Z. zhao, Y. Sun, Q. Lu, Design of nonliear robust excitation control for multimachine power systems. IEE Proc. -Gener. Transm. Distrib, Vol.143, No.3, May, 1996
3. M. A. Pai, Power System Stability, North-Holland Publishing Co., Amsterdam, 1981
4. M. A .Pai, Energy Function Analysis for Power System Stability, Kluwer Academic, Boston, 1989
5. Young-Hyun Moon, Heon-Su Ryu, Byoung-Hoon Cho, Baik Kim, Second-kind Energy Function of Power System and Its Applications, 2001 IEEE
6. MA. Pai, P. G. Murthy, On Lyapunov Functions for Power Systems with Transfer Conductance, IEEE Trans. On Automatic Control, VOLAC-18, April 1973
7. U. Gudaru, A general Lyapunov function for multimachine power systems with transfer conductance, M, J cont , VO121, pp.333-343, 1975
8. Prabha Kundur, John Paserba, Venkat Ajjarapu, Goran Andersson , Anjan Bose , Claudio Canizares, Nikos Hatziargyriou, David Hill, Alex Stankovic, Carson Taylor, Thierry Van Cutsem, and Vijay Vittal, IEEE/CIGRE Joint Task Force on Stability Terms and Definitions, Definition and Classification of Power System Stability, IEEE Transactions on power systems, 0885-8950/04,2004,page 4-5

9. LU. Q and SUN. Y. Z:The nonlinear control of power system'(Academic press,Beijing,1993)
10. R. Ortega, A. van der Schaft, B. Maschke and G. Escobar, Interconnection and damping assignment passivity-based control of port-controlled hamiltonian systems, Automatica, Vol. 38, No. 4, 2002.
11. R. Ortega, Some applications and extensions of interconnection and damping assignment passivity-based control, Proc. IFAC Workshop on Lagrangian and Hamiltonian methods in nonlinear systems, Seville, Spain, April 3-5, 2003.
12. O. Dahl, Electric Power Circuits: Theory and Applications, McGraw-Hill, NY, 1938.
13. P. C. Magnusson, The transient-energy method of calculating stability, AIEE Transactions, Vol. 66, pp. 747-755, 1947.
14. Q. Lu, Y. Z. Sun and S. Mei, Nonlinear Excitation Control of Large Synchronous Generators, Kluwer Academic Publishers, Boston, MA., 2001.
15. P. M. Anderson and A. A. Fouad, Power Systems Control and Stability, The Iowa State University Press, 1977.
16. H.D.Chiang, Study of the Existence of Energy Function for Power Systems with loss, IEEE Trans. On Circuits and Systems, Vol. CAS-36, p1423-1429, Nov. 1989
17. R. Ortega, M. Galaz, A. Astolfi,Y.Z. Sun, T. Shen,Transient Stabilization of Multimachine Power Systems with Nontrivial Transfer Conductances, February 2, 2004
18. R. Ortega, A. van der Schaft, I. Mareels and B. Maschke, Putting energy back in control, IEEE Control Systems Magazine, Vol. 21, No. 2, April 2001, pp.18-33. submitted for publication
19. H-D. Chiang, C-C. Chu and G. Cauley, Direct stability analysis of electric power systems using energy functions: theory, applications, and perspective, Proceedings of the IEEE, Vol. 83, No. 11, pp. 1497-1529, 1995.
20. F. H. J. R. Silva, L. F. C. Alberto and N. G. Bretas, Extended Lyapunove function for power systems with transmission losses, 2003 IEEE Bologna PowerTech Conference, June 23-26, Bologna, Italy.

Extremum Seeking Method and Its Applications for Operational Cost Minimaization Control of Efflunet Quality for Advanced Wastewater Treatment Process

Jo Sato, Tomokazu Ueda, Hiromitsu Ohmori

Department of System Design Engineering, Keio University, Yokohama
ohm@sd.keio.ac.jp

1 Introduction

Every society discharges waste in both solid and liquid form. The liquid part of the waste, namely the wastewater, is the result of water supply used up in the society, ending up contaminated. From the perspective of its origin, wastewater could be defined as wastage drifting down from house, office, or facilities both commercial and industrial. Subsoil water, surface water, and rainwater may join the flow depending on the circumstances.

Thus, one characteristic of the wastewater treatment process is that it undergoes massive load changes. These changes may be driven by season, weather, and often by random factors. However, since the effluent quality is subject to legal restrictions, the plant must be operated to meet the demands despite the type of load changes to come. It is also desirable if energy-conserving approaches were taken, regarding plant operation.

Activated sludge process, known for its utilization of microorganisms in advanced wastewater treatment process, also sides with the difficulty regarding plant operation. Today, still new operational methods, including plant layout and equipments to be used, are being proposed and argued in order to achieve more stability, efficiency, cost performance and so on, despite the fact that 70 years have past since the birth of activated sludge process [1].

This paper focuses on optimality of Nitrogen removing process, achieving both clean effluent quality and low operational cost. Extremum seeking approach as reported in [2] will be utilized for the upper level of controller to directly minimize the certified cost function as shown in [3].The effectiveness of the proposed method is tested on biological wastewater treatment simulator SIMBA.

2 Plant Model and Control Objective

Wastewater treatment process mainly consists of two processes: treatment of wastewater and treatment of sludge. From the treatment standpoint, the process can be divided into three parts: physical process, biological process, and chemical process. More specifically, for example, they are processes such as solid-liquid separation by precipitation, decomposition of organic matter by microorganisms, and chemical treatment such as coagulated sedimentation, respectively. This paper will focus mainly on biological process.
The so-called pre-denitrification process is one of the wastewater treatment plant configurations that enables nitrogen removal [4]. The plant consists of two biological reactors and one settler, and the process looks as shown in Fig.1. The goal of control, in the first place, is to accomplish clean effluent

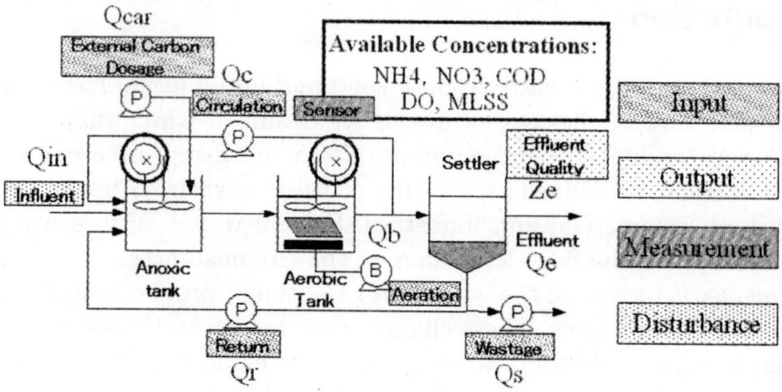

Fig. 1. Pre-Denitrification Process

quality. To be more specific, the process aims at keeping the concentrations of nitrogen and organic matter below certain level by manipulating aeration q_B, external carbon dosage Q_{car}, return flow rate Q_r, circulatory flow rate Q_c and wastage flow rate Q_s. Influent is considered as disturbance to the system, and available sensor values are limited.
Applying Activated Sludge Model No.1 (ASM1) proposed by a task group from the International Water Association (IWA) to model the process, wastewater can be divided in to 13 components as shown in Table 1 [5]. S_* denotes soluble component whereas X_* denotes particulate components. The pre-denitrification process, as shown in Fig.1, can be written as follows [6]. $\boldsymbol{z}_{a,i}$, $\boldsymbol{z}_{b,i}$ and $\boldsymbol{z}_{in}$ represents the concentration of the i-th wastewater component in anoxic tank, aerobic tank and influent, respectively. V_a, V_b and V_s denote the volume of the anoxic tank, aerobic tank and settler, respectively. The settler is modeled by applying the popular multi-layer idea, where the settler is assumed to consist of n horizontal slices with the feed into slice m, and the layers of the settler is numbered from the top to bottom [7]. Thus,

Table 1. State Variables of ASM1

Description	Symbol	Unit	
Inert organic matter	z_1	S_I	$gCOD/m^3$
Readily biodegradable substrate	z_2	S_S	$gCOD/m^3$
Inert organic matter	z_3	X_I	$gCOD/m^3$
Slowly biodegradable substrate	z_4	X_S	$gCOD/m^3$
Active heterotrophic biomass	z_5	X_{BH}	$gCOD/m^3$
Active autotrophic biomass	z_6	X_{BA}	$gCOD/m^3$
Products arising from biomass decay	z_7	X_P	$gCOD/m^3$
Oxygen	z_8	S_O	$gCOD/m^3$
Nitrate and nitrite nitrogen	z_9	S_{NO}	gN/m^3
Ammonia	z_{10}	S_{NH}	gN/m^3
Biodegradable organic nitrogen	z_{11}	S_{ND}	gN/m^3
Biodegradable organic nitrogen	z_{12}	X_{ND}	gN/m^3
Alkalinity	z_{13}	S_{ALK}	mol/m^3

the volume of each layer is $v_s = V_s/n$. $\boldsymbol{z}_{s,i,k}$ denotes the concentration of the i-th component in the k-th layer.

$$\frac{dz_{a,i}}{dt} = m_i^T \boldsymbol{r}(\boldsymbol{z_a}) + \frac{1}{V_a}\{Q_{in}z_{in,i} + Q_c z_{b,i} + Q_r z_{s,i,n} - (Q_{in} + Q_c + Q_r) z_{a,i}\} \tag{1}$$

$$\frac{dz_{b,i}}{dt} = m_i^T \boldsymbol{r}(\boldsymbol{z_b}) + \frac{Q_{in} + Q_c + Q_r}{V_b}(z_{a,i} - z_{b,i}) \tag{2}$$

$$\frac{dz_{s,i,k}}{dt} = \frac{Q_{k,in}}{v_s} z_{s,i,k,in} - \frac{Q_{k,out}}{v_s} * \psi_{s,i,k}(z_{s,k-1}, z_{s,k}, z_{s,k+1}) \tag{3}$$

$$k = 1, 2, \ldots, n$$

$$z_{in,i}(t) \geq 0, \ {}^{\forall}t \in \Re_+ \tag{4}$$

$$z_{a,i}(0) \in \Re_+, \ z_{b,i}(0) \in \Re_+, \ z_{s,i,k}(0) \in \Re_+, \tag{5}$$

$$\boldsymbol{z_a}, \boldsymbol{z_b} \in \Re^{13}, M = [m_1 \, m_2 \, \cdots \, m_{13}] \in \Re^{8\times 13} \tag{6}$$

$$\boldsymbol{r}(\boldsymbol{z_*}) = \begin{bmatrix} \rho_1(\boldsymbol{z_*}) \\ \rho_2(\boldsymbol{z_*}) \\ \vdots \\ \rho_8(\boldsymbol{z_*}) \end{bmatrix} \in \Re^8 \tag{7}$$

$m_i^T \boldsymbol{r}(\boldsymbol{z_*})$ represents the biological reactions of the microorganisms, and this is where all the nonlinearity of the system belong. Effluent quality must be controlled despite the massive load changes by influent and the nonlinearity by microorganisms. Besides the hydraulic input Q_*, carbon dosage $u_{a,2}$ is added to the right side of (1) if $i = 2$, and aeration term $u_{b,8}(K_L a)$ is added to the right side of (2) if $i = 8$.

This is a positive system where all the inputs are non-negative and all the state variables are also non-negative at any $t \in [0, \infty) = \Re_+$.

$$Q_{m,in} = Q_{in} + Q_r,\ Q_{1,out} = Q_{in} - Q_s,$$
$$Q_{n,out} = Q_r + Q_s \tag{8}$$

$$z_{s,i,k,in} = \begin{cases} z_{s,i,k+1} & \text{if } 1 \le k \le m \\ z_{b,i} & \text{if } k = m \\ z_{s,i,k-1} & \text{if } m+1 \le k \le n \end{cases} \tag{9}$$

$$\psi_{s,i,k}(\ldots) = \begin{cases} g_{s,i,k+1}(\boldsymbol{z_{s,k-1}}, \boldsymbol{z_{s,k}}, \boldsymbol{z_{s,k+1}}) \\ \qquad \text{if } i \in \{3,4,5,6,7,12\} \\ 0 \qquad \text{otherwise} \end{cases} \tag{10}$$

$$g_{s,i,k+1}(\boldsymbol{z_{s,k-1}}, \boldsymbol{z_{s,k}}, \boldsymbol{z_{s,k+1}}) \ge 0 \text{ if } z_{s,i,k} > 0 \tag{11}$$

(8)–(11) illustrates further details of (3), but it must be noted that no reaction is assumed to occur in the settler. Modeling of settler is not essential to the process since its main concern is solely on solid-liquid separation.
Let $\boldsymbol{x}$ denote the state vector of the entire plant for ease of expression.

$$\boldsymbol{x} = \begin{bmatrix} \boldsymbol{z_a} \\ \boldsymbol{z_b} \\ \boldsymbol{z_s} \end{bmatrix} \in \Re^{26+13n},\ \boldsymbol{z_a} \in \Re^{13},\ \boldsymbol{z_b} \in \Re^{13},\ \boldsymbol{z_s} \in \Re^{13n} \tag{12}$$

Table 2 summarizes the inputs subject to our manipulation, and outputs, the concentrations that must be kept below regulation level.
Guaranteeing this effluent quality, the next aim is to reduce financial cost necessary for plant operation.
(13) is the certified cost function as proposed in [3] and [4], in which each

Table 2. Control Inputs and Outputs

(a) Inputs

Description	Symbol	Unit
Aeration Coefficient	K_La	1/d
External Carbon dosage	$u_{a,2}$	g COD/m^3
Return flow rate	Q_r	m^3/d
Circulatory flow rate	Q_c	m^3/d
Wastage flow rate	Q_s	m^3/d

(b) Outputs

Description	Symbol	Unit
Readily biodegradable substrate	S_S	g COD/m^3
Ammonium	S_{NH}	g N/m^3
Nitrate and nitrite Nitrogen	S_{NO}	g N/m^3

operation and effluent quality are evaluated in terms of money.

$$y = \frac{1}{T}\int_{t_0}^{t_0+T} [(w_1 COD + w_2 BOD + w_3 SS + w_4 TN + w_5 TP)\, Q_e + w_6 Q_b + w_7 Q_r + w_8 Q_c + w_9 Q_s + w_{10} Q_{car} + w_{11} Q_{sludge}]\, d\tau \quad (13)$$

COD, BOD, SS, TN, TP denotes chemical oxigen demand, biological oxygen demand, suspended solids, total Nitrogen and total Phosphorus, respectively. These values are available online, and they can be expressed as:

$$COD = z_{e,1} + z_{e,2} + z_{e,3} + z_{e,4} + z_{e,5} + z_{e,6} + z_{e,7} \quad (14)$$

$$BOD = 0.25\left\{z_{e,2} + z_{e,4} + (1 - f_P)(z_{e,5} + z_{e,6})\right\} \quad (15)$$

$$SS = 0.75\,(z_{e,3} + z_{e,4} + z_{e,5} + z_{e,6} + z_{e,7}) \quad (16)$$

$$TN = z_{e,9} + z_{e,10} + z_{e,11} + z_{e,12} + i_{XB}(z_{e,5} + z_{e,6}) + i_{XP}(z_{e,3} + z_{e,7}) \quad (17)$$

where the subscript e indicates that the value of the state variable is taken from effluent. w_1–w_{11} are weighting factor coefficient which convert each operation and effluent quality to amount in Yen.
Cost function (13) is useful for evaluating the tradeoff between effluent quality and operational cost. Both clean effluent and low operational cost may be achieved by applying control law that minimize total cost y.

3 Cost Minimization Control

In order to accomplish the control object, extremum seeking algorithm as shown in Fig. 2 is proposed. The system consists of lower level as denoted by ① and upper level as denoted by ②. Actual input-generating scheme is governed by the lower level $u = \alpha\left(\rho, \hat{\boldsymbol{\theta}}\right)$, and the upper level determines the control parameter of the lower level to achieve minimum y based on [2].
Fundamental idea of the upper level is to minimize total cost y by acquiring gradient information on y. This is done by perturbing the system with a periodic test signal, sine wave in this case, and observing the response from the output. $\boldsymbol{H}_1$–$\boldsymbol{H}_3$ are diagonal matrices as shown in (18)–(21).

$$\zeta(t) = \begin{bmatrix} a_1 \sin \omega_1 t \\ \vdots \\ a_6 \sin \omega_6 t \end{bmatrix} \quad (18)$$

$$\boldsymbol{H}_1(s) = diag\left[\frac{s}{s+\omega_{h,1}} \cdots \frac{s}{s+\omega_{h,6}}\right] \quad (19)$$

$$\boldsymbol{H}_2(s) = diag\left[\frac{\omega_{l,1}}{s+\omega_{l,1}} \cdots \frac{\omega_{l,6}}{s+\omega_{l,6}}\right] \quad (20)$$

$$\boldsymbol{H}_3(s) = diag\left[\frac{k_1}{s} \cdots \frac{k_6}{s}\right] \quad (21)$$

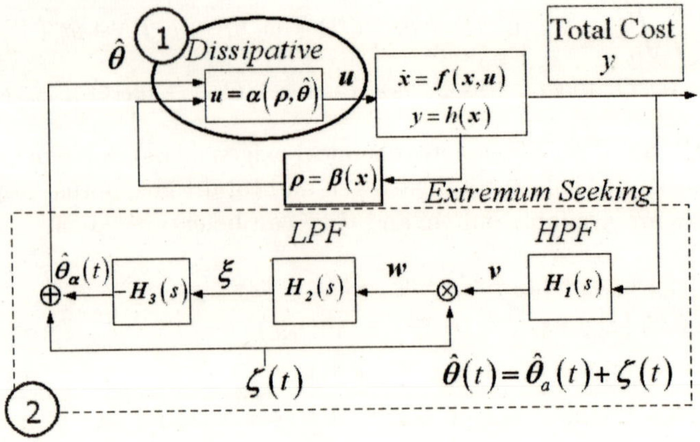

Fig. 2. Proposed Extremum Seeking Algorithm

Test signal will give output y a periodic response either in phase with the original sine wave, or out of phase depending on whether the optimum seeking parameter $\hat{\theta}_p, p = 1, \ldots, 6$ is greater or smaller than the actual optimum value θ_p^*
Thus, a group of high-pass filters $\boldsymbol{H_1}$ is employed to eliminate the "DC component" of y to see the response on output caused by $\boldsymbol{\zeta}$. Then, the "DC component" of the product of approximate sinusoids is extracted by low-pass filter $\boldsymbol{H_2}$. The sign of $\boldsymbol{\xi}$ is then used to provide the direction to the integrator $\boldsymbol{H_3}$ for moving $\hat{\theta}_p$ towards θ_p^*.
Dissipative controller as presented in [6] is employed for the lower level. The control inputs of lower level are governed by (22)–(36). ρ_* denotes the process rate of the biological reaction to be incorporated in the control law. $\beta(\boldsymbol{x})$ in the minor loop is an observer as to be also referred to in [6].

Aeration

$$u_o(\boldsymbol{z_b}) = \begin{cases} k_{o1} 2 \frac{1-Y_H}{Y_H} \rho_1(\boldsymbol{z_b}) + k_{o2} \left(\frac{4.57}{Y_A} - 2\right) \rho_3(\boldsymbol{z_b}) + \alpha_o (e_o - S_O) \\ \qquad \text{if} \quad e_o - S_O > 0 \\ k_{o1} 2 \frac{1-Y_H}{Y_H} \rho_1(\boldsymbol{z_b}) + k_{o2} \left(\frac{4.57}{Y_A} - 2\right) \rho_3(\boldsymbol{z_b}) \\ \qquad \text{otherwise} \end{cases} \tag{22}$$

$$0 < k_{o1} < 1, \quad 0 < k_{o2} < 1 \tag{23}$$

$$K_L a(\boldsymbol{z_b}) = \min\left\{ \frac{u_o(\boldsymbol{z_b})}{S_{O,sat} - z_{b,8}}, \quad K_{O,sat} \right\} \tag{24}$$

External Carbon Dosage

$$u_{car}\left(\boldsymbol{z_a}\right) = \begin{cases} k_{car}\left(1+\frac{1}{2.86}\right)\frac{1-Y_H}{Y_H}\rho_2\left(\boldsymbol{z_a}\right)+\bar{\alpha}_{car}\left(\bar{e}_{car}-\bar{S}_S\right) \\ \qquad\qquad \text{if} \ \ \bar{e}_{car}-\bar{S}_S>0 \\ k_{car}\left(1+\frac{1}{2.86}\right)\frac{1-Y_H}{Y_H}\rho_2\left(\boldsymbol{z_a}\right) \\ \qquad\qquad \text{otherwise} \end{cases} \tag{25}$$

$$0 < k_{car} < 1 \tag{26}$$

$$\bar{S}_S = \rho_1 \frac{K_{OH}+S_O}{\mu_H S_O} \tag{27}$$

$$u_{a,2} = \min\left\{u_{car}\left(\boldsymbol{z_a}\right), \ U_{car,sat}\right\} \tag{28}$$

Return Flow Rate

$$Q_r = \begin{cases} \min\left\{\bar{c}_r\left(X_{return}-X_{anoxic}, \ \ Q_{r,sat}\right\}\right) \\ \qquad \text{if} \ X_{return}-X_{anoxic}>0 \\ 0 \qquad \text{otherwise} \end{cases} \tag{29}$$

$$\bar{c}_r > 0 \tag{30}$$

$$X = X_S + X_{BH} + X_{BA} + X_P + X_{ND} + X_I \tag{31}$$

Circulatory Flow Rate

$$Q_c = \begin{cases} \min\left\{c_c\left(z_{b,9}-z_{a,9}\right), Q_{c,sat}\right\} \text{ if } \ z_{b,9}-X_{a,9}>0 \\ 0 \qquad\qquad\qquad\qquad \text{otherwise} \end{cases} \tag{32}$$

$$c_c > 0 \tag{33}$$

Wastage Flow Rate

$$Q_s = \begin{cases} \min\left\{c_s\left(X-e_{qs}\right), Q_{s,sat}, Q_{in}\right\} \text{ if } \ X-e_{qs}>0 \\ 0 \qquad\qquad\qquad\qquad \text{otherwise} \end{cases} \tag{34}$$

$$c_s > 0 \tag{35}$$

$$X = X_S + X_{BH} + X_{BA} + X_P + X_{ND} + X_I \quad \text{at the bottom of te settler} \tag{36}$$

ρ_* are process rates of biological reactions, and Y_*, K_*, μ_* are stoichiometric parameters. $0 < k_{o1} < 1, 0 < k_{o2} < 1, 0 < k_{car} < 1, \bar{c}_r > 0, c_c > 0, c_s > 0$ are tunable parameters to be tuned by the upper level. $*_{sat}$ represents the limitation of compressor, flow rate and capability of pumps.
Dissipative controller is employed for the lower level for its primary objective to not make the process unstable. Stability must be given top priority since the plant is a major part of city infrastructure.

4 Simulation

The most accepted biological wastewater treatment simulator in the field, SIMBA 4.1 by ifak system was used to confirm the effectiveness of the proposed method. Since the proposed method takes into account the trade off

Table 3. Plant Parameters

Parameter	Value	Unit
Temp.	20	
V_a	2000	m^3
V_b	4000	m^3
V_s	6000	m^3
n	3	–
m	2	–
$S_{O,sat}$	8.6	g DO/m^3

Table 4. Lower Level Parameter

Parameter	Value	Unit	Parameter	Value	Unit
$K_{O,sat}$	360	1/d	k_{o1}	0.56	–
$U_{car,sat}$	600	g COD/m^3d	k_{o2}	0.56	–
$Q_{r,sat}$	40000	m^3/d	e_o	0.6	–
$Q_{c,sat}$	100000	m^3/d	α_o	9000	–
$Q_{s,sat}$	2000	m^3/d	$\bar{c}_r$	4	–
k_{car}	0.4	–	c_c	6800	–
$\bar{e}_{car}$	15	–	c_s	0.043	–
$\bar{\alpha}_{car}$	300	–	e_{qs}	200	–
			ω_f	150	–

Table 5. Upper Level Controller Parameters

Symbol	Value	Symbol	Value	Symbol	Value
k_1	100	k_2	100	k_3	10000
a_1	0.05	a_2	0.04	a_3	0.05
ω_1	12	ω_2	15	ω_3	17
$\omega_{h,1}$	10000	$\omega_{h,2}$	10000	$\omega_{h,3}$	1000000
$\omega_{l,1}$	0.01	$\omega_{l,2}$	0.01	$\omega_{l,3}$	0.001
k_4	5000	k_5	800	k_6	200
a_4	60	a_5	0.03	a_6	0.0001
ω_4	19	ω_5	21	ω_6	23
$\omega_{h,4}$	1000000	$\omega_{h,5}$	1000000	$\omega_{h,6}$	12000
$\omega_{l,4}$	0.005	$\omega_{l,5}$	0.005	$\omega_{l,6}$	0.01

Table 6. Cost Function Parameters

Weight	w_1	w_2	w_3	w_4	w_5	w_6	w_7	w_8	w_9	w_{10}	w_{11}
Set Value	15	60	60	90	1200	0.234	0.18	0.18	0.18	10050	55

between effluent quality and operational cost, control experiment implementing just the lower level only, a control strategy that does not incorporate the idea of cost, is also conducted for comparison.

Simulation was carried out under the following parameters as shown in table 3–6. Fig. 3–7 shows the comparison of input responses between dissipative control and the proposed method. In terms of (13), they are time responses of manipulation that affect operational cost of the plant. Thus, small values taken in the responses imply retrenchment of expenditures by electricity, chemical substances, etc. Fig. 3–7 show that the proposed method has the tendency to take small manipulating values, thereby reducing the operational cost. Limiter values of each manipulation is also shown. Fig. 8–9, on the other hand, show time responses of concentrations that affect effluent quality part of the cost function (13), and they are especially the components subject to laws and regulations. *COD* response is plotted in Fig. 8 in place for readily

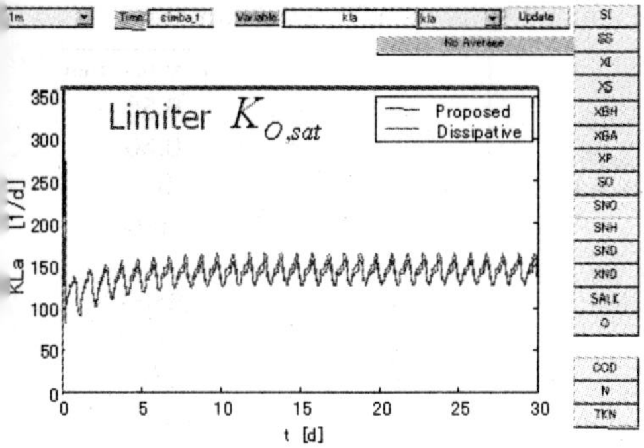

Fig. 3. Aeration

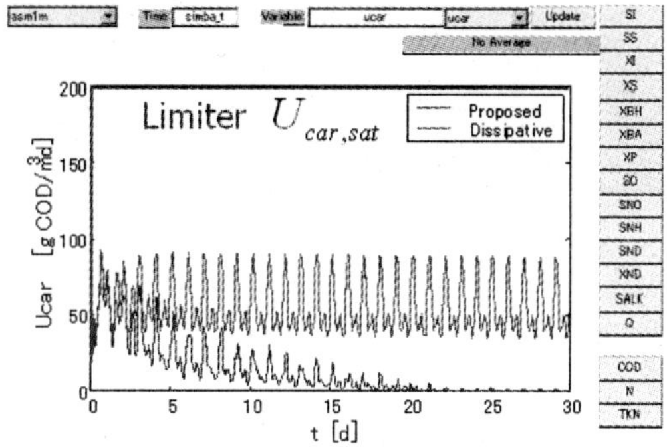

Fig. 4. External Carbon Dosage

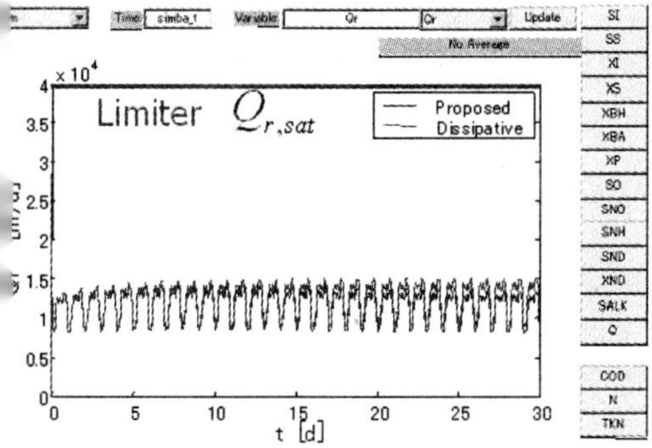

Fig. 5. Return Flow Rate

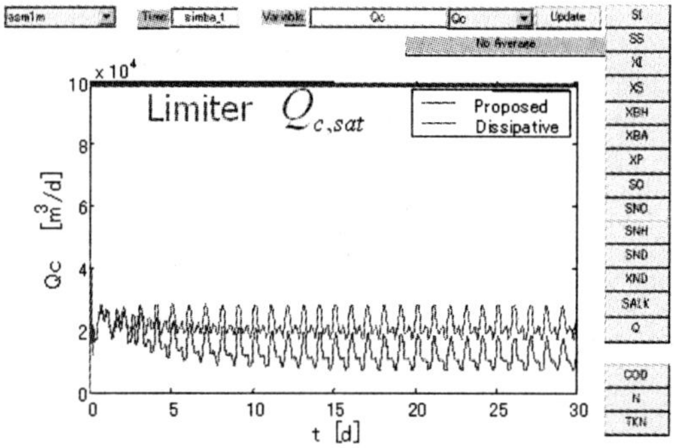

Fig. 6. Circulatory Flow Rate

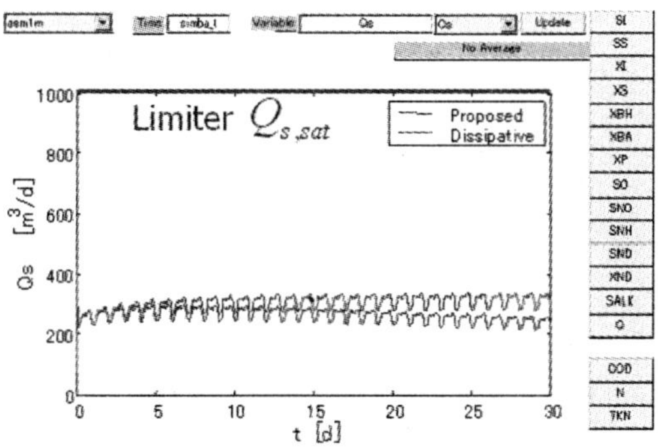

Fig. 7. Wastage Flow Rate

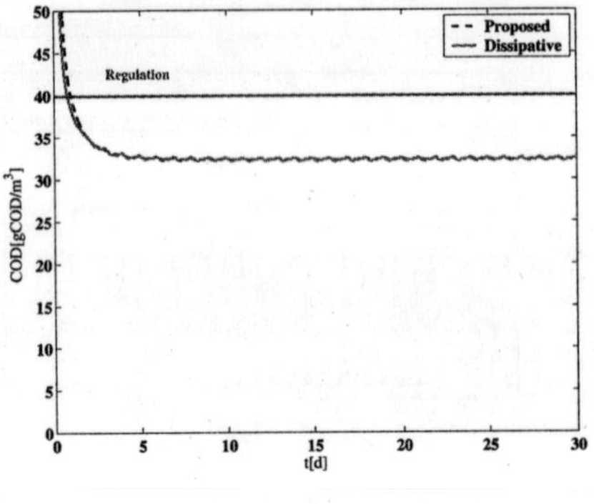

Fig. 8. COD in Effluent

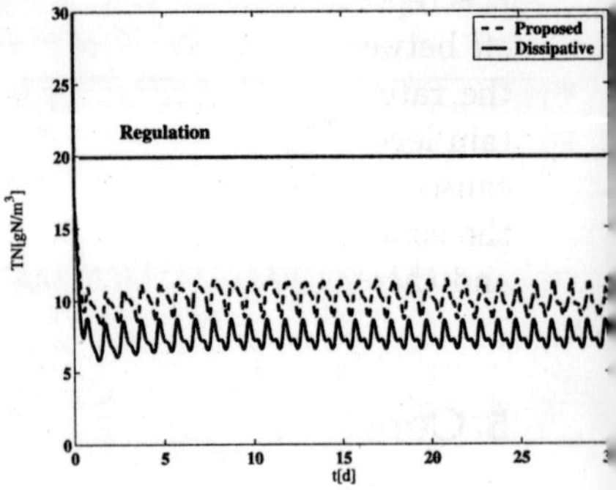

Fig. 9. Total Nitrogen in Effluent

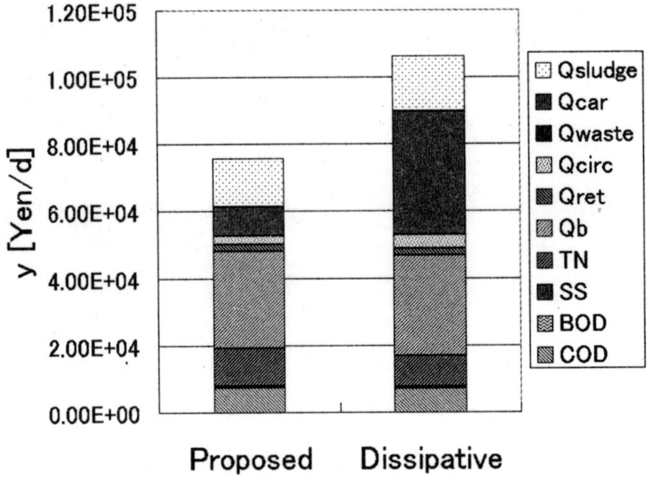

Fig. 10. Total Cost Comparison

biodegradable substrate S_S, because S_S is not directly measurable and also because COD value has strong correlation with S_S.

It can be seen that the effluent quality of the proposed method slightly falls below dissipative control for taking into consideration the aspect of cost. The oscillation observed in both responses of the two methods are caused by load changes from the influent. It can be concluded that the effluent quality is well below the regulatory level despite the load changes and the expenditures retrenchment.

The total cost of the process calculated by (13) is shown and compared in Fig. 10. The proposed method accomplished 28.7% cutback on the total cost compared to manipulation done just by the dissipative controller.

It is reported from actual plants and experimental laboratories that the trade-off between operational cost and effluent quality is done solely by adjusting the rate of external carbon dosage, once the effluent quality reached a certain level. Fig. 10 shows that the major difference in the total cost is indeed, caused by the change in external carbon dosage. It is interesting to see how the control strategy determined by the extremum seeking algorithm in Fig. 2 and the empirical strategy coincide with one another.

5 Conclusion

In this paper, a cost minimization control of effluent quality has been proposed for pre-denitrification process. Certified cost function that considers every penalty of the plant and converts them to Yen has been appointed to achieve both clean effluent quality and low operational cost. Optimization probelm then, has been solved by the upper level controller incorporating the extremum seeking algorithm to minimize total cost.
The effectiveness of the proposed method has been tested by the well accepted, biological wastewater treatment simulator SIMBA 4.1.

6 Acknowledgement

This work is supported by the Grant-in-Aid for Scientific Research (B) (Project Number:13555120) of Japan Society for the Promotion of Science, and also by the following people who have provided special research materials, intellectual stimulation, and other forms of assistance: Hiroshi Ito, Kyushu Institute of Technology; Koichi Hidaka, Tokyo Denki University; Akihiro Nagaiwa, Toshiba; and Osamu Yamanaka, Toshiba.

References

1. Metcalf and Eddy, INC. (1991), Wastewater Engineering. *McGraw-Hill Publishing Company.*
2. Krstić, M., Wang, H., (2002), Analysis and Design of Multivariable Extremum Seeking Feedback for General Nonlinear Systems. *Proc. of the American Control Conference*, 2903–2908.
3. Copp, J.B.(2001), The COST Simulation Benchmark: Description and Simulator Manual. *Office for Official Publications of the European Community.*
4. Yamanaka, O., Obara, T., Ashikaga, N., Yamamoto, K. (2004), An Optimal Control Strategy for Predenifitrication Type of Biological Wastewater Treatment Process based on A Total Cost Index – Validation of The Effectiveness of Optimal Control based on Benchmark Process – (in Japanese), *EICA*. vol.9, No.2, 91–98.

5. Henze, M., Gujer, W., Mino, T., Loosdrecht, M.(2000), Activated Sludge Models ASM1, ASM2, ASM2d and ASM3. *IWA Publishing.*
6. Ito, H.(2003), Dissipative Control with Estimation for Biological Wastewater Treatment Plants Based on Entire Nonlinear Process Models. *Proc. of the 32nd Control Theory Symposium*, 291–298.
7. Olsson, G., Newell, B. (1999), Wastewater Treatment Systems Modelling, Diagnosis and Control. *IWA Publishing.*

Sound Measuring of Motorcycle Engine with Wavelet Analysis

Masaki Suido, Seiichi Shin, and Tetsuya Tabaru

School of Information Science and Technology, The University of Tokyo, Japan
sui@axis.t.u-tokyo.ac.jp

1 Introduction

Japan is an aging society and younger engineers are not familiar with mechanisms. Therefore, retirement of skilled engineers is one of the problems in the production and maintenance of Japanese manufacturing.
Visualization is one of the keywords to cope with such a problem. Seeing is very significant as Eastern People say "One scene is superior to hundred saying" and Western people say "Seeing is believing." We can read about 300 Japanese words per minutes. It corresponds to 80 bits per second since one Japanese character consists of 2 bytes. It is too slow to write and read reports, which consist of words. Visual Sensor, eyesight, is the most powerful tool of human beings among the five senses, so that visualization is very important in engineering in all countries including the Eastern and the Western. We focus on the wavelet transform (WT) as a tool of visualization. The transform was invented by Mallat, a French engineer in 1985 as an improvement of the short time Fourier transform (STFT). A mathematical framework was developed by Daubechies, a French mathematician.
The time window which is fixed in STFT (Fig. 1) is changed along frequency in WT (Fig. 2). A width of the window is narrowed in a high frequency region and broadens in a low frequency region in WT. Therefore, the similarity of transform is restored in WT, whereas the similarity is loosed in STFT. We can get a good visualization of signals for time-frequency analysis with WT, although many transforms changing the width of the window are necessary in STFT.
We apply WT to the real field with many companies, for example, Yamatake Corporation, Matsushita Electric Industrial Co., Ltd., Taiheiyo Cement Corporation, Aisin Seiki CO., LTD., Bridgestone Corporation, Digital Electronics Corporation, and so on [1]-[10]. Moreover, Yamatake has made some tools of WT, for example, Time2wave (Fig. 3), Wavedoctor for PC, and Wavedoctor for PDA. The first one is a general tool for WT and can operate continuous WT and discrete WT. The latter two the same software, however, the

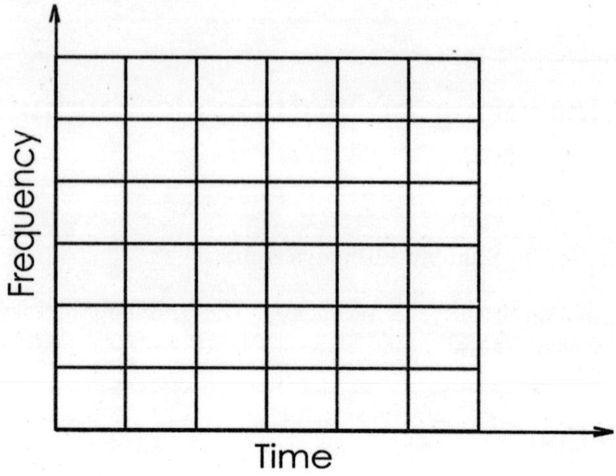

Fig. 1. The time window is fixed in STFT

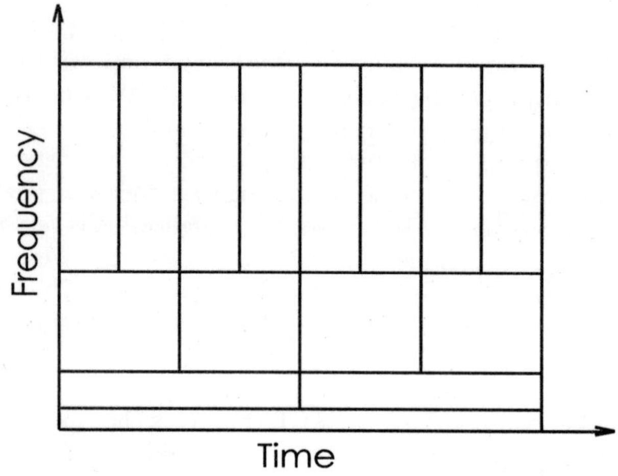

Fig. 2. The time window is changed along frequency in WT

platform is different. Wavedoctor for PC is for Windows PCs and does WT on sound files of Windows PCs. Wavedoctor for PDA runs on Windows CE machines. Engineers can sample a target sound with the PDAs which have microphones and see WT directly on the display of PDAs.

This paper presents an application of WT to discrimination of the number of cylinders of a motorcycle. A sound of an engine is sampled with an IC voice recorder and the sampled data is transformed as a visualization of the engine sound. Skilled engineers can tell the number of cylinders from the engine

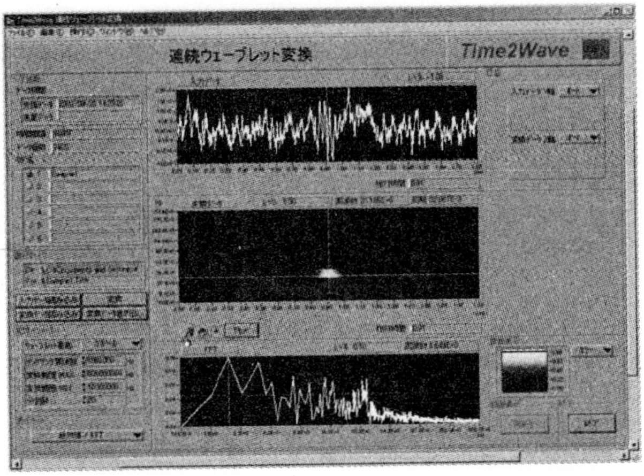

Fig. 3. Time2wave by Yamatake corporation

sound, but, the younger can not. The younger does not know what feature is significant on the sound, so the visualization is very useful to transfer the skill from the skilled to the younger. Moreover, we can develop a discrimination tool based on the visualized data. This is not so remarkable, however, is simple enough to understand the usability of WT. This work also presents a potential ability for usage of WT to maintenance of motorcycles.

2 Wavelet transform [11]

Wavelet transform $S(a,b)$ of a scalar signal $s(t)$ is defined by

$$S(a,b) = \frac{1}{\sqrt{a}} \int_{-\infty}^{\infty} g^* \left(\frac{t-b}{a} \right) s(t) dt \tag{1}$$

where b is a shift parameter, which corresponds to time and a is a scale parameter, which corresponds to an inverse of frequency. Therefore, WT is a transform from the time domain to the (a,b) domain which corresponds to the time-frequency domain.

The function $g(t)$ is a mother wavelet, which is selected based on an aim of analysis. Generally, Gabor function, $\exp(-t^2) \cdot \exp(j\omega_0 t)$, is good for the time-frequency analysis, where j is the imaginary unit and ω_0 is a parameter related to relative sensitivity between time and frequency. We set $g(t)$ as this function and ω_0 as 5, which provides a good balance of time and frequency sensitivity. Figure 4 shows a real part of Gabor function.

With Gabor function, the definition of WT becomes,

$$S(a,b) = \frac{1}{\sqrt{a}} \int_{-\infty}^{\infty} \exp \left(- \left(\frac{t-b}{a} \right)^2 \right) \exp \left(-j\omega_0 \left(\frac{t-b}{a} \right) \right) s(t) dt. \tag{2}$$

We can see from this equation, $\exp(-(\frac{t-b}{a})^2)$ corresponds to a window in STFT, however, it is a function of scale parameter a and the inverse of the parameter corresponds to the frequency in STFT. Therefore, WT can be seen a sort of STFT with a variable window on frequency.

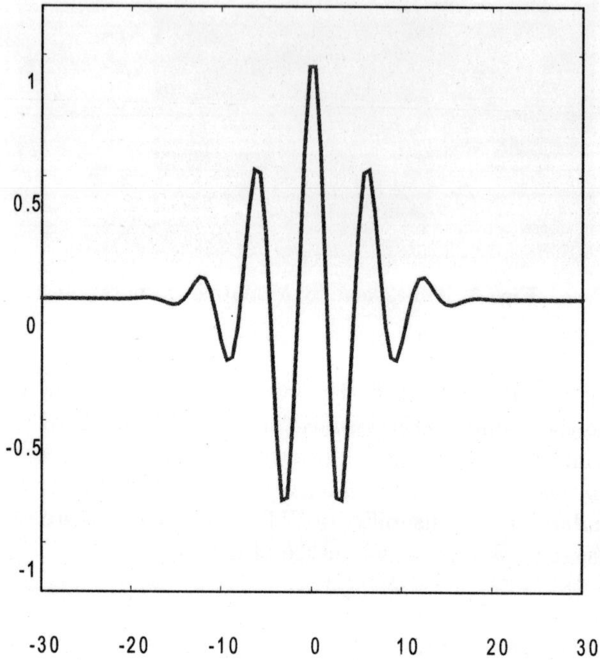

Fig. 4. The real part of Gabor function

3 Sampling system

Ordinary engines of motorcycles are categorized as single, twin, or four cylinders. Sounds of the engines are recorded by an IC recorder as shown in Fig. 5. We gathered sounds of 13 motorcycles in idling mode. They are listed in Table 1.

The sampled data (Fig. 6) is visualized by WT. Figure 7 is an example of WT of a motorcycle engine sound. We can see there are some sound data concentrated in the lower part of the plot. Figure 8 shows an extracted view of the lower part. The visualized sound data (WT) can be categorized into

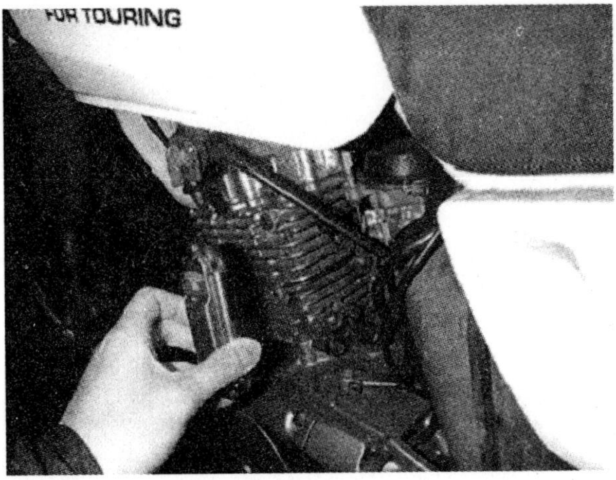

Fig. 5. The recording scene

Table 1. The motorcycles used in this paper

Model	Manufacturer	Cylinders	Cyl. Layout	Cooling	Displacement	Model Year
GPz900R	Kawasaki	4	InLine	Water	900cc	2002
CB400SF	HONDA	4	InLine	Water	400cc	2000
GSX250S	SUZUKI	4	InLine	Water	250cc	1991
FZS1000	YAMAHA	4	InLine	Water	1000cc	2001
F4S	MV AGUSTA	4	InLine	Water	750cc	2000
ZZ-R250	Kawasaki	2	Parallel	Water	250cc	1999
VT250J	HONDA	2	V-90	Water	250cc	1988
SV400S	SUZUKI	2	V-90	Water	400cc	1998
TL1000S	SUZUKI	2	V-90	Water	1000cc	1997
DL1000	SUZUKI	2	V-90	Water	1000cc	2002
CBX125C	HONDA	1	Single	Air	125cc	1987
DR125	SUZUKI	1	Single	Air	125cc	1993
XR250	HONDA	1	Single	Air	250cc	1995

three parts. The high frequency band corresponds to the cam and valve part, the middle frequency is the explosion part, and the low frequency is the exhaust noise part. We can see that explosions in the motorcycle engine are not steady on each period. In this way, we can see the inside of alive motorcycle engines by WT. Moreover, we can get a sound on a specified band with the inverse transform of WT. Skilled engineers can concentrate on a sound from a particular component of a motorcycle. Maybe they have the ability to execute WT with their brains.

By the way, the number of cylinders is related to the middle frequency part since the number of cylinders can be discriminated from the number of explosions during one cycle (revoluion) as long as four-cycle engines are used.

We suggest an algorithm to discriminate the number of cylinders in the next section.

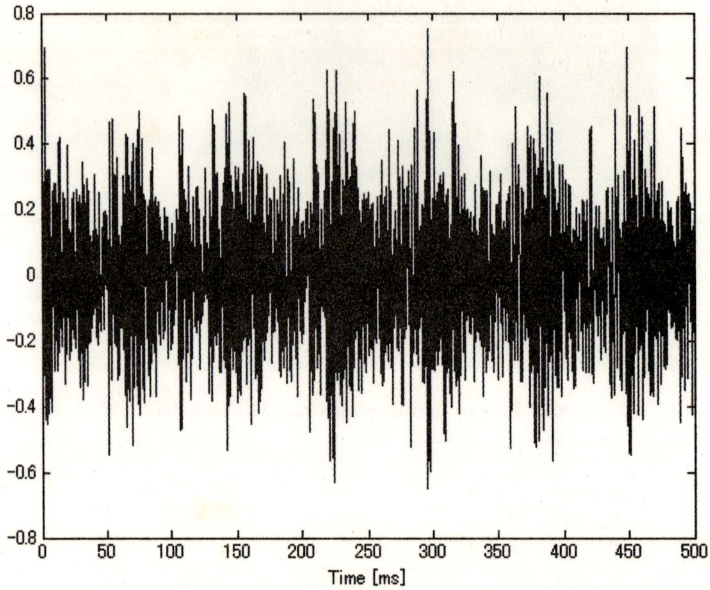

Fig. 6. The original sampled data (DR125)

4 Algorithm

Figure 9 - 11 are the brief illustrations of our method. The vector **B** is defined by

$$c(b) = \max_{a \in [a_m, a_M]} |S(a, b)| \tag{3}$$

$$\mathbf{B} = [c(1), c(2), ..., c(n)] \tag{4}$$

$$n = T/\Delta T \tag{5}$$

where a_m and a_M is the minimum and maximum values of a to be extracted, and $[a_m, a_M]$ correspond to the frequency band of the explosion sound. The variable T is the interval length to be transformed and ΔT is the sampling rate of the original data. Figure 10 is an example plot of **B**. And Fig. 11 illustrates the way to get the explosion process time. We can count the number of explosions on a period if an adequate threshold *thd* is set on **B**. This means

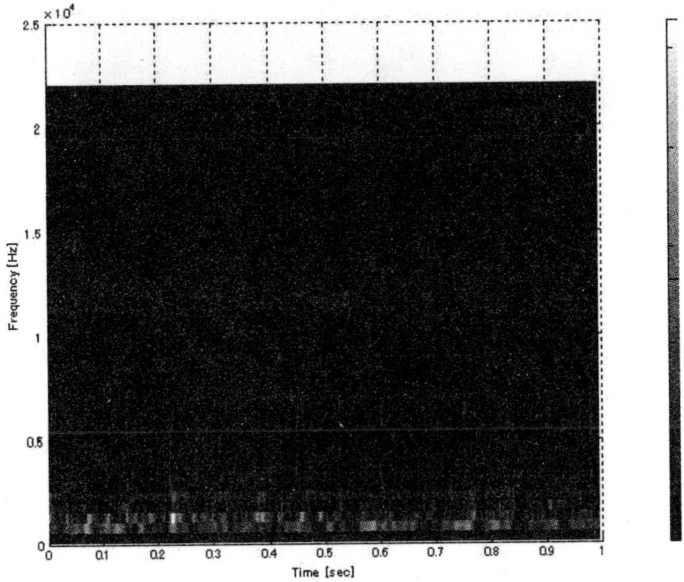

Fig. 7. WT of the original data (DR125)

we can discriminate the number of cylinders from the sampled sound of the motorcycle. Therefore, the proposed algorithm consists of the following five steps.

1. Get WT of the sampled data.
2. Extract the explosion band from the WT.
3. Calculate the vector **B** by the equation 4. (as seen in Fig. 9.)
4. Specify the time during the explosion process by comparing the **B** and the threshold *thd*.
5. Discriminate the number of cylinders from the ratio ($R(thd)$) between the explosion time and one period.

In short, we can obtain the ratio of time in the explosion process as $R(thd)$. As long as ordinary four-cycle engines with no overlap of the explosion process is used, this ratio is expected to be 1.00 in four cylinders, 0.50 in twin cylinders, and 0.25 in a single engine. However, for many reasons, $R(thd)$ won't be such an exact values. So we have to find the best discriminating values. We calculated the best threshold *thd* by checking the relation between *thd* and $R(thd)$. Table 2 gives the relation. We ploted them on *thd* and $R(thd)$ graph as shown in Fig. 12.

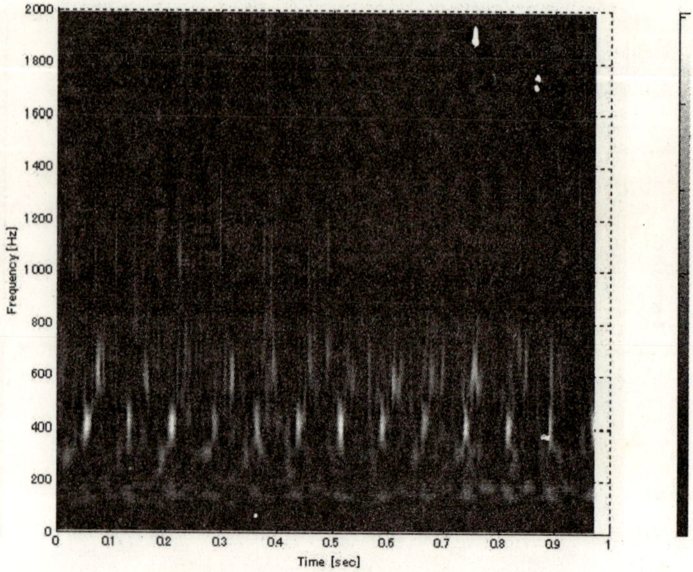

Fig. 8. Extracted WT (1-2000Hz, DR125)

Table 2. The relation between the difinition of *thd* and *R*(*thd*)

Model	cyl.	type	thd=max/2	/3	/3.5	/4
GPz900R	4	InLine	0.3047	0.6213	0.7133	0.7747
CB400SF	4	InLine	0.2350	0.6705	0.7752	0.8350
GSX250S	4	InLine	0.0534	0.3618	0.5019	0.6188
FZS1000	4	InLine	0.1620	0.4042	0.5015	0.5774
F4S	4	InLine	0.3069	0.4359	0.4804	0.5327
ZZR250	2	Parallel	0.1297	0.2294	0.2747	0.3252
VT250J	2	V-90	0.1507	0.3003	0.3601	0.4069
SV400S	2	V-90	0.1473	0.2868	0.3394	0.3816
TL1000S	2	V-90	0.1509	0.3670	0.4562	0.5255
DL1000	2	V-90	0.1641	0.2829	0.3268	0.3643
CBX125	1	Single	0.0783	0.2204	0.2978	0.3808
DR125	1	Single	0.0307	0.0888	0.1328	0.1829
XR250	1	Single	0.0534	0.1062	0.1390	0.1730

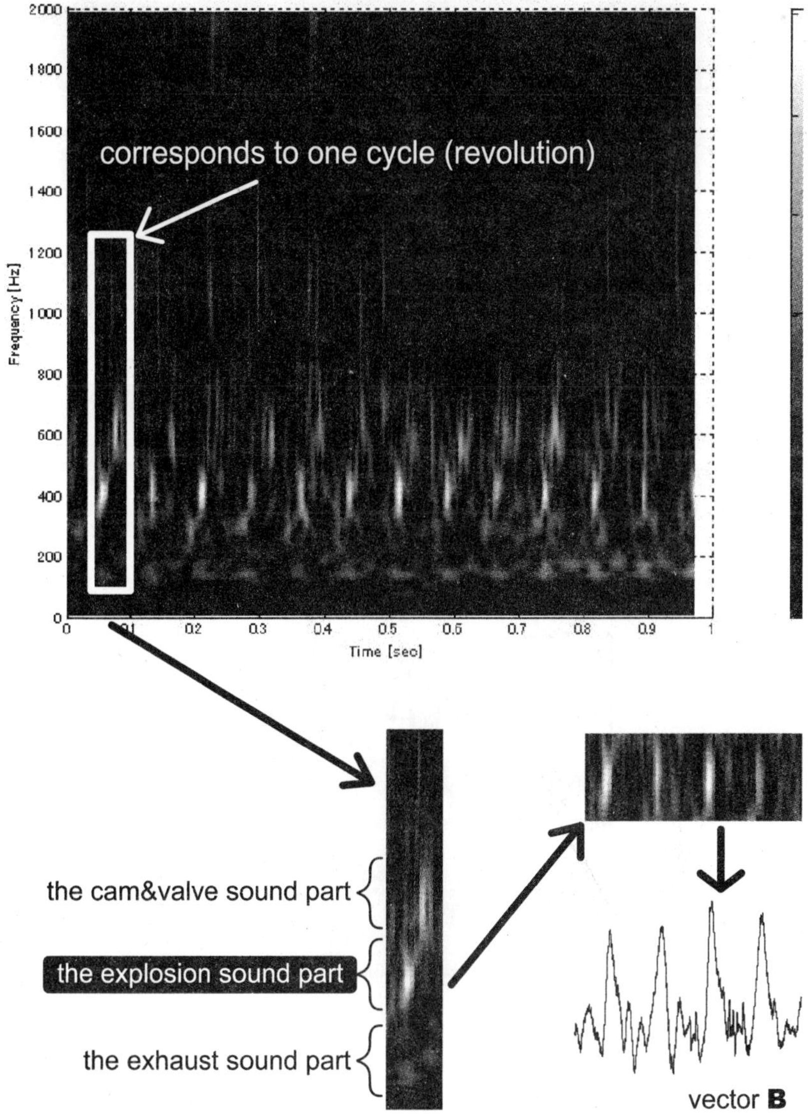

Fig. 9. The method to get **B** from the WT

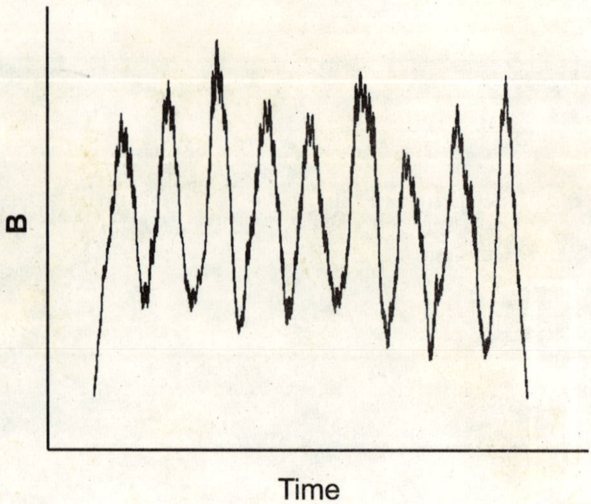

Fig. 10. An example plot of **B**

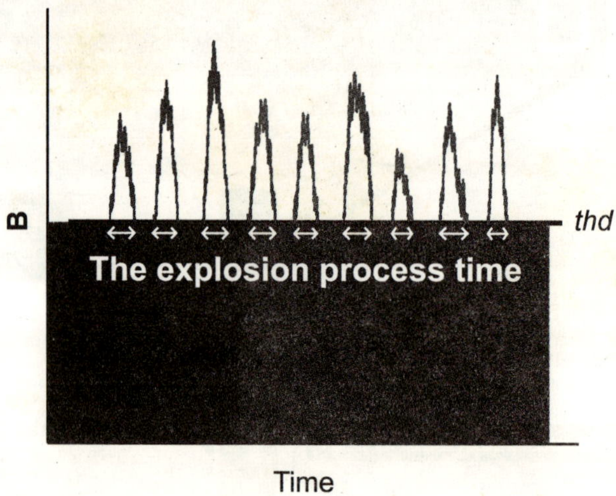

Fig. 11. The method to get the explosion process time

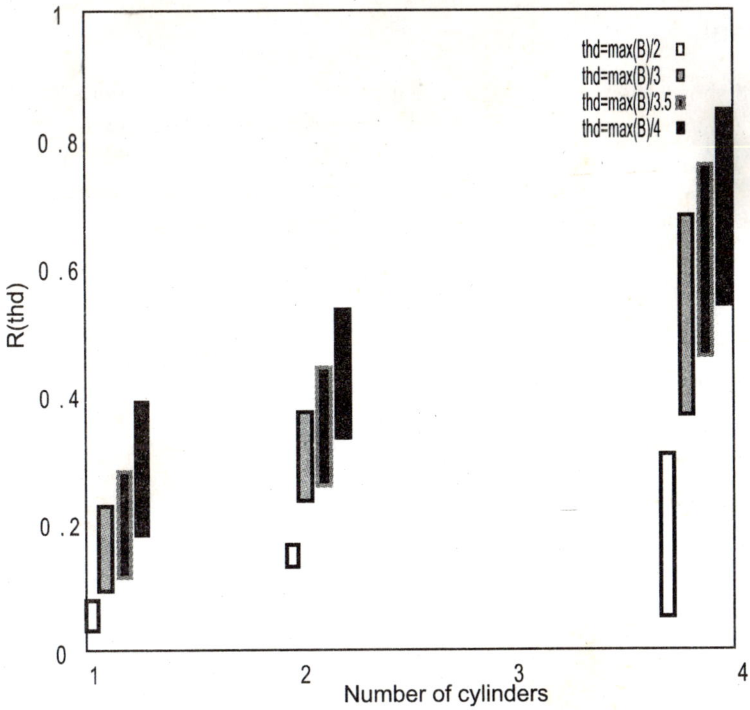

Fig. 12. The relation between the number of cylinders and $R(thd)$

5 Result

Table 3 shows the result by the algorithm proposed in the former section. The explosion time is extracted using a threshold *thd*. The threshold is set based on the maximum value because the level of the sampled sound depends on a sort of motorcycles and sampling conditions. We succeeded in discriminating 12 sorts of motorcycles with $max/4 < thd < max/3$ except for the CBX125 Custom by HONDA.

This motorcycle has an exceptional carbulating and exhausting system. Its sole cylinder has two carbulators and two in&ex manifolds each and two mufflers. Therefore there are interferences between sounds from explosion and exhaust and the proposing algorithm mistook that the engine has two cylinders, while there is only one cylinder in the engine. We can improve this problem if we use one more filter, which extracts the time of the exhaust sound and with which the interference is cancelled out.

Table 3. rate of success

R-index	single	2 cyls.	4 cyls.	failed	rate of success
max/2.0	-0.783	0.1297-0.1509	0.2350-	FZS1000,GSX250S	84.6
max/3.0	-0.2204	0.2294-0.3003	0.3618-	TL1000S	92.3
max/3.5	-0.1328	0.2747-0.4562	0.4804-	CBX125	92.3
max/4.0	-0.1829	0.3252-0.5255	0.5327-	CBX125	92.3

6 Conclusion

We proposed the method of discriminating the numbers of cylinders used in a motorcycle engine by wavelet analysis. The result of the experiment showed that the method is reasonable enough. However, in some engine which has an overlap between the frequency of explosion part and the other part, this method doesn't work. We'll get better results if we use a time-frequency filter in (a, b) domain to get the time ratio of the explosion part.

References

1. T. Tabaru, S. Shin, and T. Kitamori (1994), Identification of frequency response based on multiresolution analysis, Proc. Asian Control Conference, 27-30, July, 1994, Tokyo, Japan, pp. 413-416
2. K. Nakano, T. Tabaru, S. Shin, H. Mitsuo, Y. Toyoda, and Y. Akiyama (1997), Wavelet based identification for transfer function with dead time and its application, Proc. IFAC Symposium on System Identification, vol. 1, pp. 51-56

3. T. Tabaru and S. Shin (1997), Dead time detection based on wavelet analysis of cross correlation data, Proc. IFAC Symposium on System Identification, vol. 1, pp. 33-38
4. M. Fukuda and S. Shin (1997), Model reference learning control with a wavelet network, Proc. The second Asian Control Conferense, Seoul, Korea, 22-25, July, 1997, vol. 2, pp. 247-250
5. K. Nakano, T. Tabaru, S. Shin, and M. Satou (1997), Wavelet analysis approach to identification of systems with dead time – application to boiler process, Proc. IEEE CDC 97, pp. 3375-3380, 10-12, Dec., 1997, San Diego, USA
6. M. Fukuda and S. Shin (1998), Model Reference Learning Control with a Wavelet Network, Iterative Learning Control (Z. Bien and J. -X. Xu ed.), Kluwer Academic Publishers, pp. 211-226
7. K. Hattori, H. Kawai, T. Kawai, and S. Shin (1999), Development of airbag operating algorithm with wavelet transform, Proceedings of the 14th World Congress of IFAC (Edited by Han-Fu Chen, Dia-Zhan Cheng and Ji-Feng Zhang), vol. P, pp. 187-192
8. K. Nakano, T. Tabaru, S. Shin, and Y. Toyoda (2002), WAVELET-BASED IDENTIFICATION FOR CONTROL OF WATER-TUBE DRUM BOILER, Proc. 15th Triennial World Congress of the International Federation of Automatic Control, T-Mo-M01, Barcelona, 21-26, July, 2002
9. T. Tabaru and S. Shin (2003), Dead time measurement of closed loop system by wavelet, Preprints of the 13th IFAC Symposium on System Identification, Rotterdam, Netherlands, Aug., 2003, pp. 437-442
10. K. Seki, S. Shin, and T. Tabaru (2004), Discrimination of Normal and Studless Tires by Wavelet Sound Analysis, Proc. SICE04, Sapporo, 4-6, Aug., pp. 2312-2315
11. G. Kaiser (1994), A Friendly Guide to Wavelets, Birkhauser